6 June 2012

To Bob Wiebe—
Esteemed colleague of many decades.
Remember the Rain? Always the Wind!
Still a glorious wilderness.

Complements of the authors

Wes + Judy

The Novarupta-Katmai Eruption of 1912—Largest Eruption of the Twentieth Century: Centennial Perspectives

Wes Hildreth and Judy Fierstein

Professional Paper 1791

U.S. Department of the Interior
U.S. Geological Survey

U.S. Department of the Interior
KEN SALAZAR, Secretary

U.S. Geological Survey
Marcia K. McNutt, Director

U.S. Geological Survey, Reston, Virginia: 2012

This report and any updates to it are available online at:
http://pubs.usgs.gov/pp/1791/

For more information on the USGS—the Federal source for science about the Earth, its natural and living resources, natural hazards, and the environment—visit http://www.usgs.gov or call 1–888–ASK–USGS

For an overview of USGS information products, including maps, imagery, and publications, visit http://www.usgs.gov/pubprod

To order this and other USGS information products, visit http://store.usgs.gov

This book also for sale by the Superintendent of Documents, U.S. Government Printing Office
Internet: bookstore.gpo.gov *Phone:* toll free (866) 512-1800; DC area (202) 512-1800
Fax: (202) 512-2104 *Mail:* Stop IDCC, Washington, DC 20402-0001
ISBN 978-1-4113-3380-2

Suggested citation:
Hildreth, W., and Fierstein, J., 2012, The Novarupta-Katmai eruption of 1912—largest eruption of the twentieth century; centennial perspectives: U.S. Geological Survey Professional Paper 1791, 259 p.

Cataloging-in-publication data are on file with the Library of Congress (http://catalog.loc.gov/).

Foreword

The Novarupta-Katmai eruption in early June 1912—the most powerful in the world in the 20th century—was surpassed in size during all of recorded history only by the 1815 eruption of Tambora Volcano (Indonesia). When this voluminous eruption occurred, however, Alaska was quite different from what it is today. The total population of the Alaska Territory in 1912 was only about 50,000, living in a vast, remote region that was terra incognita to most of the world. Alaska's geology and volcanoes had been little studied, and the science of volcanology was nascent—even the early investigators at Katmai lacked the basic understanding of magmatic and volcanic phenomena that we take for granted today. Moreover, doing fieldwork—then as now—in the rugged Alaskan terrain posed daunting challenges, such as coping with the ever-present storms, icy streams, windblown dust and ash, and bears. However, such disadvantages of Alaskan field conditions are counterbalanced by the scientific bonus that the desolate, unvegetated Katmai volcanic landscape also presented exceptionally well exposed large pyroclastic deposits available for detailed investigations.

The initial information about the 1912 eruption was necessarily limited, consisting mostly of anecdotal accounts—of felt earthquakes, explosions, eruption plumes, and ashfalls—reported from places distant from the volcano. The full story of the eruption and what actually transpired at Katmai—above and below the ground surface—would only become clear after intensive scientific forensic work over subsequent decades, the results and implications of which are synthesized in this centennial volume. Wes Hildreth and Judy Fierstein are amply qualified and most experienced to undertake this synthesis, because of their own decades of substantial field and laboratory investigations on the eruption since the mid-1970s. Their encompassing synthesis is an encyclopedic and authoritative interpretation of all available data about the processes, products, and impacts of this eruption, within the context of modern scientific understanding of how volcanoes work.

With its in-depth discussion and diverse topical coverage, together with copious illustrations, this volume should be of immense utility to a wide variety of readers. For nonspecialists, the engaging narrative about the early investigations at Katmai makes for fascinating reading. For specialists in volcanology, the topics treated include stratigraphy of pyroclastic deposits, seismicity, caldera collapse, vent structure, chemical and isotopic composition of eruptive products, posteruption geomorphic evolution, and comparison of Novarupta-Katmai with other large historical eruptions. Indeed, detailed studies of the well-exposed and large-volume 1912 pyroclastic deposits have provided diagnostic insights to decipher the emplacement processes of similar deposits (ignimbrites)—prehistoric and less well preserved—in other volcanic regions, whose origins long perplexed geoscientists. In addition, the volume traces the evolution of scientific thought about the causative mechanisms for large rhyolitic eruptions through progressive advances in the field of volocanology.

Hildreth and Fierstein mention that the Novarupta-Katmai eruption occurred about two months after the RMS *Titanic* struck an iceberg on 14 April 1912 and sank, killing more than 1,500 passengers and crew. Although the huge 1912 eruption certainly had the potential to produce a massive volcanic disaster, it fortunately occurred in a sparsely populated region, causing only a single fatality (not directly eruption related) and minimal socioeconomic disruptions. Nowadays, even though the Alaska Peninsula is hardly densely populated, more people on the ground would be affected should another large eruption recur in the Katmai area. Because of the exponential increase in air traffic in the North Pacific region during recent decades, greater potential hazards are posed by possible encounters between jet aircraft and drifting clouds of volcanic ash from explosive eruptions.

The broad scope and meticulous scholarship of this volume virtually guarantee a very long shelf life. Indeed, the detailed investigations of the 1912 Novarupta-Katmai eruption, as reviewed herein—in comprehensive and compelling fashion—by Hildreth and Fierstein should be emulated at other Alaskan volcanoes to prepare the best possible assessments of potential hazards, on the ground and in the air. The future occurrence of another great explosive eruption in Alaska is not a question of if, but when, and we should be better prepared in such eventuality.

Robert I. Tilling
Senior Research Volcanologist, Emeritus
USGS Volcano Science Center

The SS *Dora*, a mail steamer of the Alaska Commercial Company, served coastal villages from Unalaska to Valdez for several decades. She was in Shelikof Strait in the early afternoon of 6 June 1912, when her crew and passengers became the first witnesses to report the great plinian eruption plume that ascended from Novarupta to the stratosphere. Photo by J.E. Thwaites, mail clerk of the *Dora*.

Contents

Figures

Tables

The Novarupta-Katmai Eruption of 1912—Largest Eruption of the Twentieth Century: Centennial Perspectives

By Wes Hildreth and Judy Fierstein

Abstract

The explosive outburst at Novarupta (Alaska) in June 1912 was the 20th century's most voluminous volcanic eruption. Marking its centennial, we illustrate and document the complex eruptive sequence, which was long misattributed to nearby Mount Katmai, and how its deposits have provided key insights about volcanic and magmatic processes. It was one of the few historical eruptions to produce a collapsed caldera, voluminous high-silica rhyolite, wide compositional zonation (51–78 percent SiO_2), banded pumice, welded tuff, and an aerosol/dust veil that depressed global temperature measurably. It emplaced a series of ash flows that filled what became the Valley of Ten Thousand Smokes, sustaining high-temperature metal-transporting fumaroles for a decade.

Three explosive episodes spanned ~60 hours, depositing ~17 km³ of fallout and 11±2 km³ of ignimbrite, together representing ~13.5 km³ of zoned magma. No observers were nearby and no aircraft were in Alaska, and so the eruption narrative was assembled from scattered villages and ship reports. Because volcanology was in its infancy and the early investigations (1915–23) were conducted under arduous expeditionary conditions, many provocative misapprehensions attended reports based on those studies. Fieldwork at Katmai was not resumed until 1953, but, since then, global advances in physical volcanology and chemical petrology have gone hand in hand with studies of the 1912 deposits, clarifying the sequence of events and processes and turning the eruption into one of the best studied in the world. To provide perspective on this century-long evolution, we describe the geologic and geographic setting of the eruption—in a remote, sparsely inhabited wilderness; we review the cultural and scientific contexts at the time of the eruption and early expeditions; and we compile a chronology of the many Katmai investigations since 1912.

Products of the eruption are described in detail, including eight layers of regionwide fallout, nine packages of ash flows, and three lava domes that followed the explosive pyroclastic episodes. Changes in the proportions of coerupting rhyolite, dacite, and andesite pumice documented for the fallout and ash-flow successions, which are locally interbedded, permit close correlation of those synchronously emplaced sequences and their varied facies. Petrological correlation of the sequence of deposits near Novarupta with ash layers at Kodiak village, 170 km downwind, where three episodes of ashfall were recorded (to the hour), provides key constraints on timing of the eruptive events.

Syneruptive collapse of a kilometer-deep caldera took place atop Mount Katmai, a stratovolcano centered 10 km east of the eruption site at Novarupta, owing to drainage of magma from beneath the Katmai edifice. Correlation of ~50 earthquakes recorded at distant seismic stations (including 14 shocks of magnitude 6.0 to 7.0) to fitful caldera collapse provides further constraints on eruption timing, because layers of nonjuvenile breccia and mud ejected from Mount Katmai during collapse pulses are intercalated with the pumice-fall layers from Novarupta. Structure of the Novarupta vent, a 2-km-wide depression backfilled by welded tuff and inferred to be funnel-shaped at depth, is described in detail, as is the 4-km-wide caldera at Mount Katmai.

Discussions are also provided concerning: (1) the impact on global climate of the great mass of sulfur-poor but halogen-rich aerosol ejected into the atmosphere by the rhyolite-dominated eruption; (2) chemical and mineralogical effects of the fumarolic acid gases; and (3) the timing of several syneruptive landslide deposits sandwiched within the pumice-fall sequence. Secondary posteruption phenomena characterized include impounded lakes, ash-rich debris flows, phreatic craters on the ignimbrite sheet, responses of glaciers to the fallout blanket and to beheading by caldera collapse, growth of new glaciers inside the caldera, and gradual filling of the caldera lake.

Structure, composition, and ages of the several andesite-dacite stratovolcanoes, closely clustered near Novarupta, all of which remain fumarolically and seismically active, are summarized. But among them only Mount Katmai extends compositionally to include basalt and rhyolite. The petrological affinities of 1912 magmas erupted at Novarupta with pre-1912 Katmai lavas are outlined, and various chemical, mineralogical, isotopic, and experimental data are assembled to construct a model of preeruptive magma storage beneath Mount Katmai.

The monograph concludes by comparing the 1912 eruption with several other well-studied large explosive eruptions, 14 of them historical and 9 prehistoric. Finally, we retrospectively review the historical difficulties in understanding what had actually taken place at Katmai in 1912 and the century of progress in volcano science that has allowed most of it to be figured out.

Campsite of Griggs expedition of 1917 on bench at north toe of Mount Cerberus. View to north-northwest down Lethe arm of the Valley of Ten Thousand Smokes. Photo courtesy of National Geographic Society.

Chapter 1

The 1912 Eruption and its Importance

The explosive eruptive sequence at Novarupta on 6–8 June 1912 was the world's most voluminous volcanic eruption of the 20th century (fig. 1), but its multifaceted significance for the developing science of volcanology involved far more than size alone. After summarizing the complex eruptive sequence, much of which was long misattributed to nearby Mount Katmai, we discuss how the 1912 events have remained scientifically important for 100 years and why the 1912 deposits continue in the 21st century to provide insights about volcanic and magmatic processes.

The first generation of posteruptive investigations of Mount Katmai, Novarupta, and the Valley of Ten Thousand Smokes was summarized in a splendidly illustrated book by Robert Fiske Griggs (1922), director of the 1915–1919 expeditions. The present volume is, in one sense, a sequel to Griggs' book, updating a century of progress in learning how volcanoes work and in understanding, in particular, the complexities and wonders of the great eruption of 1912.

Overview of the Eruption

The greatest eruption of the 20th century began at 1300 (Alaskan time) on 6 June 1912, continuing violently for ~60 hours in three discrete explosive episodes (fig. 2) that produced ~17 km^3 of fall deposits and 11±2 km^3 of ignimbrite, altogether representing ~13–14 km^3 of zoned magma. The opening episode (I) released ~70 percent of the total volume, including plinian fallout Layers A and B and the main Valley of Ten Thousand Smokes (VTTS) ignimbrite (fig. 3), which were emplaced concurrently and have matching compositional zonation that extends from 100 percent rhyolite initially through to dacite-dominant with fluctuating fractions of andesitic scoria (fig. 4). The ignimbrite, which ranges from nonwelded to eutaxitic, was constructed during the first 16 hours by accumulation of nine compositionally distinct, sequentially emplaced flow packages.

Episode II began after a break of a few hours and produced 4.8 km^3 of dacite-dominant plinian fallout (Layers C and D), accompanied proximally by several sector-confined falls and small pyroclastic density currents. After another short break, Episode III produced 3.4 km^3 of dacite-dominant plinian fallout (Layers F and G), accompanied proximally by several thin andesite-rich falls and pyroclastic density currents. All three explosive episodes issued from the 2-km-wide Novarupta depression (fig. 5), which is inferred to be a flaring funnel-shaped vent backfilled by welded Episode I ejecta and mantled by stratified deposits of Episodes II and III. Sometime after the final dust layer (Layer H) had settled, a dacite dome that plugged the Episode III vent was explosively destroyed and strewn about the depression as a block bed (Episode IV). It was subsequently replaced by a high-silica rhyolite dome (Episode V), slightly contaminated by andesite and dacite, the emplacement of which terminated the sequence.

Novarupta lies at the foot of an andesite-dacite stratocone group called Trident Volcano (figs. 5, 6), but the eruption broke out through horizontally stratified Jurassic marine sedimentary rocks. Caldera collapse at Mount Katmai, 10 km east of Novarupta, began around midnight, ~11 hours into the eruptive sequence, after escape of ~8.5 km^3 of magma. During the following 27 hours, fitful subsidence was accompanied by more than 50 earthquakes, of which ten were of magnitude 6.0 to 7.0, and by phreatic ejections of hydrothermal mud and breccia that produced several circumcaldera layers intercalated with the contemporaneous plinian pumice-fall layers from Novarupta. No juvenile material erupted from Mount Katmai during formation of its kilometer-deep, 5.5-km^3 caldera, but a small dacite dome extruded on the caldera floor sometime after the collapse. Modest earthquakes continued for months after the eruption ended.

Volcanological Significance of the 1912 Eruption

The 3-day explosive eruption at Novarupta in June 1912 was the world's most voluminous volcanic event of the 20th century. For volcanologists and petrologists, it was also one of the most provocative. Long misinterpreted and in a region long difficult to visit, Mount Katmai and the Valley of Ten Thousand Smokes (VTTS) attracted rampant speculation but rather little fieldwork by experienced volcanologists, until 1953. Since then, however, advances in understanding physical volcanology have gone hand-in-hand with numerous field studies, which have clarified the eruptive events and processes of 1912 and have turned the eruption into one of the most intensively studied in the world.

Many aspects of the 1912 eruption made it extraordinary and of enduring interest:

1. Its great volume, roughly 13 km^3 of magma, places it among the five largest in recorded history.

2. It was one of few historical eruptions during which a caldera collapsed.

3. It was one of very few historical eruptions to eject voluminous rhyolite.

4. The ashfall was global, an aerosol-dust veil was reported as far as the Mediterranean, and worldwide temperature depression was measurable.

5. Pumiceous pyroclastic flows and pumiceous fallout were of roughly equal volume and were eventually shown to have been emplaced concurrently.

6. The VTTS ignimbrite was the first large pumiceous pyroclastic-flow deposit of the scientific age to be emplaced on land.

7. It was also the first known to have produced welded tuff.

8. The eruptive products spanned a wide compositional range, with andesite, dacite, and high-silica rhyolite each representing an important proportion of the ejecta.

9. Conspicuous and abundant compositionally banded pumice provoked speculative assimilation hypotheses, because such mingling had seldom been reported previously and magma mixing had not yet been widely recognized.

10. "Ten thousand" fumaroles rose from the ignimbrite, and it took many years for observers to apprehend that the vapors owed their heat to the deposit, rather than to magma that was initially presumed to underlie all or much of the VTTS.

11. High-temperature vapor transport of Fe, Pb, Zn, Cu, Sn, Mo, Mn, and other metals in halogen-bearing fumaroles and rapid precipitation of minerals containing such metals fascinated students of ore deposition.

12. The eruption and caldera collapse were accompanied by extreme seismicity, later shown to include 14 earthquakes with magnitudes of 6.0 to 7.0.

13. C.N. Fenner (1926) published analyses of 18 rocks from the Katmai region that became a focus of petrological debate (principally with N.L. Bowen) about the processes that produce magmatic diversity, particularly the relative importance of assimilation and fractional crystallization. Because Fenner's suite included pre-Quaternary intrusive rocks as well as Pleistocene lavas and 1912 pumice, it was an inappropriate and unfortunate basis for that debate.

In later years, as more became known about the 1912 deposits, additional extraordinary aspects attracted the attention of volcanologists:

14. Lateral transport of magma to a vent some 10 km away from the site of caldera collapse raised as many questions as it solved.

15. The major ignimbrite-producing episode and the largest of three plinian pumice-fall episodes were simultaneous, apparently at odds with the then-standard model of ash-flow generation by column collapse (misinterpreted by many to imply column extinction).

16. Two subsequent plinian episodes took place many hours after the large-scale ignimbrite emplacement, contrary to the expectations of conventional eruption scenarios.

17. Dispersal of plinian fallout was strongly eastward, reflecting the stratospheric jet stream, in contrast to more distributed dispersal from comparably great eruptions of the equatorial volcanoes, Tambora in 1815 and Krakatau in 1883.

18. A high-silica rhyolite lava dome plugged the vent inside an ejecta ring composed of dacite and andesite.

19. It became clear that the VTTS ignimbrite was not emplaced en masse but by a rapid succession of dozens of flow units, divisible into at least nine compositionally distinctive, mappable packages.

20. Because the valley-filling ignimbrites were shown to have possessed neither great momentum nor high degrees of inflation during outflow, the VTTS came to be regarded as representing the low-energy extreme of a spectrum of ash-flow emplacement styles.

21. Pioneering studies of melt inclusions in quartz from the 1912 rhyolite provided insights about preeruptive magmatic conditions, helping initiate an expanding field of research on magmatic volatile contents.

22. It surprised then-conventional petrological thinking that the 1912 high-silica ("minimum-melt") rhyolitic magma contains quartz and plagioclase phenocrysts but lacks alkali feldspar.

23. A compositional gap of ~8 weight percent SiO_2 between 1912 dacite and rhyolite stimulated a variety of models for the origin of such gaps, whether by secular crystal fractionation, by separation of rhyolitic melt from crystal-rich intermediate magma, by partial melting of wall-rock granitoids, or by some independent origin of the rhyolite.

24. Application of the Fe-Ti-oxide geothermometer of Buddington and Lindsley (1964) yielded a thermal continuum for the 1912 andesite-dacite-rhyolite suite, despite the wide compositional gap, suggesting contiguity in the magma reservoir.

25. Because the 1912 vent broke out through nonvolcanic basement rocks, any hydrothermal convection induced

by the intrusive feeder beneath Novarupta would uniquely have a known age of initiation (1912).

26. Because compensatory caldera collapse took place elsewhere and the Novarupta vent was cleanly reamed with only modest marginal slumping, complex pyroclastic deposits are preserved to within meters of the plug dome, providing virtually unique exposure of the ultraproximal products of high-energy sedimentation from the margins of a plinian jet and the lower part of its convective plume.

27. Preservation of the vent structure, filled by its own ejecta, intruded only by the dome conduit, and reamed through horizontal Jurassic marine sedimentary rocks in a 3-day episode in 1912, provides an attractive target for geophysical investigation and scientific drilling; its relatively brief history offers insights about intrusion, decompression, degassing, fragmentation, and ejection of magma in a setting where the mechanical, chemical, and thermal consequences can be related to a short-lived episode.

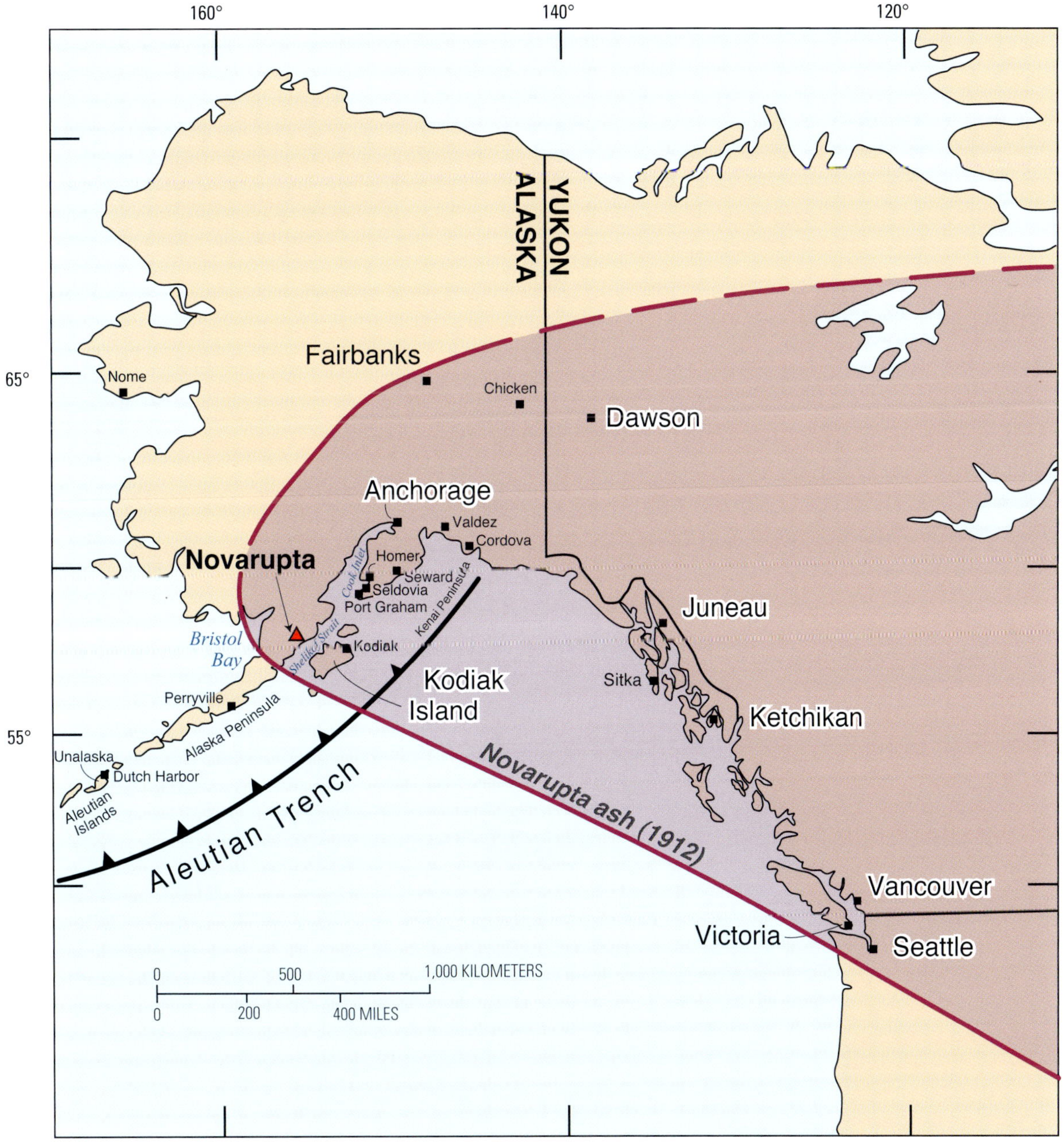

Figure 1. Map of Alaska and adjacent North America, showing location of Novarupta, several geographic names mentioned in text, and region impacted by ashfall within a few days after 1912 eruption.

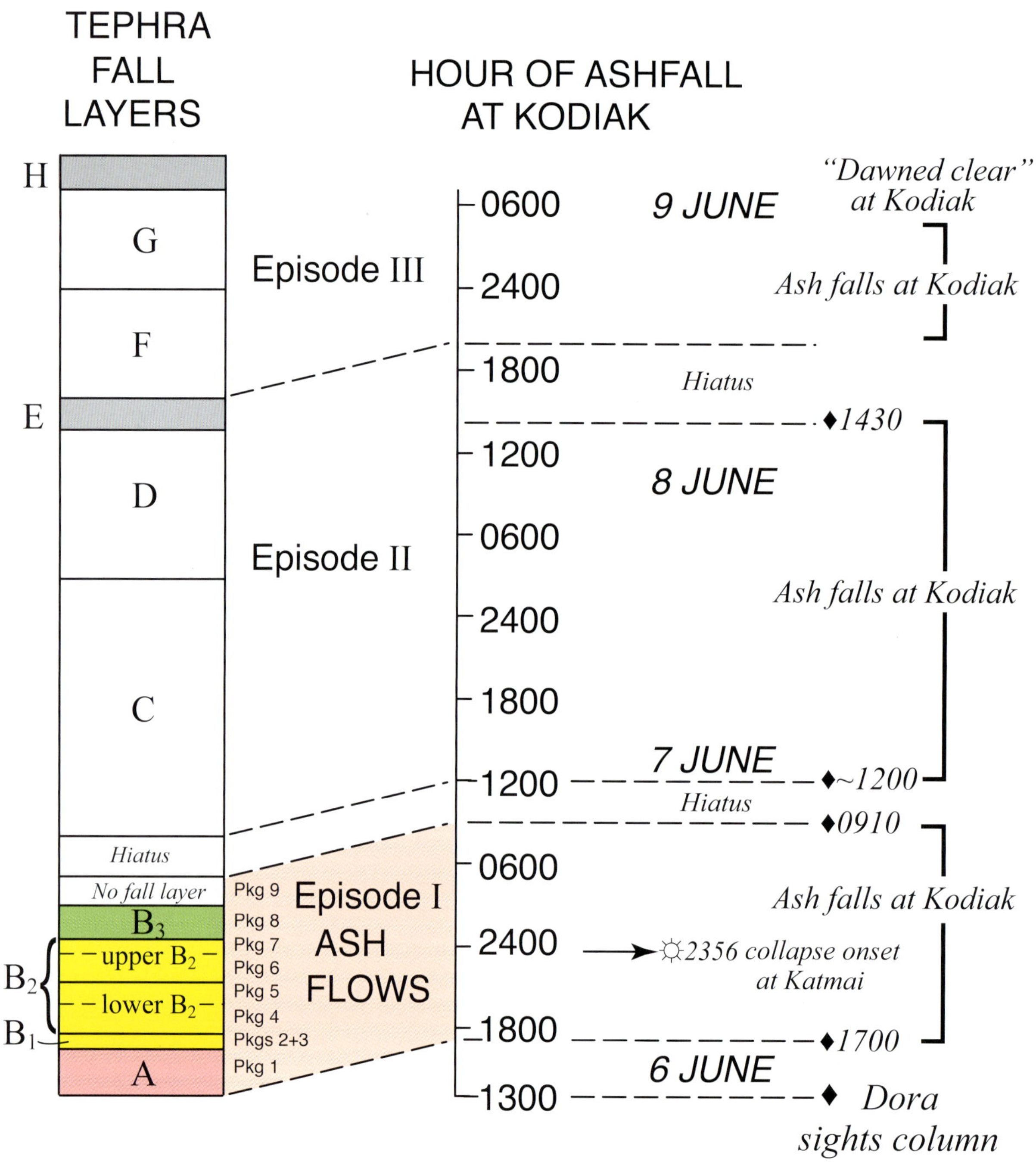

Figure 2. Generalized 1912 Novarupta eruptive sequence and chronology of ashfall recorded at Kodiak village, 170 km downwind. Four-hour offset at base represents time between initial sighting of eruption column by S.S. *Dora* and beginning of ashfall at Kodiak.

Figure 3. Aerial panorama of Valley of Ten Thousand Smokes, extending 20 km northwest from vent at Novarupta to ignimbrite terminus at base of Mount Katolinat. Baked Mountain (BM) and Broken Mountain (BR) are pumice-mantled ridges of glaciated Jurassic siltstone that separate three arms of upper valley—River Lethe, Knife Creek, and a lesser valley between them. Ignimbrite surface slopes 1°-2° downvalley and is incised by several gorges as deep as 35 m. Scarps that truncate Broken and Falling Mountains define Episode I backfilled vent depression. Just beyond tip of Falling Mountain shadow (left center), where pumice-alluvial fan enters meandering gorge, tuff dips gently back toward Novarupta, owing to intravent compaction and welding, thus defining buried west rim of vent structure. Asymmetrical ejecta ring around Novarupta lava dome consists of products of eruptive Episodes II and III. Mount Griggs, upper right, is 2,330-m-high andesitic stratocone with several postglacial eruptions and is highest peak in district, 1,750 m above valley floor. Buttress Range and Mount Katolinat are subhorizontal siltstone-sandstone strata of Jurassic Naknek Formation.

Figure 4. Representative pumices from 1912 eruption: White crystal-poor high-silica rhyolite, pale gray crystal-rich dacite, black crystal-rich andesite, and rhyolite-andesite banded pumice.

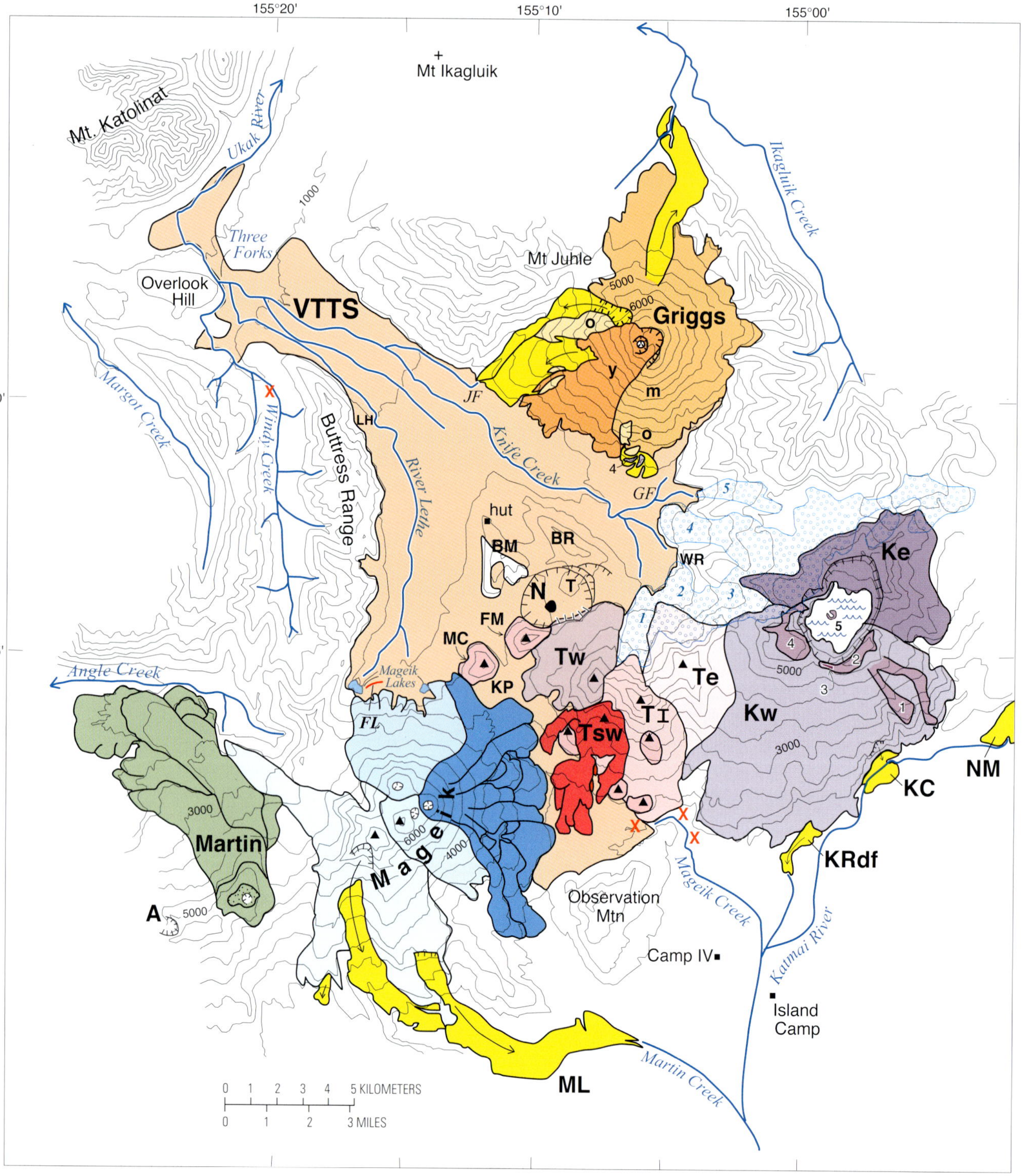
155°20'
155°10'
155°00'
58°20'
58°15'
Mt Ikagluik
Mt. Katolinat
Ukak River
Ikagluik Creek
Three Forks
Mt Juhle
Overlook Hill
VTTS
Griggs
Margot Creek
Windy Creek
Buttress Range
LH
River Lethe
Knife Creek
JF
GF
hut
BM
BR
WR
N
T
FM
MC
KP
Tw
Te
Ke
Kw
Tsw
TI
Angle Creek
Mageik Lakes
FL
Martin
Mageik
A
Observation Mtn
Mageik Creek
Camp IV
Island Camp
Katmai River
NM
KC
KRdf
ML
Martin Creek
0 1 2 3 4 5 KILOMETERS
0 1 2 3 MILES

Figure 5. Simplified geologic map of Valley of Ten Thousand Smokes (VTTS) and volcanoes of the Katmai cluster. Contour interval 500 feet (152 m). From hachured vent depression around Novarupta lava dome (N, black), Episode I ignimbrite (tan) extends northwest 20 km along VTTS, as well as across Katmai Pass (KP) for 10 km down Mageik Creek on Pacific slope. Knife Creek Glaciers (blue pattern) are numbered 1-5; other glaciers are omitted for clarity. Alagogshak volcano (A), long extinct, is indicated only by its eroded crater. Mount Mageik consists of four overlapping centers (in shades of blue); only the youngest and easternmost center (for which individual lava flows are indicated) is Holocene. Mount Martin, entirely Holocene, consists of small fragmental summit cone and several overlapping coulees. Trident group consists of three Pleistocene cones, East Trident (Te), Trident I (TI), and West Trident (Tw), as well as historical Southwest Trident (Tsw) lavas and fragmental cone of 1953–1974; several peripheral Pleistocene lava domes, comagmatic with Trident, include Mount Cerberus (MC) and Falling Mountain (FM). Mount Katmai consists of two overlapping centers, Northeast Katmai (Ke) and Southwest Katmai (Kw), both truncated by 1912 collapse of hachured Katmai caldera, which is now partly filled by lake. Five youngest eruptive units of Mount Katmai are numbered (1–5) and discussed in chapter 12: 1, leveed dacite lava flows; 2, south-rim rhyodacite lavas; 3, zoned scoria fall atop unit 2; 4, dacite agglutinate sheet on caldera rim and correlative scoria-flow deposit at Knife Creek; and 5, Horseshoe Island dacite dome. Remnants of 22.5-ka plinian rhyodacite pumice-fall deposits (and ignimbrite) in Windy and Mageik Creeks (sites indicated by red X) are related to most evolved lava of unit 2. Products of Mount Griggs are subdivided by age into older (o, middle Pleistocene), middle (m, late Pleistocene), and younger (y, postglacial) exposures. Holocene debris-avalanche deposits are in bright yellow; those emplaced in 1912 are labelled: KC, Katmai Canyon landslide; KRdf, Katmai River pumiceous debris flow; ML, Mageik Landslide; NM, Noisy Mountain landslide. Uncolored basement rocks are Jurassic Naknek Formation or Tertiary porphyritic intrusions. Miscellaneous features: BM, Baked Mountain; BR, Broken Mountain; FL, site of Fissure Lake; GF, Griggs Fork of Knife Creek; JF, Juhle Fork of Knife Creek; T, Turtle; WR, Whiskey Ridge; hut, Baked Mountain Hut, research shelter; Island Camp and Camp IV were way stations between Katmai Bay and VTTS.

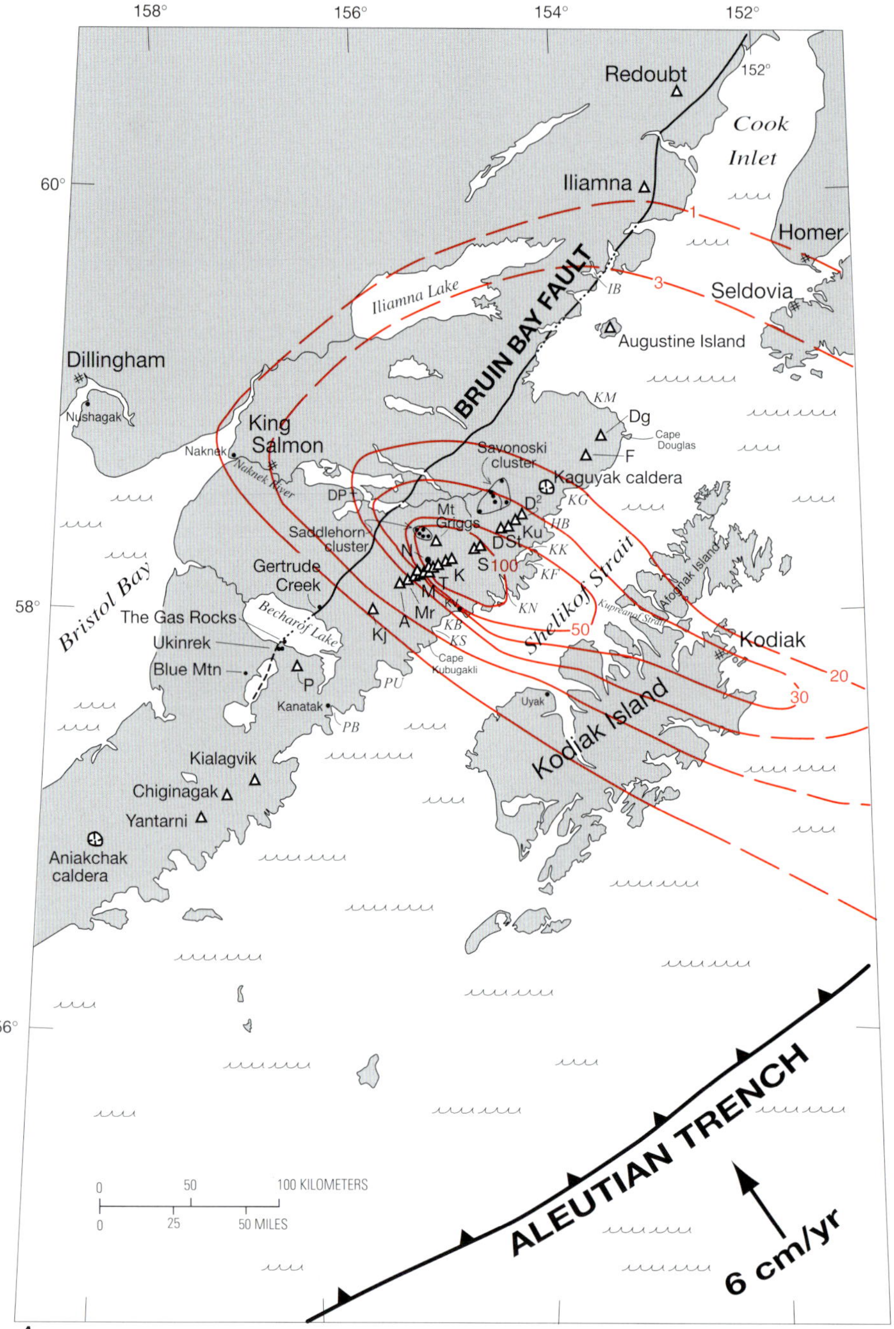

A

Figure 6. Location maps. *A*, Map of upper Alaska Peninsula and Kodiak Island group, showing many geographic locations mentioned in text and, in red, regional isopachs for total 1912 tephra-fall deposit (thickness contours in cm). Triangles indicate stratovolcanoes; hachured ovals indicate calderas. Abbreviations for volcanoes: A, Alagogshak; D, Denison; D², Devils Desk; Dg, Douglas; F, Fourpeaked; K, Katmai; Kj, Kejulik; Ku, Kukak; M, Mageik; Mr, Martin; P, Peulik; S, Snowy Mountain; St, Steller; T, Trident. Behind volcanic chain is Novarupta (N, black dot), vent site of 1912 eruption. The Savonoski and Saddlehorn clusters are rear-arc groups of small monogenetic cones, domes, and plugs described by Hildreth and others (2004). Abbreviations for other geographic features: DP, Dumpling Mountain; IB, Iliamna Bay; HB, Hallo Bay; KB, Katmai Bay; KF, Kaflia Bay; KG, Kaguyak Bay; KK, Kukak Bay; KM, Kamishak Bay; KN, Kinak Bay; KS, Kashvik Bay; KV, Katmai village; PB, Portage Bay; and PU, Puale Bay (called Cold Bay in 1912). At the Aleutian Trench, 350 km southeast of the volcanic line, the Pacific Plate subducts northwestward at 6 cm/yr.

B, Drainage system near Valley of Ten Thousand Smokes (VTTS, shaded); additional abbreviations here: B, Brooks Camp; BR, Brooks River; G, Mount Griggs; HP, Hagelbarger Pass; Kg, Kaguyak caldera; KjP, Kejulik Pass; KP, Katmai Pass; and YP, Yori Pass.

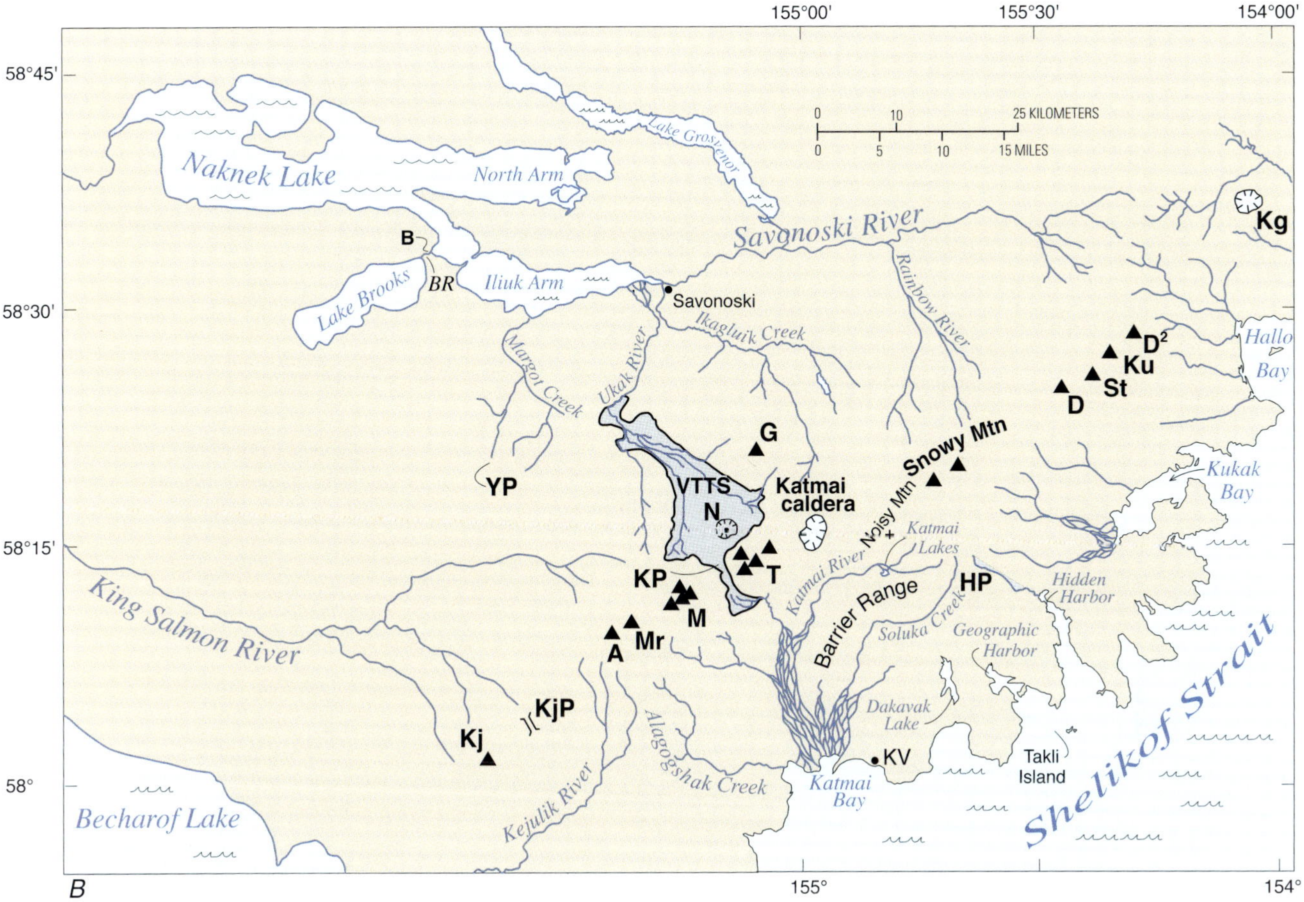

Figure 6 —Continued

Ashfall accumulation at Kodiak village, 170 km downwind from Novarupta, at the end of the 1912 eruption. Total deposit from three episodes of fallout at Kodiak during 6–8 June 1912 eventually compacted to a thickness of 25–30 cm. Photo by Wilbur J. Erskine, in charge of commerce at Kodiak.

Chapter 2

The Context of the 1912 Eruption

In this section we describe where the eruption took place—in a remote, sparsely inhabited wilderness—and we provide synopses of the cultural and scientific contexts of the eruption and its subsequent investigations. These are followed by an overview of the geologic setting of the Katmai region, which completes the background information preparatory to launching into a detailed eruption narrative.

Geographic Setting

The Alaska Peninsula is the great 750-km-long arm that dominates the map of southwestern Alaska and extends southwestward into the Aleutian Islands chain (figs. 1, 6). From Cape Douglas to Becharof Lake, the Pacific half of the upper peninsula is a highland of basement rocks standing 1,200 to 1,600 m above sea level, overlain by an ice-clad chain of active volcanoes and sculpted by active glaciers on both northwest and southeast flanks. Turbulent braided streams descend both flanks, those on the southeast side being much shorter and generally steeper than those to the northwest. The Pacific coastline is a rugged wonderland of narrow fjords and open bays with sandy beaches. The northwest half of the peninsula descends toward Bristol Bay through an ice-scoured hilly terrain of long spurs and outlying mountains of moderate relief and of large moraine-dammed lakes, merging into a broad coastal wetland of glacial and lacustrine deposits near sea level. The 1912 vent at Novarupta broke out at the head of a glacial valley, only 3 km northwest of the drainage divide along the volcanic axis, just north of Katmai Pass—one of the lowest saddles on the divide.

Novarupta is 450 km southwest of Anchorage, 250 km southwest of Homer, 170 km west-northwest of Kodiak, and 100 km east-southeast of King Salmon, which is the nearest year-round town. Most of the 200-km-long reach of the peninsula from Cape Douglas to Becharof Lake lies within Katmai National Park and Becharof National Wildlife Refuge. Despite a few seasonal fishing and tourist resorts, most of the vast area is managed as wilderness and lacks either roads or permanent habitation. Access is by small boat or by float plane from King Salmon, where there is a regional airport.

The climate ranges from maritime to arctic-alpine. There are no weather records for the mountains, but long-term observations at King Salmon (near sea level in the Bristol Bay lowlands) yield average normal low and high temperatures of 8°F and 22°F (–13°C and –6°C) in January and 46°F and 63°F (8°C and 17°C) in July. Average monthly water-equivalent precipitation is ~2.7 cm in January and 5.6 cm in July, but August is typically the wettest month, with an average of 7.5 cm; the annual average is 50 cm (far below the average of 172 cm recorded at Kodiak on the Pacific coast). Snow has fallen at King Salmon in all months except July and August and is, of course, far more frequent and severe in the mountains. Clouds and heavy morning fog are routine at King Salmon, which averages one clear day per month in summer and seven per month in January. Because the North Pacific storm track runs up the Alaska Peninsula, the region is exceptionally windy. At King Salmon, mean wind speed is ~18 km/hr every month of the year, and storm winds can reach 100-150 km/hr in the mountain passes.

Plant life in the treeless high country is limited to species resistant to cold, wind, and dessication—lichens and mosses, grasses and sedges, willows and dwarf birch, saxifrages, mountain sorrel, arnica, *Oxytropis*, and *Dryas*. In the lowlands, spruce forests are widespread, along with alder and willow thickets, grasses and sedges, fungi and mosses, and balsam poplar along streams. On the basis of three summers' collections, Cahalane (1959) identified 273 plant species in Katmai National Monument. He also listed 117 species of summer birds, common among which are hawks and eagles, owls and ravens, ducks, loons, and grebes, geese and swans, gulls and shorebirds, woodpeckers and jays, grouse, thrushes, warblers, and sparrows. In high or open treeless country, one finds the raven and gyrfalcon, ptarmigan and snow bunting, Lapland longspur and rosy finch. Rare nests of Kittlitz's murrelet were discovered in the VTTS by U.S. Geological Survey (USGS) geologists (Day, 1995).

Mammals include the wolf and wolverine, lynx and red fox, hares and lemming, ground squirrel and red squirrel, marten, mink, and weasels, porcupine and marmot, beaver, river otter, and muskrat, moose and caribou, and the preeminent brown bear. Furtive but abundant are shrews, mice, and bats. At the coast, one finds seals, sea lions, sea otters, and grey whales.

Cultural Setting: Katmai Before 1912

Native people of Eskimoan (not Aleut) linguistic, cultural, and ethnic stock had reached the upper Alaska Peninsula by ~10 ka and were distributed in a few villages and several seasonal camps on both sides of the mountain range when first contacted by Russian hunters and fur traders ~1750. The Russians reached Katmai village by 1782 and villages near today's Naknek by 1791, and in 1784 they established headquarters on Kodiak Island. Savonoski had been occupied for several centuries by Alutiiq people related to those along Shelikof Strait and apparently hostile to the Aglurmiut people who had by 1800 invaded the lower Naknek River and Bristol Bay from the north; Russian missionaries reached Savonoski by 1840 (see fig. 6). From the 1790s until the 1867 purchase of Alaska by the United States, the Russian American Company made Katmai village its principal trading post for the many small native encampments located on most of the coves and inlets along the rugged coast of Shelikof Strait. It ruthlessly organized most native men on both sides of the strait into annual hunting parties of hundreds of bidarkas (kayaks) in pursuit of sea otters and other marine mammals. The Alaska Commercial Company bought out the Russians in 1868 and had completed driving the sea otter to virtual extinction by 1900, after which the company declined. Its trading post at Katmai was neglected by 1906 and was certainly abandoned before the company sold its Kodiak-based holdings to Wilbur J. Erskine in 1911. Erskine took most of the photographs that documented the impact of the 1912 eruption in and near Kodiak.

Traditionally, native economy had been based on hunting sea mammals, supplemented by hunting, fishing, trapping, and gathering shellfish, berries, and roots. Dwellings (barabaras) were semi-subterranean and sod-roofed, and seaworthy spruce-framed sealskin bidarkas were built in great numbers. On the Naknek side of the divide, settlements were at key fishing locations, and the annual salmon run was critical for supplying fish to dry for the winter. The year-round village at Savonoski also relied on berries, bear, birds, and caribou; it was said that caribou were hunted in the valley that later became the VTTS.

By 1867, the natives were extensively Russianized culturally. Trading posts were central economically, priests were influential, chapels were built at Katmai and Savonoski, and most natives relied on Russian trade goods in exchange for furs and sea otter pelts. Unlike the Russians, the Alaska Commercial Company exercised no governmental authority, so leadership after 1868 reverted to local village elders. Depletion of sea otters by the 1890s impoverished the natives, many of whom became seasonal workers at fishing camps, salteries, and canneries, commercial enterprises that expanded rapidly after 1895. The cannery culture accelerated intermarriage among Alutiiq, Aglurmiut, Europeans, and Americans. The first cannery at Naknek opened in 1890. The town of King Salmon did not exist before construction of an Army Air Forces base there during World War II; the connecting road to Naknek was completed in the 1950s, and the base was closed in 1993.

Katmai village had a population of 457 in 1863. Its importance declined after the American takeover, but it still had 218 residents in 1880 and 132 in 1890. Savonoski had a population of 162 in 1880 but only ~90 in 1912. Influenza contributed to the population decline of the late 1800s, and the pandemic of 1918-19 again killed as many as half of the local population.

The trail over Katmai Pass from Katmai to Savonoski had been a centuries-old link between the related peoples of the Shelikof coastline and the upper Naknek drainage. It became one of several alternative routes to Nome for gold seekers during the gold rush of 1899-1900, used especially in winter when sea and river routes were frozen. The U.S. Post Office provided mail service to western Alaska by way of Katmai Pass, using dogsleds in the winters of 1900–1902 and perhaps a bit longer. Savonoski and Katmai villages were both abandoned permanently in response to the eruption of 1912, as were numerous satellite villages and seasonal camps on both sides of the divide. Hussey (1971) and Dumond (2005) provide many details of the history and prehistory of the Katmai region, and their work formed the basis for the summary just made.

Cultural Setting: The World in 1912

Alaska was little known and almost mythical to most Americans in 1912. After its purchase from Russia in 1867, the inadequately explored region had been under the jurisdiction of the military and various Federal agencies. The U.S. Army, U.S. Coast and Geodetic Survey, Western Union, U.S. Geological Survey, and the Alaska Commercial Company contributed much to its exploration. By 1912, commercial fishing, canneries, and fur trapping had become important, replacing the hunting of dwindling marine mammals. Alaska's population was ~50,000 in 1912, ~34 km^2 per person territory-wide. About 30,000 had come to Alaska in the Gold Rush of 1899–1900. Congress reorganized Alaska as an incorporated Territory in 1912, roughly halfway between purchase and statehood in 1959. The first law passed by the Territorial legislature (1913) extended voting rights to women.

Travelers to Alaska in 1912 took the train to Seattle or San Francisco, where they could arrange transport on various commercial vessels, taking two weeks to Kodiak or a month to Naknek. The SS *Dora*, a mail steamer of the Alaska Commercial Company (and later of the Alaska Steamship Company), served coastal villages from Unalaska to Valdez for several decades and sometimes ran as far south as Sitka and Seattle; she transported mail, supplies, fishing and mining equipment, passengers, and cattle, and she played a key role in reporting the 1912 eruption. In 1918, the coastal passenger liner SS *Princess Sophia* ran aground north of Juneau and was lost with all 343 aboard, Alaska's most deadly shipwreck.

The automobile was still a toy of the rich, but introduction of the Model T Ford in 1908 started its expansion into the middle class, culminating in a far more mobile society after World War I. The first modern electric traffic light was installed in 1912 (in Salt Lake City), and the Turkish Petroleum Company was founded in 1912 to exploit the oil reserves of Mesopotamia. Also in 1912, the United States issued its first postage stamp picturing an airplane, though commercial flight was still decades away. The first municipally owned streetcars began operation in San Francisco. In April, the RMS *Titanic* hit an iceberg and sank in the North Atlantic.

In politics, 1912 saw the admission of New Mexico as the 47th state and Arizona as the 48th. Also in 1912, the Massachusetts legislature enacted the first state law setting a minimum wage for women and children. Women's suffrage was adopted that year by the voters of Oregon, Arizona, Michigan, and Kansas, which thus joined Utah, Idaho, Washington, and California, where women already had the vote; it was not made national until 1920 on ratification of the 19th Amendment. In 1912, the "Bread and Roses" strike against the textile mills of Lawrence, Massachusetts, involved 20,000 workers and lasted for two months; Industrial Workers of the World (IWW) leadership appointed a strike committee with two members of every ethnic group in the mills and had its meetings translated into 25 languages. The successful strike defied assumptions of conservative unions that immigrant, largely female and ethnically divided workers could not be organized. Woodrow Wilson (Democrat) was elected president, beating Theodore Roosevelt (Progressive or "Bull Moose" Party), incumbent William Howard Taft (Republican), and Eugene V. Debs (Socialist). It was the "Progressive Era," and three of the four main candidates advocated political and social reform, antitrust legislation, and more direct democracy. Alaskans were still half a century from being allowed to vote for president.

Internationally in 1912, Sun Yat-sen was inaugurated President of China, ending 2,000 years of imperial rule, and the Japanese Emperor Meiji died, ending the transformative Meiji Era. The British National Health Insurance Act took effect, requiring contributions from all wage earners and employers, and providing free medical attention and medicine as well as unemployment benefits. The Bolshevik faction led by Lenin split from the Russian Social Democratic Labour Party and printed the first issue of *Pravda*. Montenegro declared war against Turkey, starting the First Balkan War, and Bulgarian pilots soon performed history's first aerial bombardment. The African National Congress was founded, 82 years before becoming the first postapartheid ruling party of South Africa. And the U.S. Marines invaded Cuba, Honduras, and Nicaragua (and in 1915, Haiti).

In the popular culture of 1912, Oreo Biscuits were introduced, *Tarzan, Lord of the Apes* was first published, W.C. Handy published the first "Blues" song (*Memphis Blues*), the mayor of Tokyo gave 3,000 cherry blossom trees to be planted in Washington, and the first annual "Cross-City Race" was run in San Francisco (renamed "Bay-to-Breakers" in 1965). Also in 1912, L.L. Bean opened in Freeport, Maine, and Fenway Park opened in Boston. Jim Thorpe won both the pentathlon and decathlon at the Olympic Games in Stockholm, where the host country refused to permit boxing.

In science in 1912, Alfred Wegener proposed the theory of continental drift, Max von Laue made the first observation of diffraction of X-rays by a crystal lattice, and Norman Levi Bowen joined the staff of the Geophysical Laboratory, beginning a fabled career in which he revolutionized petrology. Piltdown Man, famous "missing link" and paleontological hoax, was presented to the British Museum (but not exposed as a forgery until 1953). Robert Falcon Scott arrived on foot at the South Pole on 17 January, finding the tent of Roald Amundsen, who had reached the pole on skis 35 days earlier with five companions and 16 dogs. And the Hawaiian Volcano Observatory formally opened on the rim of Kilauea Caldera, largely on the initiative of Thomas A. Jaggar (professor of geology at MIT), who then served as its director for decades.

It can be added that during the 1915–1919 series of Katmai expeditions led by Griggs, many historic events took place elsewhere. The Mexican Revolution, the Armenian Holocaust, and World War I were ongoing. The year 1916 saw the Easter Rising in Ireland, creation of the U.S. National Park Service, election of the first woman to Congress, arrest of Emma Goldman for lecturing on birth control, and appointment of Louis Brandeis to the U.S. Supreme Court. The year 1917 saw U.S. intervention in World War I, the Balfour Declaration, and the Bolshevik Revolution. The year 1918 saw the Spanish Flu pandemic and the Armistice that stopped the First World War. The year 1919 saw the enactment of Prohibition in the United States, founding of the Fascist Party by Mussolini, the Paris Peace Conference, founding of the League of Nations, and the Versailles Treaty, terms of which ensured resurgence of populist nationalism in Germany, Yugoslavia, Iraq, and Palestine.

Scientific Context of the Early Katmai Investigations

The many publications of Griggs and Fenner made the 1912 eruption famous, and their provocative interpretations attracted widespread skepticism and commentary. Difficulty of access, however, dissuaded scientific reinvestigation for another four decades. Their failure to understand fully what they observed is attributable in part to the unusual nature of the eruption, in part to botanist Griggs' lack of geological experience, and in part to Fenner's restless imagination, his compulsion to challenge orthodoxy, and his wistful hope that field observations could qualify or supplement the logic of Bowen's petrological experimentation. (Bowen and Fenner were colleagues and rivals at the Geophysical Laboratory, which had been established by the Carnegie Institution in 1905.) Nonetheless, the insights and interpretations made by Griggs and Fenner, both mainstream scientists, were rooted in the scientific milieu, conventions, and ideas of their time. It may thus be useful to provide some background on the state of volcanological and petrological understanding during the decade of their field expeditions.

Microscopic petrography had long been sophisticated, and by 1912 bulk major-element analysis of igneous rocks was a thriving field. But little was known about magma sources, and awareness of plate tectonics and subduction lay a half century in the future (Yoder, 1993).

Influential books on igneous geology and petrology were published within a decade of the eruption by Alfred Harker (1909), Cambridge; Joseph Paxson Iddings (1909, 1913), USGS, University of Chicago; and Reginald Aldworth Daly (1914), Harvard, all of whom were widely travelled experts. It was well known by then that most magmas are multicomponent aluminosilicate solutions, characterized by freezing-point depression, wide temperature ranges of crystallization, and orders of mineral crystallization that reflect pressure, temperature, and volatile content as well as bulk composition. Bowen's experimental work was just beginning in the decade of the eruption, however, so many details of magmatic petrology remained nebulous and interpretations widely divergent.

It was nonetheless already appreciated (by most petrologists) that dissolved water lowers magmatic viscosity and the temperature range of crystallization, that water is the dominant volatile component in most magmas, that much of the volatile content of a magma is lost on eruption, and that volatile retention under pressure at depth requires that plutonic crystallization extend to lower temperature than typical for extrusive equivalents. Although most geologists thought by then that silicic magmas crystallize at lower temperatures than mafic ones, the ambiguity of pioneering melting experiments and skepticism about inferences drawn from simple experimental systems left many unconvinced. Not until 1930 was it proven that dry granite can melt at lower temperature than dry basalt, that its lower temperature range of crystallization is not solely dependent on magmatic volatile content, and that superheat is thus not required for basalt to remelt solid granite (Day, 1931).

Many had long inferred that basalts represent parental magmas globally, but only in 1916 did Arthur Holmes call basalts partial melts of peridotite, and not until 1928 did Bowen confirm that experimentally. Flux melting had been suggested but with no real notion of a volatile source nor any suspicion of subduction. Decompression melting was understood in principle, but presumption of a rigid earth prevented its acceptance until long after Holmes had proposed convective mantle upwelling in 1928.

Magmatic differentiation, the processes by which the great diversity of igneous rock composition originates, was then a key focus of research and debate. Iddings (1892) had summarized numerous potential alternative processes responsible for magmatic diversity.

Thoughts on more evolved magmas developed broadly along three lines—independent primary origins, assimilation of crustal rocks by basalt, or derivation from basaltic parents by liquid or crystal fractionation. Darwin made a strong case for the latter as early as 1844, suggesting both crystal settling and "straining" out the interstitial melt from crystal mush. Mixing of magmas was proposed by Bunsen as early as 1851 and highlighted again by Harker (1909), but few had accepted it as common or important (Wilcox, 1999) by the time Fenner and Griggs neglected to invoke it for the banded pumice at Katmai. Several petrologists had proposed, nonetheless, that intermediate magmas in general were deep hybrids of independently generated mafic and silicic parents, although Iddings (1914) turned this around by arguing that the rhyolitic and basaltic extremes both differentiate from primitive intermediate parents, which he presumed fundamental to each region. Regional "petrographic provinces" had been discussed for decades by many petrologists, notably emphasizing distinctions between alkaline and subalkaline suites.

Experimental petrology made great strides during the decade of the Katmai expeditions, principally at the Geophysical Laboratory, where phase-equilibria techniques were then being developed (Yoder, 2004). Bowen (1913) completed the phase diagram for plagioclase solid solutions and for numerous magmatically important binary and ternary systems during the following decade. Already in 1915, he argued that basalt is parental to most igneous rocks, that crystal fractionation controls differentiation, that assimilation is unimportant, that the Soret effect is negligible, and that experimental evidence was against a significant role for liquid immiscibility (Bowen, 1915). At the time, most petrologists remained skeptical or reluctant to generalize his experimental inferences to igneous rock suites, particularly to voluminous continental granitoids (Holmes, 1916). In the communitywide debate that ensued in 1916, Fenner listened but remained silent (Yoder, 1992). Nonetheless, Fenner participated in the first experimental study of a hydrous system at elevated pressures and temperatures (Morey and Fenner, 1917), the first conducted above the critical point of water; they showed that mineral solubility increases with temperature, that water strongly lowers the melting point and viscosity of silicate magma, that crystallization of non-volatile-bearing phases raises system pressure, and that compressed vapor at high temperature readily transports chemical constituents of magmas. Such results certainly influenced Fenner's later interpretations at Katmai.

Seismology had made major advances by 1912, but there was only one seismograph operating in Alaska at the time of the eruption. Although at least 30 Milne seismographs were by then distributed in Europe, Asia, Japan, Hawaii, North Africa, and North America (Abe, 1992), those closest to Katmai were at Sitka (Alaska) and Victoria (British Columbia). By 1912, studies of seismic-wave propagation had shown that most of the deep earth is solid, undermining age-old speculations about molten shells or global reservoirs of magma in the crust and mantle. Mohorovicic had discovered the seismic discontinuity defining the base of the crust in 1909.

As for volcanology in 1912, a fundamental distinction had been drawn between fissure and central-vent eruptions and among Hawaiian, Strombolian, Vulcanian, and Peléan eruption styles, and the importance of gas release for volcanic behavior was widely appreciated. It was generally acknowledged that vesiculation, fragmentation, and production of pyroclastic deposits owed to exsolution of water-dominated gas during shallow magma ascent. Harker emphasized the role of elevated viscosity for eruption of plugs, domes, and spines, and it was widely known that incandescent clouds (including pyroclastic density currents) are sometimes associated with such viscous extrusions. But there was little appreciation yet for the scale, abundance, nature, origin, or lithologic variability of ignimbrites.

Large-scale stoping and assimilation of the sunken blocks at depth, as advocated by Daly (1914), was an idea adopted by Griggs for caldera formation at Mount Katmai. In the perennial debate about the relative importance of magmatic differentiation versus assimilation, Daly had strongly favored the latter. He nonetheless postulated that differentiation accompanies and continues long after main-stage assimilation, and he accepted roles for fractional crystallization, assimilation of felsic wall-rock melts that ascend buoyantly within a chamber, expulsion of silicic melts from crystal mush, gaseous transport of some components, and liquid immiscibilty. Despite his broad perspectives, like others in his time, Daly greatly underestimated the abundance of silicic volcanic rocks, and he thus proposed that most volcanic rocks are mafic and most intrusive rocks granitoid. Noting that some large continental volcanoes are mafic, along with most oceanic volcanoes and most fissure-fed flood lavas, Daly inferred that basalt is the universal primitive magma and everything else derivative. Daly was also an advocate of the importance of sills and thus of lateral transport of magmas, perhaps in this respect influencing Fenner's interpretation of the VTTS.

Pyroclastic density currents (pyroclastic surges and pyroclastic flows) were poorly understood during the early investigations at Katmai. Principal comparators at the time were the small-volume eruptions of May 1902—andesitic scoria flows generated by marginal collapse of a vertical column from the open crater of the Soufrière of St. Vincent (Anderson and Flett, 1903) and the nuée ardente generated by a sector-confined, laterally directed blast from an andesite dome in the crater of Mont Pelée, Martinique (Lacroix, 1904). At Krakatau, a more appropriate comparator for the 1912 eruption, most of the 1883 flow deposits were underwater, and prevalent notions about the 1883 eruption (Verbeek, 1885) emphasized only its fall deposits, as was then likewise true for the 79 C.E. eruption of Vesuvius. Not until the 1930s did the volcanological community begin to appreciate the scale and eruptive style of ignimbrites and their abundance in continental volcanic belts worldwide.

Calderas also remained poorly understood, and there was no consensus on a collapse origin before the overviews by Reck (1928, 1936) and Williams (1941). Although Fouqué (1879) had made a strong case for subsidence of Santorini caldera, as had Verbeek (1885) for Krakatau, many still interpreted calderas as explosively excavated craters well into the 20th century. As late as 1914, Iddings wrote that Krakatau had "blown up" in 1883. Tambora was a remote jungle island, little investigated. Clough and others (1909) had shown Glen Coe to be a subsided cylinder within a steep ring fault with felsic ring dikes, and they had compared it to the Askja caldera of similar size. For Mount Katmai, Griggs and Fenner thought partly in terms of explosive shattering of the edifice, although Griggs also speculated about magmatic stoping. Fenner's recognition that the lithic ejecta were far from sufficient to match Katmai's caldera volume may have helped lead to his radical hypothesis that the missing mass was melted, dissolved, or assimilated by the 1912 rhyolite magma.

Exothermic chemical reactions postulated to attend gas exsolution and shallow oxidation of magmas had a history dating back to Newton, so Fenner was hardly alone in proposing such reactions as the source of the excess heat that should be required for the 1912 rhyolitic magma to have assimilated solid andesite. Just earlier, Day and Shepherd (1913) had argued that such reactions made Kilauean magmas hotter, and Iddings (1914) agreed that they could be a "source of enormous volcanic energy." Even Verbeek had suggested that "internal fusion," wholesale remelting of the interior of the volcanic edifice, may have prepared Krakatau for caldera collapse.

Students of ore deposits and metallogenesis were by that decade generally agreed that most metallic ores were deposited by fluids, whether deep or shallow or at high or low temperature, that were heated by and in part emanated from igneous intrusions (Lindgren, 1913). Debate centered on what fractions of the fluids were meteoric versus magmatic and what fractions of the metals were recycled versus magmatically juvenile. Such uncertainties rendered the work of Allen and Zies on the VTTS fumarolic gases and precipitates of exceptional interest.

Institutionally, no national or state equivalent of the National Science Foundation existed in 1912, and university research budgets were lean or lacking. The Geophysical Laboratory was established in 1905 and the Hawaiian Volcano Observatory in 1911-12, organized and funded by exceptional individuals. The USGS had for 30 years supported research in geophysics and geochemistry, but on a modest scale without secure Congressional funding. Perret's monitoring of eruptions at Pelée and Vesuvius were largely self-supported. It fell to the Research Committee of the National Geographic Society to sponsor Martin's trip to Katmai in 1912 and the Katmai expeditions led by Griggs in 1915–1919.

Geologic Setting of the Katmai Volcano Cluster

The Quaternary volcanic chain of Katmai National Park is the most tightly spaced line of stratovolcanoes in Alaska (figs. 5, 6). Mount Katmai lies midway along a 95-km-long linear reach of the arc that extends from Devils Desk to Kejulik Volcano. Crater-to-crater spacing between adjacent (commonly contiguous) edifices here is typically 5 km or less. The frontal volcanoes of the arc segment are constructed along the preexisting rangecrest (the Shelikof Strait-Bristol Bay drainage divide), where the prevolcanic basement reaches elevations of 1,200 to 1,600 m. Because the volcanic summits reach 1,830 to 2,290 m and lie in a region of high precipitation, all centers on the main chain are extensively ice-covered, as (to a lesser degree) is 2,330-m Mount Griggs, highest peak in the region but 12 km northwest of the frontal axis. Several rear-arc volcanoes lie 11 to 22 km behind the front (Hildreth and others, 2004), all of them monogenetic except Mount Griggs.

Beneath the Katmai volcanic cluster and adjacent to the VTTS, exposed basement rocks belong exclusively to the Late Jurassic Naknek Formation, which here averages 1,700–2,000 m thick (Detterman and others, 1996). The Naknek Formation strata consist here of subhorizontal to gently warped marine siltstone, arkosic sandstone, and lesser lithic wacke and conglomerate that represent marine shelf, submarine fan-delta, and channelized prodelta slope facies of a forearc basin. Ranging in age from Oxfordian to Tithonian, they are thought to have been deposited synorogenically during crustal shortening and exhumation of the adjacent Jurassic batholith (Mullen, 1987; Trop and others, 2005). Although underlain regionally by an additional 3,500 m of Triassic and Jurassic sedimentary strata (Detterman and others, 1996), the Naknek Formation supplied all the nonvolcanic lithic fragments ejected at Novarupta in 1912, including sparse granitoid cobbles from its conglomerate members. Between Naknek Lake and Shelikof Strait, the Naknek strata are intruded by about 20 dioritic to granodioritic Tertiary plutons (Riehle and others,1993).

The Mesozoic sedimentary rocks upon which the volcanic centers are built belong to a tectonostratigraphic package commonly called the Peninsular terrane, which is thought to have originated far to the south and to have been added to southern Alaska in middle or late Mesozoic time (Plafker and others,1994). The Mesozoic strata are altogether as thick as 8,500 m and range in age from Late Triassic to Late Cretaceous; their base is nowhere exposed, although an islet of Permian limestone crops out near Puale Bay (Detterman and others, 1987, 1996). Tertiary volcanic-arc rocks have been eroded from the highland beneath the Katmai volcanic chain, but they are preserved outboard along Shelikof Strait as an 800-m-thick section (Riehle and others, 1993) ~30 km wide that extends 60 km northeast from the Barrier Range at Katmai River to beyond Kukak Bay.

Crustal structure in the Katmai region is poorly known, and the nature of the basement under the subhorizontal Mesozoic marine section south of the Bruin Bay fault (fig. 6) remains particularly uncertain. Alaska Peninsula crust is thought to be 30–36 km thick farther to the northeast (Moore and others, 1991), as well as near the southwestern extremity of the peninsula (Fliedner and Klemperer, 2000). Although the forearc and the subarc lower crust may be unusually mafic, having atypically high seismic velocities for normal continental crust (Fliedner and Klemperer, 2000), the many Tertiary granitoid plutons in the Katmai district indicate that much of the subarc crust here has become effectively continental.

The Aleutian trench (5–6 km below sea level) lies 350 km southeast of the Katmai volcanic front. Eocene seafloor is now being subducted there, nearly orthogonally, at a convergence rate of 6 cm/yr. The inclined seismic zone is well defined and 20 to 30 km thick (Kienle and others, 1983; Page and others, 1991). It dips only about 10° NNW for some 300 km, then steepens to about 45° beneath the present-day arc, and persists behind the arc front to depths as great as 200 km. The top of the seismic zone lies about 100 km beneath the arc front near Mount Katmai, but it shallows northeastward to ~85 km beneath Devils Desk, and it may deepen slightly along the front to the southwest (Kienle and others, 1983).

Chapter 3

1912 Eruption Narrative: What was Actually Recorded in 1912

Before discussing the history of investigation and interpretation of the 1912 eruption and our own detailed characterization of the deposits and eruptive processes, it is appropriate to summarize what was actually reported by observers and instruments in 1912. All times cited are local Alaskan time in 1912. We call attention to what appear to be uncertainties, discrepancies, or yarns. Primary sources are largely Perry (1912) and Martin (1913), who interviewed and quoted many observers.

Chronology of Observations in 1912

1 June

Seismicity began? Martin (1913) stated (without citing his source) that earthquakes were felt at Katmai village "for at least five days prior to the eruption." As there were no seismic instruments within 1,000 km of Katmai in 1912, no earthquakes were recorded instrumentally until the first of many at 0508 on 6 June.

4 June

Severe earthquakes were felt at Kanatak, Uyak, Nushagak, and Katmai village. The only two families (six people) present at Katmai village evacuated, paddling down the coast toward Cold Bay (now called Puale Bay). It was inferred that their departure was prompted by the earthquakes. By June, most Katmai residents were away at Kaflia Bay and Cold Bay, fishing either privately or as employees of commercial fish camps. Most were probably at Cold Bay, explaining why the six evacuees headed that way two days before the eruption.

5 June

Severe earthquakes were felt at Kanatak, Uyak, and Nushagak.

6 June

Morning.—Explosions were reported, but only at Nushagak and Seldovia. At Seldovia they later became louder and more frequent, continuing all day. At least eight earthquakes (M_s 4.8–5.5) were instrumentally recorded remotely (and relocated by Abe, 1992) in the interval between 0508 and 1229, before the first sighting of the eruption plume at 1300.

1300.—A great explosion and earthquake were reported at Cold (Puale) Bay. The mail steamer *Dora* first viewed the eruption cloud at 1300, immediately after hearing a detonation. Captain C.B. McMullen reported that the *Dora* had left Uyak at 0830 and steamed up Shelikof Strait "in fine clear weather," seeing no plume until entering Kupreanof Strait at 1300, when he sighted a "heavy cloud of smoke" over Mount Katmai 55 miles (88 km) astern (Martin, 1913). First seen was an immense column ~1 km in diameter that was soon obscured by the dark ash cloud expanding above it. The ash cloud was overhead by 1500 (or 1600), ashfall began on the ship ~1800, and the vessel was enveloped in darkness soon after 1830. The ash raised the temperature on deck perceptibly, and the heavy, sultry, dust-filled air became stifling. Rather than risk landing at Kodiak, McMullen headed out to sea. He said that thunder and lightning had begun early in the afternoon and that helpless "birds of all species kept falling on the deck," floundering and crying wildly.

1500.—A great explosion was heard for hundreds of miles around, as far as Juneau and Fairbanks. Detonations were first noted at 1900 at Cordova but as early as 1430 at Chicken, where they continued until 0600 on 9 June, sometimes in quick succession, sometimes at longer intervals. Griggs wrote that the three most violent of many reported explosions took place at 1500 and 2300 on 6 June and at 2240 on 7 June. The most distant reports were from Dawson (1,050 km) and Juneau (1,200 km). No explosions, however, were recorded at Kodiak, even during the interval before arrival of the ash cloud in late afternoon, after which thunder might easily have been confused with detonations.

1530.—Ashfall began at Uyak.

1600.—Captain K.W. Perry, aboard U.S. Revenue Service cutter *Manning*, already moored at Kodiak in clear weather, "observed a peculiar-looking cloud slowly rising to the southward and westward." Distant thunder was heard soon after.

1700.—Light ash began falling at Kodiak and grew more severe by 1800, gradually increasing. Perry (1912) wrote that the "cloud bank had spread past the zenith when I observed another bank to the northward, and the two met about 30 degrees above the northern and eastern horizon." There may thus initially have been two separate ash clouds spreading downwind at different altitudes on slightly different azimuths.

1900.—"A black night had settled down" at Kodiak, although long before sunset (Perry, 1912). Thunder and lightning had become frequent and at times very intense (unnerving to local people because normally rare in southern Alaska). Ball lightning was reported near the ground. Wireless apparatus on the *Manning* and at Woody Island naval wireless station nearby were rendered useless by the electrical conditions (severe static). Thunder and lightning diminished about midnight but continued intermittently until 8 June, when the fallout ceased. Sometime during the ashfall, the Woody Island station was destroyed by fire, presumed to have been caused by lightning.

2356.—The first major earthquake (M_s 6.5) was recorded, following the first 11 hours of the eruption, during which at least eight instrumentally recorded events (M_s 4.8–5.8) had taken place (1325 to 2253 on 6 June). This was soon followed at 0035 and 0223 (7 June) by earthquakes of magnitude (M_s) 6.3 and 6.1, respectively.

The two native families who evacuated Katmai village on 4 June were camped ashore somewhere near Cape Kubugakli at the start of the eruption. When they arrived at Cold Bay, people there saw rice-size pumice in their boat, and they were reported (third-hand) to have seen "the top of Katmai Mountain blown off." If taken literally, this seems unlikely, because of the viewpoint from their campsite, the ashfall dispersal from Novarupta directly across Mount Katmai, and the timing of incremental collapse-related seismicity.

A fourth-hand report that reached George Martin months later said that native residents of Savonoski had paddled away at the start of the eruption and, on reaching Naknek, reported "the upper half of Katmai Mountain gone." The timing of their trip is poorly known, as is the location from which they might have seen Mount Katmai, which is not visible from Savonoski. How much of this testimony was later embellishment has never been clear. At Naknek in 1918, Paul Hagelbarger interviewed a Savonoski leader, who said that, in response to the earthquakes, he had been returning to Savonoski from a camp farther up the Ukak River when the eruption began. He spoke of hot ashfall and darkness while paddling down the lake to Naknek. His statement, later set in quotes by Griggs, that "fire came down the trail from Katmai" is likely to have been an inflated yarn. Decades later, his wife, who was born and raised in Savonoski, said most of the residents were already away working at the canneries; only a few remained at Savonoski, and she did not confirm that her husband, Petr Kayagvak ("American Pete"), had been among them. In 1923, Fenner met a prospector who said he had been at Savonoski when the eruption started, earthquakes were first felt, explosions were first heard, dust and darkness enveloped them, and the natives paddled away—all alleged to happen on the same morning. Inconsistencies with the regional record, conflict with other testimony, and conflation of all these phenomena to a single hour ("immediately") make his story appear to be a latter-day concoction. In 1975, a National Park Service ranger interviewed a 68-year-old Perryville man, who had lived as a child at Katmai village and had been at Kaflia Bay during the eruption; he stated that the Savonoski people (who included his relatives) had evacuated because of the earthquakes and that "they all move down to Naknek. They all ran away. Nobody left in Savonoski during that eruption" (National Park Service, 1975).

7 June

0300.— Ashfall began at Port Graham and Seldovia at the southwest end of the Kenai Peninsula. Light fallout continued there all day but without darkness, the main dispersal sector being farther south.

Dawn.—Ash was still falling at Kodiak. Out at sea, the *Dora* "cleared the black smoke" at 0430, "emerging into a fiery red haze, which turned into yellow," and by 0600 ashfall ceased and the "horizon was perfectly clear from west to north" (Martin, 1913), although the ash cloud still filled the sky to the south and southwest.

Morning.—No ashfall had yet been reported (upwind) at Cold Bay, but "increasing noise of thunder and explosions, together with the growing violence of the earthquakes" drew attention to the great ash cloud to the northeast. Father Patelin at Kanatak (on Portage Bay) nonetheless reported dustfall in the morning and earthquakes nearly all day. Instrumentally recorded earthquakes before noon included a M_s 6.5 event at 0825 and two others, M_s 5.1 and 5.8.

0910.—After gradually diminishing since 0300, ashfall ceased at Kodiak. About 5 inches (13 cm) of ash had fallen there, forming the lower gray, coarsest layer (of three), fouling all the streams and wells, and prompting the *Manning* and another ship to furnish water to the local inhabitants.

1000.—People from Kodiak took shelter on the *Manning* and in a warehouse on the wharf, the number rising the next day to 835; others arrived later from vessels, a salmon cannery, and Woody Island.

1200.—Ashfall resumed at Kodiak, increasing steadily in density until it was pitch dark by 1400, remaining so for the next 24 hours. Perry (1912) wrote that the ash now was "fine dust and flakes of a yellowish color," and that sulphurous fumes were at times in the air. Avalanches of ash on nearby hills could be heard and "sent forth clouds of suffocating dust." Crews worked vigorously with shovels and fire mains to clear the ship's decks.

Noon.—Ashfall started near Cordova.

Evening.—At Kanatak, Patelin reported earthquakes so severe that barabaras were abandoned for tents and that the strongest earthquake yet was felt sometime after 2200, accompanied by heavy rumbling, rockslides all around, and a strong glare of light from the volcano. After midnight, there was a big noise like thunder, after which he went to sleep.

2240.—Severe earthquakes were felt at Cold Bay. At Iliamna Bay, earthquakes persisted all night. The instrumentally recorded earthquakes relocated by Abe (1992) include events of M_s 6.0 at 2011, M_s 6.3 at 2049, M_s 7.0 at 2136, and M_s 6.8 at 2248, as well as 11 others between M_s 5.4 and 5.9

scattered between 1247 and 2029 on 7 June. These were soon followed by shocks of M_s 6.2 at 0029 and M_s 6.6 at 0300 on 8 June, after which there was an apparent seismic lull of ~18 hours until the next instrumentally recorded earthquake (M_s 5.4) at 2057 on 8 June.

8 June

0330.—The *Dora* reentered a "bank of volcanic ash" en route to Seward, having stopped under clear conditions at Seldovia at 2000 and Homer at 2300 the previous evening; by 0700 there was complete darkness again, which continued until 1600, when the ship emerged from the ashfall into "brilliant clear weather" (Martin, 1913).

Morning.—Two hours of darkness were reported in lower Cook Inlet, and ashfall continued all day at Seldovia, much heavier than the light ashfall there on 7 June. It was still dark at Kodiak.

1430.—Ashfall decreased at Kodiak, skies assumed a reddish color, and objects became dimly visible. Perry (1912) wrote that "seismic disturbances were still felt." Ashfall "almost ceased" by late afternoon, and to this lull Martin attributed the top of the second ash layer, 4.5 inches (11.5 cm) thick and both browner and finer grained than the basal layer.

Evening.—Ashfall resumed at Kodiak, depositing the third and final ash layer, 1.5 inches (4 cm) thick, very fine grained, and light gray, and continued until an unrecorded hour that night. Following an 18-hour seismic lull, earthquakes resumed (less energetically) with instrumentally recorded events of M_s 5.4 at 2057, M_s 5.1 at 2145, and M_s 5.6 at 2221.

9 June

0300.—A "dense cloud of dust came slowly" into Seldovia from the south. Ash fell from 0500 until nearly noon, causing a few hours of partial darkness. Accompanied by sulphurous fumes, fallout was heavier at Seldovia than on the previous days, "about ¾ inch accumulating on the ground." Intermittent explosions (or thunder?) continued to be heard there from the 9th through the 14th of June. Ashfall persisted at Seldovia until 13 June.

Dawn.—Ashfall had ceased at Kodiak and did not resume. Perry (1912) wrote that during "the 9th, 10th, and 11th ...some substance was held in suspension, and at times the most unpleasant and strangling gases filled the air, making it difficult to breathe." Wilbur J. Erskine, Alaska Commercial Company successor in charge of Kodiak commerce, took many photographs during the ensuing week—of the ash cover, slides and fans of remobilized wet ash, and collapsed or dislodged buildings.

Morning.—An "illuminated" funnel-shaped cloud was seen from Naknek and Iliamna, rising straight up before spreading into cloud banks at altitude. Only three instrumentally recorded earthquakes on 9 June were relocated to the Katmai area by Abe (1992)—events of M_s 6.0 at 0714, M_s 5.7 at 1138, and M_s 5.7 at 1210.

Evening.—Anne Olsen, an elderly refugee already suffering from tuberculosis and nephritis, died aboard the *Manning* in Kodiak harbor, the only known death attributable to the 1912 eruption (Hussey, 1971).

10 June

Three instrumentally recorded earthquakes took place here on 10 June—events of M_s 5.1 at 0142, M_s 4.8 at 0200, and M_s 6.9 at 0606. Light ashfall was noted at noon at Tatoosh Island and also that day at Olga (both in Washington state).

11 June

Dustfall was noted at Dawson, Vancouver, Victoria, and Puget Sound. Atmospheric effects (haze, smoke, red twilights) had already been observed downwind, beginning in British Columbia on 6 June, Wisconsin on 8 June, Virginia on 10 June, Algeria on 19 June, California on 21 June, and at various European locations starting 20–27 June (Kimball, 1913). Volz (1975a, b) later determined that mean atmospheric residence time of the aerosol was 9–12 months and that the main fraction that persisted for a year in the atmosphere was at altitudes of 15–20 km. In Seldovia and Cordova, acid rain caused painful burns, killed vegetation, and corroded metal on buildings; it tarnished brass and ruined linens on clotheslines as far away as Victoria and Puget Sound. Only one earthquake (M_s 6.1) was recorded at Katmai on 11 June (Abe, 1992).

12 June

About 2 cm of ash accumulated at Naknek, where (owing to wind direction) skies had remained clear during the eruption (Dumond, 2005). This and a second ashfall episode at Naknek on 15 June probably resulted from westward spreading of fine ash held aloft, perhaps augmented by ash associated with dome emplacement at Novarupta or Katmai caldera, phreatic plumes from the Katmai caldera floor or through the VTTS ignimbrite, or convective remobilization from ash-covered surfaces. Aeolian remobilization of fines from the VTTS during dry spells continues, a century later, to loft occasional dust clouds to elevations of thousands of meters. In 1912, dustfall was noted as far west as Nushagak later in the summer.

Summer

After the eruption, earthquakes were felt at Cold (Puale) Bay on 50 of the next 70 days through mid-August. According to Martin, they were "heaviest" on 8, 11, 13, 17, 21, 22, and 28 June and on 4, 16, 21, 23, 24, 30, and 31 July. The most severe shocks felt there were said to be on 11 and 21 June and 30 July; the only one instrumentally recorded remotely, however, was that of 17 June, a M_s 6.2 event at 0128. This was the final seismic event Abe (1992) was able to relocate to the Katmai area from the instrumental record. Dust remained in the air at Cold Bay until 24 June, at times falling in sufficient amount to be seen on the ground. Griggs quoted the diary of C.L. Boudry, cannery agent at Cold Bay, who wrote of smoke, fumes, tremor, and shaking lasting all summer and an occasional roar or red clouds over the volcano. The earthquake at 1145 on 21 July was "hard enough to throw hat off table" and lasted 1–1.5 minutes. The same day

Boudry wrote of "smoke from new crater like steam escaping or else light yellow" and "not a steady flow but puff every 2 or 3 minutes." He wrote (on 23 November) that he had tried three times to go to the volcano but was turned back each time by smoke and acid fumes that burned his hands and eyes. He added that dogs in Cold Bay were blinded by the acid.

12 August

From Takli Island, Martin (1913) saw "an immense column of steam ascending through the ordinary clouds" above the position of Mount Katmai (which was itself always in cloud during his expedition). While anchored there, he heard "an almost continuous roar."

15 August

At the mouth of Katmai River, Martin (1913) saw the whole valley north of Katmai village "enveloped in a blue haze," which by noon became so thick that visibility was limited. Acid rain stung the eyes and tarnished brass and silver. There were no earthquakes nor sounds from the volcanoes. Although even some of the trees had been killed at Katmai village, Martin encountered three dogs there that had survived the eruption and the summer-long desolation.

17 August

A sharp earthquake at Naknek upset lamps on the table.

20 September

Another strong earthquake was felt by Boudry at Cold Bay.

Environmental Impact of the Eruption

Beneath and adjacent to the ignimbrite in the VTTS and Mageik Creek, all living things were killed. Over a vast area, especially downwind to the southeast, pumiceous fallout blanketed shrubs and lesser vegetation, killing some of it right away. Trees, in general, were not killed by the fallout directly, but many subsequently died when partially buried by pumice-rich alluvium, especially along the Katmai and Ukak Rivers (figs. 5, 6). Where fallout was less than ~1 m thick, many plant species were recovering well by 1915, as reported by Griggs (1917, 1918d) along lower Katmai River and Soluka Creek (figs. 5, 6), as well as on Kodiak Island.

Intertidal organisms like barnacles, mussels, and kelp were killed from Katmai Bay to Afognak Island, in part by the abrasive action of floating pumice (Rigg, 1914). Land mammals were blinded by ash and acid rain and slowly starved, while birds and insects were exterminated outright. People on Kodiak Island suffered sore eyes and respiratory distress, and a few buildings collapsed (Griggs, 1918a, 1922); water was temporarily undrinkable. Seasonal settlements along the Katmai coast were evacuated and permanently abandoned, as were the villages of Katmai and Savonoski. Salmon populations dwindled in the principal fallout sector because of starvation, suffocation, and failure to spawn in ash-choked streams; because most salmon return to spawn in their fifth year, the run was lowest in 1916–1917, but it recovered well after 1920 (Evermann, 1914; Eicher and Rounsefell, 1957). Acid aerosol remained perceptible in many Alaskan settlements for months, and the acid concentration in the 1913 ice layer on the Greenland icecap remained two-thirds as elevated as in that of 1912 (Hammer and others, 1980).

Truly remarkable is the fact that the greatest eruption of the 20th century, as far as we know, killed no humans directly. A woman already suffering from tuberculosis and nephritis passed away aboard the *Manning* at Kodiak on 9 June (Hussey, 1971), her condition probably exacerbated by ash-induced respiratory distress. Fewer than 200 people were displaced from their homes by the eruption, as the resident populations of the Savonoski area and the Katmai coast had each diminished to less than 100 by 1912 (Dumond, 2005). These impacts can be compared to those of the 1991 eruption of Pinatubo, which was comparable in size and style to that of 1912—several hundred people dead and injured, mostly from roof collapse or lahars; >8,000 houses destroyed and ~75,000 damaged; 50,000 people displaced and the livelihoods of two million (from 364 villages) severely impacted (Newhall and Punongbayan, 1996).

Recovery of Flora and Fauna

The ashfall generally did not kill underground parts of plants, so where fallout was thin or eroded, antecedent plants recovered quickly. True colonization of sterile inorganic ash surfaces, whether fallout or ignimbrite, was a far slower and site-restricted process. Around fumaroles in the VTTS, mosses and algae were conspicuous pioneers in 1917–1919, but they died of dessication or sandblast after extinction of the fumaroles. In 1919, most primary ash surfaces remained barren, and the most conspicuous pioneers on disturbed surfaces were lupines on hillslopes and grasses and sedges on ash-choked alluvial flats (Griggs, 1922).

When he returned in 1930, Griggs (1933) found that little had changed except that extensive black mats of liverworts had pioneered some ashy surfaces. Cahalane (1959) reported, however, that most such mats had disappeared by 1953. Of major interest to Griggs was how soil forms from inorganic substrates and how nitrogen-fixation begins. Laboratory study of ash samples brought back in 1930 contained no fungi or actinomycetes, total bacteria counts very low relative to fertile soils, and no bacteria known to fix nitrogen, even beneath the liverwort mats; it was concluded that if any fixation of nitrogen was taking place in the VTTS ash, it could not be attributed to the microflora (Smith and Griggs, 1932).

As of 2012, Griggs' (1933) prediction of rapid revegetation in years to come has not been realized in the VTTS or on fallout-mantled terrain proximal to Novarupta. Farther than several kilometers from Novarupta, revegetation within the heavy fallout sector is unevenly distributed, depending principally on local erosion of the ash mantle, local exposure of pre-1912 substrate, and local stabilization of surfaces subject to aeolian or alluvial remobilization of ash and pumice.

In the VTTS, the principal obstacle to revegetation is persistent windscour and sandblast on the wide-open barren surface of the ignimbrite and on slopes mantled by coarse loose pumice. Along protected gullies and depressions, however, especially where sandy-silty substrate is available (either reworked ash or primary Layer H ashfall), many pioneering plant species have established a tenuous foothold. In 1953–54, Cahalane (1959) counted 35 such species in the upper VTTS, principally willows, sedges and grasses, wood rushes, horsetail, fireweed, mountain sorrel, snow pearlwort, and arctic poppy. In the upper VTTS in 1954, Cahalane surveyed six plots, each about 32 m by 16 m across, their corners marked by aluminum pipes, as a basis for long-term monitoring of revegetation. He photographed, described, and mapped the 1954 vegetation in all of them, three each in the Lethe and Knife Creek arms of the VTTS, but little followup work has been done. In the 1960s, Cameron (1970) sampled ashy surface "soil" at 10 sites in and adjacent to the VTTS; laboratory studies showed that the valley floor remained virtually sterile, with very low contents of organic matter, bacteria, algae, fungi, and nitrogen. As in the microbial ecology of extreme deserts, the limitation against establishing algal-lichen crusts in the VTTS is not moisture but, rather, porosity and instability of the substrate. And always the wind!

Mammals remain rare in the central and upper VTTS. Ground squirrels, hares, red fox, and mice are seen around the margins, and brown bears, wolverines, and moose occasionally cross Katmai Pass, typically in a hurry. By 1986, a porcupine had reached Baked Mountain Hut, and a mouse and a hare were seen at Novarupta, all far from food (and probably from survival).

C.N. Fenner, then 53, wading Mageik Creek en route to climbing the south slope of Mount Katmai to its caldera rim on 23 July 1923. Earlier that month, he and Charles Yori had ascended its west slope and became the only people ever to descend to the caldera floor, where the shallow lake seen in 1916–19 had temporarily been replaced by mudflats. Photo by Charles Yori.

Chapter 4

History of Investigations and Interpretations

We now summarize the 90-odd years of geological and geophysical research concerning the 1912 eruption and the volcanic system that gave rise to it. The chronological account that follows documents the succession of investigations, observations, and interpretations that bear upon the eruption itself and upon its products and the processes that affected them before, during, and after the eruption.

Chronology of Geologic Investigations at Katmai

1898

Josiah Edward Spurr (USGS party leader) left Savonoski village on 14 October, camped two nights in the valley of the later VTTS, crossed Katmai Pass on 16 October, and arrived at Katmai village on 17 October. Although there had been a few written accounts of crossing the pass between 1880 and 1898, Spurr's general description of the geography and pre-eruption conditions were the first by a scientific observer. He noted hot springs at the south toe of Trident and introduced the name Naknek Formation for the principal basement rocks along the VTTS and Naknek Lake. Spurr was accompanied by Oscar Rohn and George Hartman (camp hands) and topographer William Schuyler Post, who made the first topographic map of the route. They were picked up by the steamer *Dora* at Katmai Bay on 31 October.

1908

A topographic sketch, Coast Survey Chart Number 8555, of the Mount Katmai area was made by the U.S. Coast and Geodetic Survey, from a triangulation station at Cape Kubugakli, 15 km southwest of Katmai Bay. It was reproduced in part by Griggs (1922, p. 270–275) and by Fenner (1923, p. 8). There is no known photograph of Mount Katmai earlier than that taken by Griggs (1918e) in 1915.

1912

Eruptions at Novarupta 6–8 June. Initiation of the eruption column was witnessed from the *Dora* on Shelikof Strait in clear weather. Ashfall was reported across North America and as far as the Mediterranean. It was the world's most voluminous eruption of the 20th century and the largest since that of Tambora in 1815.

George Curtis Martin (USGS), sent by the National Geographic Society to investigate the eruption, arrived at Kodiak in early July. Aboard a ship transporting refugees from mainland villages, he contacted most witnesses and compiled comprehensive accounts of the eruption impact, from Cape Douglas to Pualc (then Cold) Bay and on Afognak and Kodiak Islands. Martin cruised the coast closest to Mount Katmai for two weeks in August and visited the site of Katmai village on 13–16 August. The volcano remained in cloud all summer, but "an immense column of steam ascending through the ordinary clouds" was seen above Mount Katmai (Martin, 1913). Martin compiled a regional isopach map of total fallout and described three main layers of distal ashfall that fell during discrete intervals. He observed great fields of floating pumice.

1913

George B. Rigg (U.S. Bureau of Soils) cruised the coasts of Shelikof Strait in June and July, assessing the eruption's impact on kelp and intertidal algae (mainly abrasion by floating pumice); he observed fields of pumice still floating as far as 241 km southwest of Katmai Bay.

Robert F. Griggs (Ohio State University botanist, attached to Riggs' kelp expedition) studied the impact of ash burial upon onshore vegetation, mainly near Kodiak and Takli Bay (Griggs, 1918a, d).

William A. Hesse of Cordova and Mel A. Horner of Seldovia, hiking from Katmai village, failed in a June-July attempt to reach and photograph the eruption site. After defeat on the ash-covered approach to Mount Katmai, they went west to Martin Creek and obtained the first photograph (reproduced on p. 85 of Griggs, 1922) of a previously unmapped volcano, which Griggs later named Mount Martin.

1915

Griggs returned to Kodiak in June, sponsored by the National Geographic Society (NGS) with a $4,000 grant. He had been recruited by the NGS "out of the blue" in early 1915 on the basis of his 1913 revegetation studies (Griggs, 1961). Surprised by rapid recovery of antecedent vegetation, Griggs chose to move his investigation of revegetation processes to more deeply buried areas closer to the volcano. Accompanied by Bentley B. Fulton (entomologist) and Lucius G. Folsom (Kodiak schoolteacher), Griggs landed at Katmai Bay 11 July 1915; they hiked for a week up the east side of Katmai River, crossing ash-choked Soluka Creek with difficulty, naming the Barrier Range (which blocked their view of Mount Katmai), but failing to cross Katmai River (fig. 6) for lack of a rope. Camped 3 km below the mouth of Katmai Canyon, near the base of Mount Katmai, Griggs took the first photograph of beheaded Mount Katmai, noted the Katmai River debris flow (beyond the river), named Trident Volcano and Mount Martin, and noted steam plumes on Martin, Trident, and Mageik volcanoes. The party was picked up at Katmai Bay and returned to Kodiak the first week in August.

1916

Griggs, with Folsom, Donovan B. Church (photographer, Ohio State University student), and Walter Metrokin (packer), returned in July, again hiking up the east side of Katmai River. They based at what they called Island Camp (location shown in fig. 5) in the braided stream and (this trip carrying a rope) waded it below the mouth of Katmai Canyon. Their first ascent to the caldera rim on 19 July was up the south slope via the Katmai River debris flow, which Griggs recognized to postdate the stratified fallout and to contain uncharred wood. The second ascent was on 30 July up the southwest slope from the mouth of Mageik Creek canyon to the south-rim notch. Griggs reported a milky blue lake in the caldera with a horseshoe-shaped island dome, a light-colored boiling upwelling, and many vigorous fumaroles all around the caldera floor margins. There were "two large bright-yellow spots of sulphur" low on the north wall (Griggs, 1922), but there was still no ice on the pair of benches where intracaldera glaciers later accumulated. On 31 July, Griggs and Folsom traversed Observation Mountain and ascended upper Mageik Creek to Katmai Pass (figs. 5, 6). Lured on by fumaroles in the Pass, they discovered and named the "Valley of Ten Thousand Smokes" (VTTS) and hiked toward the lava dome (later named Novarupta). Griggs estimated that ~1,000 of the VTTS fumarolic plumes were more than 500 feet high. He misinterpreted the thin (by then fumarole-free) ignimbrite along Mageik Creek as a mudflow and the prominent crag ("tooth") of rhyodacite lava in the caldera's south-rim notch as a basaltic neck, but he correctly interpreted the Novarupta dome as a plug extruded within an ejecta ring. Lacking supplies for adequate further exploration, they returned to Island Camp, thence to Katmai Bay, waiting there 10 days for a boat to return them on 14 August to Kodiak (where Griggs' family had awaited him for 5 weeks). Griggs began to plan and lobby for a major expedition the following year.

1917

Griggs returned, with a well-supplied party of ten (and a $12,000 grant from the NGS), basing for the first time at the west side of Katmai Bay on 15 June, relaying supplies up the west side of Katmai River via Martin and Mageik Creeks to the VTTS, and establishing three intermediate camps. Griggs crossed Katmai Pass with the packer Metrokin on 4 July and moved the whole party on 8 July to a camp on the bench north of Mount Cerberus (MC on fig. 5), where they remained well into August. Besides returnees Griggs, Folsom, Church, and Metrokin, the party included John Wesley Shipley (chemist), Clarence F. Maynard (topographer), James S. Hine (zoologist), Andrean Yagashof (packer), and Jasper D. Sayre and Paul R. Hagelbarger (botanical assistants from Ohio State University). Because Griggs, oddly, did not include a geologist, little geology was accomplished. In addition to general exploration of the VTTS, gas samples were collected by Shipley for analysis; a fine topographic map was surveyed by Maynard; Fissure Lake (FL on fig. 5) was named and photographed; and Katmai Canyon, Katmai Lakes (there were three at that time), and the Mageik Landslide on Martin Creek were investigated. Maynard, Griggs, and probably others climbed to the caldera rim from Camp IV (fig. 5) near the mouth of Mageik Creek canyon. The party was picked up by boat on 29 August and returned to Kodiak on 30 August.

Based on his 1917 observations and samples, Shipley wrote a (long overlooked) paper, published in 1920, about the fumarolic gases and the composition and mineralogy of their deposits. Unlike Griggs and Fenner, who speculated about hundreds of magmatic vents beneath the VTTS and who waffled about specific sources of the ignimbrite, Shipley stated flatly that the valley-filling flow deposit was "ejected by Novarupta volcano."

1918

Griggs' plan for another large expedition was postponed owing to U.S. entry into World War I. Instead, he sent Sayre and Hagelbarger to pioneer a new route to the VTTS. After reaching Kodiak on 22 May, they shipped aboard the *Dora* via Dutch Harbor to Naknek, arriving 6 June. Ascent of Naknek River by motor boat, troubled by rapids, took them from 9 June until 19 June. After climbing Dumpling Mountain (shown on fig. 6) for reconnaissance on 23 and 25 June, they reached the head of Naknek Lake on 26 June and hiked up the Ukak River to the VTTS on 27–29 June. They walked from their base camp across Katmai Pass all the way to Katmai Bay and back (1–8 July). From 13 July until 5 August, they recorded temperatures of VTTS fumaroles with pyrometers, measuring 86 vents hotter than 190°C and several hundred near the boiling point; 37 exceeded 290°C, 9 exceeded 390°C, and the hottest measured was 432°C. A mercury thermometer carried in 1917 had soon broken. Sayre and Hagelbarger noted a pre-1912 pumice-rich debris-flow deposit along lower Windy Creek (fig. 5), near the site where the 1912 ignimbrite had dammed what they called "Windy Lake" (which lasted only a few years). From 5 August to 17 August, they surveyed the Naknek Lake system and Naknek River while en route back to Naknek; eventually they shipped to San Francisco, arriving 22 September.

Katmai National Monument, to be managed by the newly established National Park Service (NPS), was created by presidential proclamation (under the Antiquities Act) on 24 September 1918, in response to promotion by Griggs (1918b) and officers of the National Geographic Society. It may have helped that Secretary of the Interior Franklin K. Lane was on the NGS Board of Managers. The first "tourist" was Rodney L. Glisan, Portland (Oregon) attorney and a founding director of the American Alpine Club, who accompanied the Griggs expedition in 1919.

1919

Griggs, with a $35,000 appropriation by the Research Committee of the National Geographic Society, invited participation of Geophysical Laboratory scientists and outfitted an expedition of 19 men (fig. 7). Sayre returned via Naknek, assisted by three of Griggs' students (A.J. Basinger, Richard E. Helt, and Harry E. Jacob), and hired Naknek Packing Company to transport the bulk of expedition supplies by motor boats (23 May to 2 June) to a base camp at the head of Naknek Lake. Griggs and a few others landed at Katmai Bay from Kodiak in late May, reestablished several relay camps via Katmai Pass, and reached the VTTS by 1 June. Along with returnees Folsom, Hine, and Paul Hagelbarger, Griggs was accompanied by Clarence Norman Fenner (petrologist; fig. 8), Emanuel George Zies (fig. 9) and Eugene Thomas Allen (chemists), Emery C. Kolb (motion picture photographer), Julius Stone (boatman), Frank I. Jones (color photographer), and five assistants—Ralph Hagelbarger, William L. Henning, August E. Miller, H.M. Wallace, and Charles Yori. Griggs, Stone, and Miller joined Sayre at base camp near Savonoski on 9 June. New camps were set up at the northeast edge of the VTTS ("Ukak Camp") and on the exposed bench at the southwest toe of Baked Mountain (fig. 5).

The Geophysical Laboratory contingent arrived at Katmai Bay on 13 June. Allen and Zies crossed Katmai Pass 16 June, reached Savonoski Camp 19 June, set up a field laboratory there 20–26 June for gas analysis and microscopy, and studied fumaroles in the lower VTTS from Ukak Camp from 27 June to 28 July. They moved to Baked Mountain Camp on 29 July to work on fumaroles in the upper VTTS but lost 16 of 25 days in August to stormy weather, leaving the VTTS frustrated on 25 August and reaching Katmai Bay on 29 August. Fenner spent the period 30 July to 12 August

Figure 7. Participants in 1919 Katmai expedition, at Kodiak. From left, Richard Helt, Jasper Sayre, Paul Hagelbarger, Emery Kolb, Charles Yori, Robert Fiske Griggs, Lucius Folsom, Clarence Norman Fenner, Eugene Thomas Allen (swatting mosquito), August Miller, Emanuel George Zies, J.S. Hine, H.M. Wallace, A.J. Basinger, Julius Stone, and William Henning. Members not in photograph were Ralph Hagelbarger, Harry E. Jacob, and Frank Jones. Roles of each were mentioned by Griggs (1922). Photo courtesy of National Geographic Society.

at Camp IV, studying the south side of Mount Katmai and ascending to the caldera rim. Expedition accomplishments included measuring several fumarole temperatures in the range 500°C–645°C; collection of 18 gas samples and numerous samples of rocks and fumarolic incrustations; ascent to the crater rim of Mount Martin by Yori, Fenner, and Griggs on 16 August from Baked Mountain camp; and further topographic surveys of the Naknek Lake system and of a potential tourist route from Geographic Harbor across Hagelbarger Pass (fig. 6). Fenner noted that stream gorges incised into the ignimbrite were already nearly 100 feet deep. Griggs noted slackening, cooling, or disappearance of many 1917 fumaroles and conversion of many former steam vents in the lower VTTS to boiling springs. Huts were built near Katmai Bay, Savonoski, and Brooks River (fig. 6). A movie outfit of two actors, an actress, and three packers arrived via Kodiak and Katmai Bay and spent a July week filming in the upper VTTS. The wives of Folsom, Kolb, and Griggs (Laura Amelia Tressel) came over from Kodiak to the VTTS in July and August (Griggs, 1961). Several participants, including Mrs. Griggs, Fenner, Hine, Basinger, and Sayre, reached the caldera rim from Camp IV. Between 26 August and 7 September, most were shuttled from Katmai Bay to Kodiak in three trips, but a few others returned via Naknek. Ever the promoter, as the fumaroles faded, Griggs (1921) increasingly advocated tourism, scenery, salmon, and wildlife.

1920

Fenner rejected Griggs' idea that the 1912 ignimbrite was a "great hot mud flow" and proposed instead that it had been a hot dry "ash- or sand-flow" analogous to *nuées ardentes* (Fenner, 1920, 1923). He argued, however, that fissure vents for the ejecta (and for the subsequently persistent fumaroles) issued from a sill that extended under the whole VTTS. Novarupta, he suggested, was similar to the spine of Pelée, representing stiff degassed magma later extruded from just one of the supposedly numerous earlier fissure vents. For Katmai caldera, he saw that lithic ejecta represented only a small fraction of the missing volume, and he granted that subsidence might have played a part, but he radically attributed

Figure 8. Clarence Norman Fenner at Katmai, summer 1923, beside Yori's pack. Fenner marked his 53rd birthday at Katmai that summer when he and Yori spent two months alone together and altogether three months in the Katmai region. Photo by Charles Yori.

Figure 9. Emanuel George Zies, 17 August 1919, at site of storm-wrecked camp on western bench of Baked Mountain. Photo courtesy of National Geographic Society.

formation of the crater principally to assimilation of older rocks by superheated rhyolitic magma, thermally aided by exothermic gas reactions.

1922

Publication of Griggs' well-illustrated book "The Valley of Ten Thousand Smokes" by the National Geographic Society.

1923

Fenner (then 53) and Yori shipped to Kanatak on Portage Bay in early June, accompanying a USGS party, which lent them two packers, a dozen horses, and at least two dogs for transport to the VTTS. Via Kejulik Pass and Yori Pass (fig. 6), the packtrain reached the VTTS on 23 June and left them alone for two months at a well-stocked camp at the north toe of Mount Cerberus. They made the first ascent from the VTTS side of the range to the Katmai caldera rim on 6 July, traversing high from Novarupta across three Knife Creek Glaciers to the west-rim notch. Descending a snow-covered talus chute from the notch, they became the only people ever to stand on the caldera floor, where the shallow lake seen in 1916–19 had been replaced by mudflats. They photographed boiling mudpots, a large mud geyser, and myriad fumaroles around the base of the wall (Fenner, 1930), and they sampled the horseshoe-island dacite lava dome. On 14 July they climbed to the crater rim of Mount Mageik in poor weather. They camped for a week at Katmai River and climbed to the south rim of Katmai caldera on 23 July, accompanying a party of at least 11 tourists that included several women. They reclimbed Mageik on 4 August, when they were able to descend to its crater floor, site of roaring fumaroles and small pools of mud but at that time no crater lake. Yori made the first ascent of Knife Peak (now Mount Griggs) on 20 August. Fenner made many regional observations on intrusive rocks and undeformed Mesozoic sedimentary rocks that dominate the arc basement in this part of the Alaska Peninsula. Fissure Lake (FL on fig. 5) had by then drained, but a glacier still occupied the later site of West Mageik Lake. The packtrain picked them up at their Cerberus camp on 31 August.

Hundreds of the 1919 fumaroles were either dead or had cooled to near the boiling point, and most had vanished from the upper Knife Creek arm. In their photographs, ~50 steam plumes survived in the upper Lethe arm and nearly as many in the middle VTTS but only a few downvalley from Three Forks. Four discrete plumes were active along the western bench of Baked Mountain. The floor of Novarupta basin was steaming densely as far west as what we now recognize as the vent margin north of the nose of Falling Mountain; the Novarupta dome was steaming vigorously but the Turtle only weakly in a few places. Around Novarupta and in the upper Lethe arm, they found a few surviving superheated fumaroles, and what they identified as the hottest remaining one was along the arcuate graben ~1.2 km northeast of the dome.

The USGS party spent two months preparing a reconnaissance geologic and topographic map from Mount Peulik to Savonoski (fig. 6), using and naming Yori Pass. The party consisted of Walter R. Smith (geologist), R.K. Lynt (topographic engineer), C.S. Franklin (recorder), three camp hands, and 15 horses. Near Savonoski on 28 August, they met another USGS party, consisting of Kirtley F. Mather (geologist), R.H. Sargent (topographic engineer), and four assistants, that had landed at Iliamna Bay off Cook Inlet on 16 June and likewise reconnoitered the area from Kamishak Bay to the Savonoski drainage (fig. 6). After retrieving Fenner and Yori, the two parties together reached Kanatak on 9 September, whence they all shipped to Kodiak.

1923–1929

Allen and Zies published a series of papers describing the composition of the fumarolic gases and the morphology, mineralogy, and chemistry of the vents and deposits. They tabulated 106 temperature measurements for fumaroles in all parts of the VTTS, noting that the hottest were in the central VTTS, with lower temperature vents distally and around the margins. They reported that many fumaroles measured in 1918 had declined in temperature within a year and that hundreds had died out completely. They estimated that less than 10 percent of stream runoff onto the ignimbrite was emerging at Three Forks, leading them to conclude that most of the fumarolic steam originated as surface waters infiltrating the hot tuff, vaporizing, and thus cooling the VTTS rapidly. The temperature of the glacier-fed River Lethe was 80-85°C at Three Forks in 1919.

1929

Bernard R. Hubbard, S.J., "the Glacier Priest," after failing to reach Mount Katmai from Kukak Bay in summer 1928, was dropped at Katmai Bay in late June (Hubbard, 1932, 1935). With four college boys and two dogs, he climbed to the caldera's south rim from the warm springs on Mageik Creek and also spent several days camped in the VTTS at the north toe of Mount Cerberus. The "milky turquoise blue" caldera lake had been restored and was by then deep enough that the Horseshoe Island was submerged. Hubbard reported a "whitish-blue boil" in the center of the lake, sporadic patches and streaks of "vivid yellow," and several dozen fumaroles around the lakeshore. Ice had begun to accumulate on both intracaldera slump-block benches, and Hubbard and Frank Klatt descended a snow slope to the southern one. In the VTTS, only a few hundred fumaroles remained active, nearly all near or below the boiling point; none were left in the lower valley, but the Novarupta lava dome was still steaming vigorously. The party was picked up at Katmai Bay in late July.

1930

Griggs made his final visit to the VTTS, again via Katmai Bay, camping in the upper valley 3–9 July. He sprained his ankle badly in the VTTS and limped painfully all the way back to Lagoon Camp at Katmai Bay (Griggs, 1961). Of a few hundred fumaroles left, none were found to be above the boiling point, but he noted that some were still active on the face of Falling Mountain. He concluded that liverworts (not mosses, bacteria, or algae) were the principal pioneering plants on barren ash-covered surfaces. In a report to the Department of the Interior (22 November 1930), Griggs advocated expansion of the Monument as a wildlife preserve. On 24 April 1931, a presidential proclamation doubled the size of Katmai National Monument.

Griggs was accompanied in 1930 by his 18-year-old son David Tressel Griggs (1911–1974). The trek helped inspire a career in experimental geophysics (at Harvard and UCLA), in which he pioneered studies of rock deformation, hydrolytic weakening, crystal dislocation, creep, oriented fabrics, earthquake mechanisms, and the role of mantle convection in mountain building. After the elder Griggs died in 1962, David climbed Mount Griggs to place his parents' ashes on top in the summer of 1963. His own son added David's ashes up there in 1975.

1932

Hubbard returned via Katmai Bay in March with two lads and a dogsled team. He made a brief trip into the VTTS and climbed to the caldera rim from the south, reporting that the "emerald lake" was unfrozen and that floating icebergs melted quickly. The icebergs calved from a "cascading glacier," which may suggest that the narrow ice tongue on the west wall had originated by then. After waiting for 11 days at Katmai Bay, they were picked up on 16 April.

1934

Hubbard returned via Katmai Bay in September with six lads and four dogs, again climbing to the south rim of Mount Katmai and visiting the VTTS and Novarupta. The caldera lake remained blue and shallow, with no island; Hubbard (1935) mentioned rockfalls and waterfalls but no fumaroles on the caldera walls, and he said the intracaldera glaciers had grown and were calving into the lake. He reported that only "a few feeble" fumaroles were left (compared to hundreds in 1929), mostly in the upper Lethe arm and in Novarupta basin—especially adjacent to the lava dome. None survived in the lower or central VTTS. Most conspicuous was a pair of big steamers on the bench alongside Baked Mountain (down from four in 1923), but the corresponding bench along the west side of the Lethe arm lacked active fumaroles. Many fissures in the upper VTTS that had been open in 1929 were by 1934 filled with windblown pumice and sand. Hubbard noted how deeply the 1912 deposits along Mageik Creek and Katmai Pass had been incised since his 1929 trip. He mentioned the odor of H_2S along Mageik Creek between Trident and Observation Mountain.

1940

Frank T. Been (superintendent of Mount McKinley National Park) and Victor H. Cahalane (Biological Survey, then a bureau in the U.S. Department of Agriculture) spent September reconnoitering by boat and plane the natural resources and the rampant poaching in Katmai National Monument. The National Park Service had never yet had any protective or interpretive personnel based in the Monument. They reported that only about ten fumaroles were still active.

1950

For the first time, the National Park Service stationed a ranger in the Monument, near the mouth of Brooks River (fig. 6) for the summer. By arrangement with the NPS, Northern Consolidated Airlines initiated its first two fly-in fishing camps for tourists.

1951

The Alaska Recreation Survey, organized by the National Park Service, flew from Kodiak around Katmai National Monument for several days in June. The party consisted of George L. Collins, Adolph Murie, Thomas Williams, and Lowell Sumner. They reported seven steam plumes near Baked and Broken Mountains, fumaroles in the crater of Mount Martin and high on Knife Peak, and a "jade-green" color for the caldera lake at Mount Katmai. Their recommendations helped lead to the Katmai Project.

1953

Katmai Project: The National Park Service proposed to the Office of Naval Research comprehensive investigation of Katmai National Monument, in part to help prepare a master plan for management and interpretation. Studies would include geography, climate, glaciers, coastal geomorphology, flora, fauna, entomology, soils, volcanic activity, regional geology, mineral resources, and archaeology. Scientists from the USGS, Public Health Service, and five universities eventually took part. Eruption of Trident, beginning in February, helped draw the attention of the military, which provided rations, rifles, radios, field equipment and clothing, air transport, and even a helicopter. Beginning 19 June, field camps were outfitted on upper Knife Creek (VTTS), Kukak Bay, Brooks Lake, and near the site of Savonoski. Project headquarters on the Air Force Base at King Salmon was open all summer, closing for the season on 7 October.

Trident began erupting explosively on 15 February. Several andesite-dacite lava flows were emplaced between 1953 and 1960, and numerous explosive events persisted as late as 1974, building a new fragmental cone. There were many observations during 1953 (Snyder, 1954), but only occasionally thereafter.

Howel Williams, Garniss Curtis, Ronald W. Kistler, and Rolf Werner Juhle (fig. 10) spent the last two weeks of June, camped on upper Knife Creek, reconnoitering the VTTS, Trident, and Mount Katmai. Correcting interpretations of Fenner and Griggs, they concluded that 1912 vents were not beneath the VTTS but at its head, that 1912 andesitic and rhyolitic magmas erupted simultaneously and mingled, that andesitic magma did not originate by fusion of older rocks by the rhyolitic magma, and that most fumaroles issued from the valley-filling ignimbrite sheet itself, not from an imagined sill or batholith beneath the valley. Williams introduced the place names for the Turtle and Whiskey Ridge (fig. 6). Poking fun at Fenner's ideas, Curtis, Williams, and Juhle (1954) wrote that "No evidence was observed to support the hypothesis of a still under the valley floor."

Juhle and Kistler (then a student at Johns Hopkins, later at Berkeley and the USGS) stayed on (fig. 10), conducting regional geologic mapping, until Juhle was lost into one of the torrent-filled gorges in the lower VTTS on 31 August (Donnay, 1956).

1954

Curtis returned to the Knife Creek campsite in summer 1954, assisted by Jack Sheehan. Fenner and Griggs had assumed that most of the stratified fallout was ejected from Mount Katmai, even though they realized that the thickest and coarsest ejecta surrounded Novarupta. They believed that some subordinate fraction of the fallout had erupted at or near Novarupta, but Curtis keyed on the fact that they had never developed criteria to distinguish products of the two postulated vents. So Curtis subdivided the stratified fallout into eight principal layers, measured 30 sections between Novarupta and Mount Katmai, and showed that all eight thickened toward Novarupta, their true source vent. He was first to propose hydraulic interconnection between the collapsed caldera and the main vent at Novarupta, 10 km away. As for the ignimbrite, Curtis recognized that it had come from Novarupta basin, and he noted its gross compositional zonation, from rhyolite-rich distally to rhyolite-poor in the upper VTTS. Observing blocks of eutaxitic welded tuff ejected phreatically from deep in the ignimbrite and noting the differential compaction across the upper VTTS, Curtis constructed profiles for nearby streams and concluded by comparison that valley-filling ignimbrite in the VTTS might be more than 200 m thick.

A. Samuel Keller and Hillard N. Reiser (USGS) devoted two months of fieldwork to preparing a geologic reconnaissance map of Katmai National Monument (Keller and Reiser, 1959), in collaboration with the NPS as a contribution to the Katmai Project. During July and August, eight field camps were put in for them by float plane supplied by the Air Force, at sites that included Iliuk Arm of Naknek Lake, Geographic Harbor, and Kinak, Kukak, Hallo, and Kaguyak Bays.

Figure 10. Werner Juhle, Howel Williams, and Ronald Kistler during Katmai Project, summer 1953. Photo by Garniss Curtis.

Emphasis was on basement rocks, but knowledge of the volcanic rocks was also extended.

1959

The Board of Geographic Names, on the initiative of the National Geographic Society, changed Knife Peak volcano to Mount Griggs.

1962

Pierre Bordet, G. Marinelli, M. Mittempergher, and Haroun Tazieff, European volcanologists, visited the VTTS in June after a scientific conference in Tokyo (Bordet and others, 1963). Despite other misconceptions, they generally agreed with the 1953 observations of Williams and Curtis, inferring the mixing of separate silicic and intermediate magmas, which they postulated to have been stored respectively beneath Novarupta and Katmai. They were the first to explicitly apply to the VTTS deposit the word "ignimbrite" (which had originated in 1932; Marshall, 1935). Griggs had called it a "hot mud flow" and Fenner a "sand flow." Williams and Curtis had considerable experience with ignimbrites, but Curtis chose to call it a "tuff flow" and Williams a "glowing avalanche deposit."

The National Park Service constructed a dirt road from Brooks River (fig. 6) to Overlook Hill near the VTTS, making access far easier.

1963

Robert W. Decker and Peter L.Ward of Dartmouth College, supported by the Geophysical Institute (University of Alaska), conducted a pilot project in July and August, proving the feasibility of seismology and gravity studies in the VTTS (Decker, 1964). Seismometers installed near Brooks River (fig. 6) and Overlook Hill (fig. 5) recorded 82 small earthquakes in 17 days. Gravity was measured at 35 stations, by boat along the lakes, on foot up the VTTS and through Katmai Pass to Mageik Creek, and by float plane near Dakavak Lake (fig. 6); their results defined a negative Bouguer anomaly of –35 mgal centered on Katmai Pass and flanked by positive anomalies on both sides of the volcanic axis. On 6–7 August, Ward and Decker climbed the new Trident cone, reporting a 150-m-wide crater with abundant fumaroles, most of them hotter than 200°C and many on the crater rim encrusted with sulfur. They sketched a map of the 1953–1960 lava flows below the cone.

1965

Baked Mountain Hut (fig. 5; two contiguous cabins plus storage and generator sheds) was constructed for the Geophysical Institute (GI), University of Alaska, with Air Force helicopter support via Brooks Camp (fig. 6). The hut was built on a pumice-mantled wind-buffeted bench devoid of surface water, 4 km northwest of Novarupta, fairly central to the upper VTTS but an uphill trudge from all directions. Ensuing geophysical research was undertaken during summers of 1965 to 1969.

Eduard Berg and Jürgen Kienle (GI), supported by NSF and the Air Force Office of Scientific Research, measured ~75 new gravity stations along five helicopter-assisted profiles across the volcanic axis, extending the along-axis negative anomaly recognized in 1963 (Berg and others, 1967). The Bouguer gravity minimum centered on Katmai Pass was recalculated to reach –42 mgal.

Dipak Kumar Ray and Robert B. Forbes (University of Alaska) sampled Trident lavas and ejecta of the 1953–1963 eruptions for Ray's petrographic and chemical Ph.D. dissertation (1967). They reported little change in activity or configuration of the Trident cone since 1963. All four lava flows were still steaming.

In the VTTS, weak fumaroles (near or below boiling) remained only on Novarupta dome, on the west rim of its ejecta ring, along faults on and near the Turtle, and on the western bench of Baked Mountain. Fumarolic plumes were reported from Mageik, Martin, and the upper west flank of Mount Griggs.

Seismometers were installed for various intervals during summers of 1965–1967 near Brooks Camp, Overlook Hill, Katmai Canyon, Baked Mountain Hut, Savonoski, and Kodiak (figs. 5, 6). Participants included Berg, Kienle, Tosimatu Matumoto (Lamont), Ward (by then at Lamont), and Susumu Kubota (University of Tokyo). Hundreds of small earthquakes were recorded, nearly all shallower than 10 km, mostly magnitude 0–3, and the majority close to the volcanic axis from Mount Martin to Snowy Mountain. Screening of S-waves and attenuation of high-frequency P-waves on certain azimuths were interpreted as suggesting magma storage beneath the volcanic axis (Berg and others, 1967; Kubota and Berg, 1967; Matumoto and Ward, 1967; Ward and Matumoto, 1967; Matumoto, 1971).

1966

Ray returned to Trident in August, reporting no changes other than continuing growth of a plug in the crater, from ~3 m high in 1965 to ~15 m high in 1966.

Kienle (1970) measured 85 gravity stations along three profiles across the central VTTS in the summers of 1966 and 1967. Uncertainties concerning density contrasts among basement rocks, variably welded ignimbrite, and unknown thicknesses of concealed pre-1912 valley fill led him to present two contrasting models. One model gave maximum tuff thicknesses of ~70 m; the second gave 140–160 m. In 1991, after the great range in VTTS ignimbrite density (ρ=1.0 to 2.1) was explained to him, he elaborated the second model, affirming the thickness estimate Curtis had based on stream profiles.

1967–1969

Attempts to determine thickness of the ignimbrite in the VTTS by seismic refraction profiles were made with a sledgehammer by Matumoto and M.L. Sbar (Lamont). Greatest calculated thickness in the lower Lethe arm was ~70 m; in Novarupta basin, it was thought thicker, but results were too complex to interpret (Sbar and Matumoto, 1972). In 1969, Larry Gedney, Charles Matteson, and Forbes (all University of Alaska) repeated such an experiment using explosives; their maximum inferred thickness was only ~50 m, in the lower Lethe arm (Gedney and others, 1970). Interpretation of such experiments was bedeviled by uncertainties about unexposed (and apparently unsuspected) welding zones in the ignimbrite and by subjacent glacial and alluvial deposits. Kienle (1991) later based his reinterpretations of those refraction profiles on new density data to give recalculated thicknesses of 120–167 m (in the middle VTTS).

1975

Roman J. Motyka (GI) descended the Yori-Fenner chute to Katmai caldera lake in early July and, using an inflatable raft, recorded water temperatures at several sites that averaged 5.5°C at depths of 10–60 m. Measured pH was 2.5–3.0. Odor of H_2S was strong, and a mid-lake yellowish bubbling upwelling contained particulate sulfur. Water samples were taken for chemical analysis. He had already surveyed the lake from the caldera's west rim in August 1974. It was found to have deepened by 182 m between 1923 and drafting of the USGS topographic map in 1951, and by an additional 47 m from 1951 to 1975. Motyka returned to the lake in August 1977 and found that its level had risen by an additional 6.5 m in two years. In 1977 he measured temperatures of 20.5°C at the lake floor and 8.8°C at depths of 10–60 m, and he noted that the lake had not frozen in winter 1975, although it had been photographed as largely frozen over in March 1967 (possibly for the first time).

1976

Wes Hildreth (then a Berkeley Ph.D. student working on ignimbrites, later at the USGS) spent five August weeks in the VTTS, assisted by Larry Jager and joined for a week by Daniel Kosco (U.C. Berkeley), and brought back 150 kg of samples for petrologic study. A single chimney fumarole (49°C) survived on the western bench of Baked Mountain; numerous centimeter-scale orifices with wispy odorless discharge (as hot as 88°C) were located on the Turtle and the northwest ejecta ring of Novarupta; and odorless vapors rose weakly from several spots atop the dome itself. They climbed to the caldera rim twice from the VTTS, observing three glaciers calving icebergs into a pale blue lake with a prominent yellowish upwelling in the middle. Depth of Knife Creek gorge south of Three Forks was measured by rope to be 33 m. Pale gray dacite, rich in crystal clots, was recognized as a magmatic product distinguishable from coerupted andesite and rhyolite. They saw that proportions of the three pumice types changed greatly through the emplacement sequence, which could be defined by numerous apparent flow units in the ignimbrite (some subtle, some compositionally abrupt). Nonwelded in the lower VTTS, exposures of the tuff were found to be widely indurated in the upper VTTS, variously by sintering, partial welding, and vapor-phase crystallization. Phreatic craters had ejected welded, even eutaxitic, blocks from unexposed depths in the tuff. "High sand mark" benches alongside the VTTS were recognized to reflect differential compaction of welded tuff, with lesser contributions by marginal run-up during inflated emplacement and locally by axial drainaway. The eruption-initiating fallout layer of rhyolite pumice was found in Mageik Creek but nowhere under the ignimbrite on the VTTS side of the volcanic line. A pre-1912 debris-flow deposit rich in dacite pumice was sampled at Windy Creek. Relatively low-energy emplacement of the ignimbrite in the lower VTTS was inferred from the way it banked against and flowed around the modest cross-valley moraine near Three Forks. Familiar with zoned ignimbrites elsewhere, Hildreth envisaged that the entire rhyolite-dacite-andesite mixture could have issued from a continuous magma body.

1978

Kosco and David A. Johnston (USGS postdoc) spent two weeks in the VTTS, sampling fumarolic gases at Trident, Mageik, Griggs, and Novarupta dome. Kosco collected lavas from the volcanoes for his Ph.D. work at Berkeley, and Johnston began investigation of the stratified fallout. Mageik crater lake was measured at 70°C with pH ~1.

1979

Hildreth, Johnston, and Anita Grunder spent 22 June until 22 July on foot in the VTTS, following a 10-day helicopter-supported USGS field trip along the Alaska Peninsula, led by Thomas P. Miller and Robert L. Smith. Gas samples were again collected from Trident, Mageik, and Griggs. Temperatures measured for sulfurous fumaroles were 97°C on Trident, 108°C on Griggs, 172°C for a roaring jet on the north shore of Mageik's crater lake, and 135°C for another issuing from talus on the northeast wall. The yellow-green lake itself was 71°C, roiling vigorously, and fringed with yellow sulfurous spindrift. Terry E.C. Keith, assisted by Peter Shearer (both USGS), spent 10 days with the party and initiated new studies of the (now-fossil) fumarolic deposits and the chemistry of spring waters. Pumice-type proportions were counted throughout the ash-flow and fallout deposits in order to document the complex compositional zonation of the 1912 eruptive sequence. Thin tephra layers in and near the lower VTTS were correlated with the meter-scale layers defined by Curtis in the area proximal to Novarupta. The basal rhyolitic pumice-fall layer was confirmed beneath the ignimbrite in lower Mageik Creek.

1980

The Monument was expanded and redesignated "Katmai National Park and Preserve," much of it a Wilderness Area, under the Alaska National Interest Lands Conservation Act (ANILCA).

Hildreth, stunned by the death of Johnston at Mount St. Helens on 18 May, returned to the VTTS for the month of August, assisted by Judith Ellen Fierstein (USGS, on her first backpack trip). A tent was pitched for 10 days on the south fork of upper Knife Creek, to facilitate study of tuff-glacier relations, phreatic craters, and disrupted deposits along the Griggs fork (fig. 5) of Knife Creek. They began systematic inventory of maximum sizes of pumices and lithics in all 1912 subunits. Logging of the complex proximal pyroclastic fall and flow deposits around Novarupta basin was initiated. They mapped phreatic, fluvial, and lake deposits in upper Knife Creek, near West Mageik Lake, and along the trough of former Fissure Lake. They also mapped a flood-margin track down the northeast side of the Knife Creek arm of the VTTS and ash-dominant debris-flow deposits near the VTTS terminus. A 3-day hike to Mageik Creek and Katmai River confirmed that the Katmai River debris-flow deposit (which overlies the stratified 1912 fallout) had engulfed uncharred trees (as noted by Griggs) and is thus a remobilized secondary deposit, not a hot pyroclastic-flow deposit from Mount Katmai (as interpreted by Fenner and Curtis). The Lethe Hills (fig. 5) and the "high sand mark" and its fissured hingeline were investigated (see chapter 6).

By microbeam analysis at Berkeley and Menlo Park, Hildreth showed that the zoned 1912 eruption products recorded a preeruptive continuum in Fe-Ti-oxide temperatures (800°C–990°C), in spite of the wide bulk-compositional gap between the crystal-poor rhyolite and crystal-rich dacite. This was interpreted as support for a unitary zoned magma reservoir rather than mixing of magmas from two or more separate reservoirs.

1982

Hildreth and Fierstein returned to the VTTS from 11 July to 9 August, joined for a week by Christina Neal (Arizona State University, later USGS) and by Terry Keith and Michael Thompson (USGS) for two weeks. Fierstein began sieving pyroclastic samples at Baked Mountain Hut for grain size and componentry studies. Recording at the outcrop the maximum sizes of pumices and lithics continued for all 1912 subunits. Climbs of West Trident and Falling Mountain and traverses of the spurs of Baked Mountain involved logging the complex proximal pyroclastic-flow and fallout sequence. Sampling of thick plane-laminated ignimbrite east of Three Forks addressed the issue of distinguishing vent-derived flow units from those developed in the moving mass by internal shear, pulsatory remobilization, or incremental deposition. Sites of several 1915–1919 photographs were reoccupied to provide a record of geomorphic changes.

Keith and Thompson continued work on fumarolic deposits and spring waters. The whole party crossed Katmai Pass and sampled seeps and springs that issue from beneath the 1953 and 1958 Trident lava flows; waters were near-neutral and ranged from 8°C to 42°C at their orifices.

1983

Hildreth (1983) published an updated interpretation of the 1912 eruptive sequence, integrating chemical petrology and microbeam mineralogy of the zoned eruptive products with advances in understanding pyroclastic processes, outflow and welding of ignimbrites, and structure of the backfilled vent area in Novarupta basin.

1985

A Katmai Working Group was assembled, largely on the initiative of John Eichelberger (Sandia National Laboratories), to propose that the U.S. Continental Scientific Drilling Program (CSDP) should support research drilling into the 1912 ignimbrite and Novarupta vent. Grappling with ambivalence between his scientific and wilderness values, Hildreth joined the effort and sold the project to the CSDP at their Rapid City meeting in June 1986, advocating that the vent was the best thermal target to be drilled because it was "young, hot, and simple" (Eichelberger and Hildreth, 1986). A formal proposal was developed as an interagency DOE-NSF-USGS effort, in need of authorization by the NPS (Eichelberger and others, 1987). There ensued meetings with the NPS and environmental societies, as well as numerous workshops that culminated in Geophysical Expeditions to Novarupta in 1989 and 1990 (Eichelberger and others, 1990, 1991). A Science Plan was prepared with Eichelberger, Hildreth, and James J. Papike (1989) as Principal Investigators, and an Operations Plan was finalized in May 1992. The NPS contracted with Dames and Moore, Inc., to prepare an Environmental Impact Statement, a 350-page draft version of which was reviewed by Hildreth and others in July 1993. Resistance mounted, and in 1994 the proposal to drill was withdrawn, apparently owing to complex politics that involved an order-of-magnitude escalation of estimated costs, fears about the perceived precedent of drilling in a Wilderness, and an announced requirement that authorization would need Congressional legislation.

1986

Hildreth and Fierstein returned to the VTTS from 18 July until 18 August, joined by Keith and Gail Mahood (Stanford) who together continued work on fumarolic deposits. The emphasis this season was on documenting lithology and mass fraction of lithic ejecta in all 1912 subunits, as input for modelling the volume and shape of the backfilled vent structure beneath and around Novarupta. Special attention was given to the Novarupta ejecta ring and stratified deposits draping the walls of Novarupta basin and all sides of Falling Mountain. The party climbed to the Katmai caldera rim, recognizing there a thick 1912 explosion breccia banked against ice on the west rim. Soil atop a meter-thick pre-1912 dacite pumice-fall deposit on the north slope of Mageik was collected and later radiocarbon dated at 2,140±60 ^{14}C yr B.P. In the course of regional geologic mapping, Robert L. Detterman and James R. Riehle arrived at Novarupta by helicopter to consult about the lithology, thickness, and stratigraphy of the Naknek Formation and deeper Mesozoic strata concealed beneath the 1912 vent.

1987

Fierstein hired a boat for a 9-day cruise along Shelikof Strait in June, measuring thickness and componentry of 1912 fallout layers at many locations on Kodiak and Afognak Islands.

Hildreth and Fierstein returned to the VTTS from 2 July to 8 August, continuing detailed stratigraphic analysis and componentry of 1912 fallout and the sequence of ash flows. They climbed Whiskey Ridge, hiked the crest of the Buttress Range, and measured many sections on spurs of Baked Mountain and on the distal Ukak lobe of the ignimbrite. Two sorties were made down Mageik Creek, where they camped for a week and logged many complete fallout sections on slopes of Trident and Observation Mountain. The ash-flow sequence was mapped along the axis of Katmai Pass and Mageik Creek, where it is intercalated with rhyolitic and dacitic fall units, and where Layer B was shown to be its contemporaneous zoned equivalent. They later logged fallout sections atop Dumpling Mountain near Brooks Camp, by boat along Iliuk Arm, and by car from Naknek to Naknek Lake (fig. 6). Hildreth took USGS Director Dallas Peck on a helicopter tour of Novarupta and the VTTS, explaining the goals of the proposed scientific drilling initiative.

Terry Keith, assisted by Rick Hutchinson (NPS, Yellowstone), spent 10 July days working on fumarolic deposits and spring waters.

Lawrence G. Kodosky (University of Alaska) sampled fresh and fumarolically altered ejecta at 127 sites within Novarupta basin and analyzed them for elemental Hg content. It was hoped that the distribution of non-sulfide Hg, precipitated from vapor and adsorbed on clay-size particles, might help delineate structures that reflect configuration of the backfilled vent—generally inferred to be funnel shaped.

Peter Wallman, assisted by Larry Mastin (both then Stanford graduate students), began a structural study of Novarupta basin as part of his Ph.D. dissertation.

DeAnne Pinney camped for a month in Windy Creek, examining pre-1912 tephra and secondary deposits as a University of Alaska thesis project (Pinney and Beget, 1991).

Peter Ward deployed an array of 10 seismometers in September (enlarged to 14 in August 1988) on both sides of the volcanic chain; signals were telemetered to King Salmon. He located 1,900 earthquakes by December 1990 (mostly magnitude 1–2 and at 3–10 km depth), and he observed that P-wave delays were greatest near Katmai Pass (a formerly narrow defile filled by thick 1912 ignimbrite), coinciding there with the regional Bouguer gravity low. Although both P and S waves were attenuated along some paths that crossed the volcanic axis, S-wave screening like that noted in 1965–67 was not found. Most earthquakes occurred in four clusters—near Martin and Mageik, near Trident and Novarupta, near Mount Katmai, and northwest of Snowy Mountain.

1988

Fierstein hired a boat for a second week-long cruise along Shelikof Strait in July, measuring thickness and componentry of 1912 fallout layers at many locations along the Katmai coast from Kaguyak to Kashvik Bay (fig. 6). Jacob B. Lowenstern (Stanford, later USGS) joined Wallmann to study the rhyolite sills that intrude the Naknek Formation near West Mageik Lake (Lowenstern and others, 1991).

1989

The Geophysical Expedition party (fig. 11) established camp in an arcuate graben just east of the Turtle; starting 31 July, three spacious fabric shelters and a ton of equipment and supplies were shuttled from King Salmon by helicopter. Surface studies were intended to help define the shape of the backfilled pyroclastic vent structure and conditions within it, preparatory to the proposed drilling project. Bench marks were established, and a 150-point grid was surveyed, spanning the vent region. Gravity and magnetic field strengths were measured at all grid points and the highly variable heat flow at selected ones. Wispy fumaroles were found to be as hot as 90°C. Electrical resistivity was measured along two profiles, and seismic refraction was recorded along lines west of Novarupta. A low-altitude aeromagnetic survey was flown by helicopter. The camp was dismantled 14 September. Participants included John Eichelberger, Vicki McConnell, Sanford Ballard, and Allan R. Sattler (all Sandia National Laboratories); James Papike (South Dakota School of Mines); John Power, Betsy Yount, John Paskievitch, Wayne Colony, G. DuBois, Peter Ward, Eugene Iwatsubo, and Jack Kleinman (all USGS); Paul Kasameyer, W. Daily, and M. Wilt (all Lawrence Livermore National Laboratory); David Pollard and Peter Wallmann (Stanford); and Chris Nye, Tim Redfield, Chris Larsen, and David B. Stone (all University of Alaska).

Hildreth and Maura Hanning (USGS) were taken from the camp on 2 August by helicopter to the south toe of Mount Katmai, just west of the mouth of Katmai Canyon, where they pitched a tent for two weeks. Climbing all over the south and southwest slopes of Mount Katmai and twice spending several hours on the caldera rim, they logged and sampled numerous fallout sections and the Katmai River debris flow. It was verified that no initial ash is present on either side of the volcanic axis that might have preceded onset of the rhyolitic plinian column that started the 1912 eruption. The big landslide that had blocked Katmai River canyon (and breached as a flood, possibly in 1915 according to Griggs, 1922) was shown to be syneruptive, intercalated within the sequence of dacite-rich fallout layers. Several layers of lithic-bearing stinking mud (and coarse breccia equivalents near the caldera rim) were logged, interbedded widely in the pumiceous fallout sequence from Novarupta but ejected from Mount Katmai by hydrothermal explosions during its caldera collapse. After a 2-day return hike via Katmai Pass, they joined 12 colleagues at the Geophysical Camp near Novarupta on 15 August and later hiked out to Three Forks Overlook.

Terry Keith, Mike Thompson, and Rick Hutchinson returned for two weeks in late August to continue work on fumarolic deposits and spring waters.

Julie M. Palais and Haraldur Sigurdsson published estimated masses of acid aerosols (H_2SO_4, HCl, HF) released to the atmosphere by numerous large eruptions (including that of 1912), based on a petrologic method that compares microbeam concentrations of such components in melt inclusions sealed in phenocrysts with those retained by the matrix glass of inflated pumice or dispersed ash. Unfortunately, their 1912 samples (supplied by J. R. Riehle) were dacitic fallout from 17 km northeast of Novarupta and did not include the volumetrically dominant rhyolite, which was dispersed southeastward. Accordingly, their whole-eruption values for HF are underestimated and for H_2SO_4 overestimated. Nonetheless, the acid fallout for 1912 measured in Greenland ice cores is less than a third that for 1783 (Hammer and others, 1980), which was produced by the basaltic Laki eruption in Iceland, despite the similar volumes and latitudes of the two eruptions; the silicic Katmai ejecta had far lower sulfur concentrations than the basalts at Laki.

1990

A second Geophysical Expedition, continuing surface studies of the Novarupta vent area, was based at Baked Mountain Hut 7–26 July, supplied by helicopter from King Salmon. Many gravity and magnetic stations were remeasured to assess any ongoing changes, and electrical, magnetic, and aeromagnetic surveys were extended beyond those of 1989. New bench marks completed a geodetic net encompassing the vent region (Kleinman and Iwatsubo, 1991). Samples were collected across 1912 fumarolic deposits and later analyzed for 50 elements to estimate mass gains and losses during alteration of primary pyroclastic material. Keith measured temperatures as high as 30°C for mid-VTTS springs issuing in a gorge of Knife Creek. The Hut was renovated. Participants included returnees Eichelberger, Papike, Hildreth, Fierstein, McConnell, Keith, Kasameyer, Kleinman, Paskievitch, and Redfield, as well as Andrew M. Goodliffe (GI), Chris D'Sa (University of Iowa), Don Felske (Lawrence Livermore National Laboratory), and Christina Neal, Game McGimsey, and Susan Walker (all USGS).

Hildreth and Fierstein were dropped by helicopter on 9 July on the southeast slope of Trident, where they pitched a tent for 10 days against a glacier front as a windbreak. They climbed high on East Trident and Trident I, reached all the Trident saddles, and measured many Novarupta fallout

Figure 11. Participants in 1989 Novarupta Geophysical Expedition. From left, Wayne Colony, Chris Nye, Gene Iwatsubo, Wes Hildreth, Vicki McConnell, John Power, Maura Hanning, Jack Kleinman, Tim Redfield, Jim Papike, Sandy Ballard, Chris Larsen, and John Eichelberger. Slope of Broken Mountain at left and Mount Griggs in distance. Photo by Paul Kasameyer.

sections with Katmai-derived mud layers intercalated. The 500-m-wide Trident fumarole bowl (at the southeast toe of Trident I, first seen by Griggs in 1916) had ~80 sites of gas emission and a strong H_2S odor. After camp was moved to Mageik Creek basin north of Observation Mountain, detailed stratigraphy between there and Katmai Pass showed (1) that during the initial all-rhyolite phase of the 1912 eruption, thin ignimbrites and plinian fallout alternated; and (2) that emplacement of the subsequent compositionally heterogeneous main VTTS ignimbrite was contemporaneous with fallout of compositionally equivalent plinian Layer B. Fierstein's region-wide stratigraphy and componentry had by now conclusively demonstrated contemporaneity and compositional equivalence of pyroclastic falls and pyroclastic flows throughout most of the 1912 eruptive sequence, contrary to the then-standard model whereby column collapse was presumed to replace the one by the other. She submitted this work as a thesis and published it in 1992 in the Bulletin of Volcanology.

Remnants of a late Pleistocene rhyodacitic plinian pumice fall and ignimbrite were mapped and sampled along Mageik Creek. Hildreth and Fierstein then met Colin J.N. Wilson (invited colleague, then at Cambridge University) at Baked Mountain Hut on 21 July, and they spent 9 days together examining most parts of the VTTS, engaged in vigorous discussion of pyroclastic-flow dynamics.

1991

Hildreth and Fierstein led a 12-day wilderness backpack trip for 12 members (fig. 12) of the Commission on Explosive Volcanism (International Association of Volcanology and Chemistry of the Earth's Interior, IAVCEI). All parts of the VTTS were visited on foot, 12–23 July, as well as Katmai Pass, Trident, and the Katmai caldera rim, and many aspects of 1912 eruption processes were discussed. Participants included Hans-Ulrich Schmincke (Germany, then Secretary General of IAVCEI), Bruce Houghton (New Zealand), Mari Sumita (Japan), Jocelyn McPhie (Australia), Mauro Rosi (Italy), Stephen Self (UK), Armin Freundt (Germany), Robert L. Christiansen (USA), Paul van den Bogaard (Netherlands/Germany), and Tim Druitt (UK/France). Key issues included (1) discriminating

Figure 12. Participants in 1991 Commission on Explosive Volcanism workshop at Valley of Ten Thousand Smokes. From left, Jocelyn McPhie, Judy Fierstein, Tim Druitt, Mari Sumita, Hans-Ulrich Schmincke, Steve Self, Paul van den Bogaard, Bob Christiansen, Mauro Rosi, Bruce Houghton, and Armin Freundt. View down valley from near Baked Mountain Hut. Mount Katolinat on central skyline. Note British short-pants colonial tradition. Photo by Wes Hildreth.

discrete vent-derived flow units from layers deposited discontinuously by sedimentation at the base of flows or developed by internal laminar shear within them; (2) processes of emplacement of high-energy proximal deposits and their relationship to valley-filling ignimbrite of inferred low velocity; and (3) origin of the Lethe Hills (fig. 5) in the mid-VTTS.

Hildreth and Fierstein returned alone (26 July to 3 August) to their tent at the toe of Knife Creek Glacier 3 and spent a week (roped up) visiting the caldera rim and the glacial north slopes of Trident and Mount Katmai. Hydrothermal explosion breccia and mud layers on the west rim were logged and sampled in a snowstorm. Pyroclastic flows rich in dacite pumice blocks that terminated Episode I were shown to be partially welded, and thus not secondary (as some had argued), and to have issued directly from Novarupta (not as a laharic slurry remobilized back down the Knife Creek Glaciers).

Keith sampled the mid-VTTS thermal springs in Knife Creek gorge when surface waters were largely frozen, landing with David Keith and Game McGimsey by fixed-wing aircraft on 15 March. Discharge was only slightly lower than in June 1990 and water temperatures were similar, but the springs had been ~10ºC cooler in July and August of 1987–1989 when diluted by cold infiltration. In 1992, Keith and coauthors published a data-rich paper that compared chemical compositions of (1) major VTTS creeks, (2) springs and streams outside the 1912 deposits, (3) the mid-VTTS thermal springs, and (4) thermal springs that emerge from beneath the 1953–1958 Trident lava flows.

Keith and Papike published several papers (discussed in chapter 9) on physical and chemical features of VTTS fossil fumaroles, addressing contrasting alteration patterns of rhyolitic versus dacitic pumiceous protoliths, chemical gains and losses, mineralogical evolution during high-temperature vapor-phase processes and subsequent cooling, retrograde reactions, argillic alteration, and weathering. Within Novarupta basin, Keith mapped locations of dozens of warm areas and wispy fumaroles that ranged from 25ºC to 90ºC during her 1979–1991 investigations.

A special issue of Geophysical Research Letters collected 15 papers that presented diverse results of investigations by the 1989 and 1990 Geophysical Expeditions.

1992

Fierstein and Hildreth (1992) published a detailed analysis of the 1912 eruptive sequence, defining three explosive episodes and documenting synchroneity of compositionally equivalent fall and flow units, lateral equivalence of high-energy proximal ignimbrite (HEPI) and valley-filling ignimbrite (VFI), progressive changes in rhyolite/dacite/andesite compositional proportions in successive emplacement units, and correlation of fall units from the VTTS all the way to Kodiak.

Transported by helicopter from King Salmon via Kejulik Pass on 7 July, Fierstein and Hildreth spent a week camped at the southeast toe of Mount Katmai, where the Noisy Mountain Glacier fork descends to Katmai River. The south and east slopes of Mount Katmai were mapped and sampled on foot, and the landslide that dammed Katmai Canyon (1912–1915) was shown, by its intercalation in the fallout sequence, to have been emplaced on the first day of eruption, midway through Episode I. They returned to King Salmon via Katmai Pass on 13 July. Fierstein met Colin Wilson in King Salmon and returned to the VTTS, together working on details of ignimbrite emplacement until 2 August.

1993

The USGS published 1:250,000-scale geologic maps of the Mount Katmai quadrangle (Becharof Lake to Cape Douglas) by James R. Riehle, Robert L. Detterman, and associates, based on 130 days of helicopter-supported fieldwork, 1983–1987.

Robert B. Symonds and Beatrice E. Ritchie (USGS) sampled fumarolic gases at Trident and Griggs, on foot in July.

Jacob B. Lowenstern (Stanford, later USGS) identified Cu-rich bubbles in melt inclusions within phenocrysts from 1912 rhyolite. These were inferred to reflect entrapment during phenocryst growth of a fluid phase that had coexisted with the silicate melt before eruption. By FTIR, he also found the melt inclusions (in quartz and orthopyroxene) to contain 3.5 to 4.7 weight percent H_2O. Its inferred vapor saturation indicates that the CO_2-poor high-silica-rhyolite melt had been stored little deeper than 4 km before its 1912 outburst.

1994

Hildreth, Fierstein, Colin Wilson, and Bruce Houghton flew by helicopter to Baked Mountain Hut from King Salmon on 4 July; they were joined by Paul Wallace (University of Chicago) until 17 July. HEPI were logged in detail all around the upper VTTS—on spurs of Baked and Broken Mountains, in gulches south of Novarupta, inside its ejecta ring, and atop the Buttress Range crest, Mount Cerberus, and Whiskey Ridge. Dozens of layers in compositionally distinctive packages at each section were field sieved; coarse blocks were weighed at the outcrop, multi-kilogram samples sieved at the Hut, and remainders finer than 8 mm bagged for drying and sieving in the lab. Largest pumice and lithic sizes and component proportions were recorded for each layer. Brown pumice blocks atop the Turtle were collected to determine affinity with either dacite (normally pale gray) or andesite (normally black); chemical analysis later showed them to be both (61.8–65.6% SiO_2). Numerous pumice clasts were collected from every major layer for vesicularity and density determinations in the lab. A helicopter took out 300 kg of samples, and the party hiked out on 25 July.

Symonds and Ritchie again sampled fumarolic gases at Southwest Trident in mid-July but were prevented from doing so on Mount Griggs by exceptionally severe weather.

1995

Hildreth, Fierstein, and Houghton returned to Baked Mountain Hut by helicopter from King Salmon for the week of 21–27 July. Logging, sampling, and sieving of HEPI deposits continued on West Trident, Whiskey Ridge, and the several deep gulches south of Novarupta and adjacent to Falling Mountain. More pumice clasts were taken for vesicularity and density determinations from the stratified fallout on Baked Mountain and the south slope of Mount Katmai. Houghton recognized that, within ~1 km of Novarupta, the topmost unit of 1912 regional fallout (Layer H) was overlain by a field of scattered blocks of dacite lava, products of an extrusive dome that had been explosively destroyed before extrusion of the rhyolitic Novarupta dome. Cirque headwalls on the north faces of Trident were photographed and sketched by helicopter. Between helicopter lifts, 1912 fallout sections were logged and sampled high on the east, north, and west sides of Mount Katmai, while Hildreth was shuttled to numerous landing sites on the edifice, including most notches around the caldera rim, to collect Katmai lava flows. On the southwest rim, a dacite agglutinate sheet was found to be >60 m thick and to drape inboard into a pre-1912 crater. On the south rim, a thick scoria fall deposit, partly agglutinated and zoned from dacite to andesite, was found to overlie a 50-m-thick rhyolite lava flow (later dated at 22.5 ka).

John Power, Elliot Endo, Art Jolly, and John Paskievitch (Alaska Volcano Observatory, AVO) reoccupied the five-station geodetic network centered on Novarupta and undertook simultaneous 12–15 hour observations with five GPS receivers. They detected no deformation in the vent area, relative to EDM and GPS surveys in 1990 and 1993 by Jack Kleinman and Eugene Iwatsubo, who had installed the stations in 1989–90 (Kleinman and Iwatsubo, 1991; Kleinman and others, 1997). Five seismic stations from Ward's 1987–1991 network were reactivated and upgraded.

Symonds and Ritchie returned and sampled fumarolic gases at Mount Griggs on foot and at Mageik crater and the southeast flank of Trident with helicopter assistance.

1996

Hildreth and Fierstein were asked by AVO to prepare a geologic map and hazards assessment of the whole Katmai volcano cluster (Fierstein and Hildreth, 2001), extending by helicopter their previous Novarupta-Katmai-Trident work, which had largely been on foot. Assisted by Michelle Coombs (University of Alaska) and David Tucker (Western Washington University), they set up three successive tent camps at the heads of the east and southeast forks of Angle Creek and of the east fork of Kejulik River (fig. 6). The eroded edifice of Alagogshak Volcano was recognized and mapped, along with parts of Martin and Mageik. Coombs sampled 1953–1974 eruptive products of Trident, which became a focus of her petrologic Ph.D. work, and Fierstein began her study of the pre-1912 tephra layers of the Katmai region. Fieldwork extended from 14 July to 1 August.

John Paskievitch and Art Jolly spent 14–28 July extending the regional seismic monitoring array by helicopter; they enclosed five existing and six new seismic stations in fibreglass huts (for protection from wind, snow, and bears).

1997

Hildreth, Fierstein, Wilson, Houghton, and Coombs flew by helicopter to Baked Mountain Hut from King Salmon on 16 July, basing there until 15 August. Detailed attention was devoted to logging, field sieving, and correlating proximal pyroclastic fall and flow layers among the many gulches south and east of Novarupta. Hildreth and Fierstein tent camped for a few days in lower Mageik Creek, where complex pre-1912 valley-fill deposits were logged. They spent many days by helicopter advancing the regional tephra and geologic mapping efforts; progress was made on Mageik, Martin, and Griggs Volcanoes by landing at numerous high points on all three. The valleys of Alagogshak, Martin, Soluka, Ikagluik, Windy, and Margot Creeks were cruised for tephra layers, as were Katmai Canyon, Noisy Mountain, and the Ukak River; the sites of Savonoski and Katmai villages were visited. Before departure on 15 August, shorelines of Iliuk Arm were cruised by boat for tephras and glacial deposits.

Symonds and Ritchie again collected fumarolic gases at Griggs, Trident, and Mageik. Since their earlier visits, highest recorded temperatures at Griggs had declined from 105°C to 99°C and at Trident from 97°C to 94°C; the highest at Mageik crater lake remained in the range 167–173°C. All had cooled slightly relative to Johnston's measurements in 1979.

Paskievitch, Jolly, and Tim Plucinski restored the array of seismic stations, several of which had been down from April until July.

Using satellite-based synthetic aperture radar interferometry (InSAR), Zhong Lu (USGS) and associates detected several centimeters of inflation that took place between 1993 and 1995 within a 7 km^2 area that included the still-fumarolic Southwest Trident cone (Lu and others, 1997). No further inflation was detected there subsequently through at least 2010.

1998

Hildreth, Fierstein, Wilson, Houghton, and Coombs returned by helicopter to Baked Mountain Hut from King Salmon on 14 July, basing there until 31 July. Logging and sieving proximal 1912 deposits continued, as did the regional search for pre-1912 tephras, this year in Angle and Margot Creeks and Rainbow River. With ample helicopter support, mapping and sampling of Griggs, Trident, Mageik, and Martin Volcanoes was completed, Mount Katmai nearly so, and Snowy Mountain was reconnoitered. Rear-arc lavas were collected near Savonoski River, Rainbow River, Mount Juhle, and Gertrude Creek (fig. 6). A reconnaissance fly-around at Kaguyak caldera and landings on its two intracaldera domes assessed logistics for a major effort there the following year (Fierstein and Hildreth, 2008). After others departed, Fierstein, Wilson, and Hildreth tent-camped near Three Forks Overlook from 31 July until 8 August, logging in detail distal 1912 deposits, pre-1912 tephras, and glacial deposits.

Paskievitch, Jolly, Plucinski, and Seth Moran spent two July weeks replacing batteries and installing seven new seismic stations, bringing the Katmai seismometer array to

18 stations and a few repeaters. Signals were telemetered in real time via line-of-sight radio transmission to King Salmon, thence via commercial phone circuits to AVO in Anchorage and Fairbanks.

Symonds and Ritchie again sampled fumarolic gases on Griggs, Trident, and Mageik, all of which had maximum temperatures similar to those of 1997. They installed a lake-monitoring package in Mageik crater lake, intended to record temperature, pH, conductivity, and lake level every two hours for the following year. At the time, lake water pH was 1.6 and temperature 72°C.

1999

Fierstein and David Tucker picked up a Zodiac inflatable boat in Anchorage, met Hildreth in King Salmon on 12 July, and were dropped by two Cessna float planes at the islet in Kaguyak caldera lake. After photographing and sketching the walls from the islet, the three sampled all lavas and fragmental deposits exposed on the walls and rim, by Zodiac and on foot. Soundings of lake depths were made and lake waters collected for analysis. From 22 July until 2 August, the outer slopes of the Kaguyak edifice, the ignimbrite outflow, and tephras exposed in coastal bluffs were sampled by helicopter. Additional sampling was done at Snowy Mountain, Devils Desk, and the several Savonoski volcanic cones.

Houghton, Wilson, and Julia Eve Hammer (Brown University petrology postdoc) arrived at King Salmon on 3 August and were flown to Baked Mountain Hut (fig. 13). Houghton and Hammer did further work on Novarupta proximal deposits until 14 August, and Fierstein and Wilson worked on the valley-filling ignimbrite in the upper and lower VTTS until 23 August. Numerous pyroclastic samples were sieved at the

Figure 13. Participants in 1999 expedition to Novarupta. From left, Julia Hammer, Bruce Houghton, Colin Wilson, Judy Fierstein, and Jim Sink. Pumice-covered slopes and ice-filled crater of Mount Griggs at upper right. Photo by Wes Hildreth.

Hut, and the sub-8-mm fractions were flown out for lab work by helicopter in several duffle bags. All four hiked out to Three Forks Overlook.

Paskievitch replaced batteries and performed system checks for the Katmai seismic array. Several stations had been inoperative during the winter months. The network became a mainstay of Alaska Volcano Observatory operations, locating 1,000 to 2,000 earthquakes per year in the Katmai region (Mount Steller to Mount Martin) for the whole decade through 2010. Nearly all have been in the M_L range –0.5 to 2.5 and shallower than 10 km, most of them clustered in the same four areas identified by Ward in 1987–1991—Martin-Mageik, Trident-Novarupta, Mount Katmai, and northwest of Snowy Mountain. Rare events have been M_L 2.8 to 4.5.

2000

Fierstein and Hildreth spent a week (16–23 July) doing helicopter geology from King Salmon, principally along Angle Creek, on the Savonoski cones, and on the Rainbow River cone and debris-avalanche deposit. Local problems concerning the pyroclastic deposits were reexamined along Mageik Creek, Windy Creek, the Griggs fork of Knife Creek, and near Three Forks.

Michelle Coombs and associates published her petrological dissertation work on lavas and ejecta of the 1953–1974 Trident eruptions, collected during her 1996–1998 fieldwork. Eruptive products were shown to range from 55.8 to 65.5 percent SiO_2 and from ~1,000°C (andesite) to 890°C (dacite) in Fe-Ti-oxide temperature. H_2O contents of melt inclusions in phenocrysts and phase-equilibria experiments indicated that pre-1953 magma storage had been as shallow as 3 km and that mixing of ascending andesite into resident dacite may have taken place just before the 1953 eruption.

Hildreth and Fierstein published an overview article that summarized the eruptive sequence, caldera collapse chronology, and 1912 magma storage model, establishing the compositional affinity of the 1912 suite to Mount Katmai.

2001

Fierstein and Hildreth spent 12 days (28 July to 9 August) doing helicopter geology from King Salmon, principally along Margot Creek and the Griggs fork, on Snowy Mountain and Mount Denison, and on the several lava remnants between Mounts Ikagluik and Juhle. Spot checks were made on Mageik and Griggs, and several rim stops around Katmai caldera. The helicopter was shared with John Paskievitch, who was installing new batteries and servicing the far-flung seismometer network, so, on flightless days, Fierstein and Hildreth investigated glacial and tephra deposits by truck between Naknek and Lake Camp. A boat was hired to examine exposures along bluffs of the Naknek River, where they learned why Sayre and Hagelbarger had a hard time ascending the rapids in 1918. While Hildreth recuperated from an injury, Fierstein devoted a day to reconnoitering Ukinrek Maars, The Gas Rocks, and Blue Mountain, all on the southwest side of Becharof Lake.

Coombs ran phase-equilibria experiments on 1912 rhyolite in the lab of James E. Gardner (University of Alaska) and Hammer on 1912 dacite and andesite in the lab of Malcolm J. Rutherford (Brown University). Results suggested preeruptive storage of the rhyolite in the depth range 1.6 to 4.4 km, and of the zoned dacite-andesite continuum at 4 to 5 km. Uncertainties include quartz stability, the effects of ranges in magma temperature, and the influence of preeruptive CO_2 and SO_2 contents of the fluid phase on pressure inferences.

2003

Fierstein, Wilson, and Hildreth arrived at King Salmon on 16 July and remained in the area until 15 August. Fierstein and Wilson based much of the time at Baked Mountain Hut, completing mapping of the compositionally distinctive packages that make up the 1912 ignimbrite in the VTTS, Katmai Pass, and Mageik Creek. Nine sequential packages of flow units, each with different proportions of rhyolite, dacite, and andesite pumice, were mapped, many with sharp mutual contacts. Pumice proportions and local intercalation of flow and fall deposits were used to link ignimbrite formation and coeval fall deposition. The compositional zonation permitted documentation of how a major ignimbrite sheet is incrementally assembled, as published by Fierstein and Wilson (2005).

Houghton and Nancy Adams (University of Hawaii) based at Baked Mountain Hut between 27 July and 10 August and concentrated on collecting pumice suites from upper Knife Creek, Novarupta basin, and the south slope of Mount Katmai; they also mapped in detail the Episode IV dacite block bed in Novarupta basin. Hildreth devoted 2 days to Mount Denison, and he flew to the Katmai caldera rim on three different days, variously with Fierstein, Adams and Houghton, or Andrew Calvert (USGS). Calvert visited 21–25 July and spent a day with Hildreth at Kaguyak caldera, scouting for samples suitable for Ar dating. Hildreth also sampled most of the Tertiary granitoid plutons in the Katmai region. Fierstein and Hildreth flew together to Kaguyak for a day and, on another day, beyond Kaguyak to reconnoiter and sample Fourpeaked and Douglas Volcanoes. They sampled all units at The Gas Rocks and many at the Blue Mountain dacite dome complex, and Hildreth and Houghton returned a week later to complete the sampling and mapping of Blue Mountain (Hildreth and others, 2007). Families and friends of Hildreth and Fierstein gathered at Brooks Lodge 13–15 August for a walk to Three Forks that celebrated completion of the project; participants included Garniss Curtis and Ian Carmichael, Berkeley professors who had encouraged Hildreth's rookie trip to the VTTS, a month on foot in 1976. It was the 50th anniversary of Curtis' first trip to the VTTS in 1953 (fig. 14).

A volcanological geologic map of the Katmai volcanoes and VTTS was published by Hildreth and Fierstein (USGS Map I-2778).

Seth Moran (AVO) determined first-motion fault-plane solutions for 140 earthquakes (1995–2001) and inverted the solutions for stress tensors. He was thus able to discriminate

(1) seismicity related to regional tectonics (near Snowy Mountain and off the volcanic axis) from (2) seismicity along and near the axis that he attributed variously to hydrothermal convection, edifice loading, or activation of fractures that may have been created during the many 1912 earthquakes of magnitude 6.0–7.0 (Moran, 2003).

2004

Houghton, Wilson, Fierstein, and Hildreth published a stratigraphic and dynamic analysis of the 1912 ultraproximal deposits of Episodes II and III around Novarupta. This was the most comprehensive characterization yet undertaken of complex proximal deposition during a plinian eruption because such deposits are seldom both preserved and exposed. Simultaneous with the buoyant plinian plume, sedimentation from margins of the jet and lower plume produced local pyroclastic flows and sector-confined fallout wedges that intercalated only proximally with the region-wide plinian pumice-fall layers. Relative to the regional falls, most such proximal deposits are enriched in andesitic and banded pumice and in denser lithic fragments, reflecting lateral heterogeneities in particle concentration and componentry around the periphery of the emerging jet. It was inferred that a concentrated annular zone surrounding the jet core was more susceptible to fountaining and collapse because of its higher density and lower velocity. Confluence of dacite and andesite magmas and the shift from lateral conduit flow to vertical flow beneath the vent were implicated as contributing to the heterogeneities and thus to jet-margin instabilities. Four concurrent transport regimes were identified for the pyroclasts ejected.

Figure 14. Garniss Hearfield Curtis (right) with the authors near Three Forks in August 2003, on 50th anniversary of his first expedition to the Valley of Ten Thousand Smokes. Curtis always wore tennis or running shoes during fieldwork. Photo by Julie Donnelly-Nolan.

2004–2010

AVO continued to amass seismic data from its 18-station array, serviced annually by John Paskievitch. Jolly and associates applied traveltime inversion tomography to >1,000 local earthquakes (1996–2001) to produce a 3-dimensional P-wave-velocity structural model for the Katmai volcano cluster (Jolly and McNutt, 1999; Jolly and others, 2007). They confirmed a Mageik-to-Trident low-velocity zone centered on Katmai Pass (extending to depths of at least 8 km) and demonstrated absence of such an anomaly beneath Griggs and Snowy Mountain Volcanoes. Stephanie Prejean (AVO) and associates relocated >7,000 earthquakes using the double-difference method to sharpen images of seismic clusters beneath Martin and Mageik volcanoes and planar structures near Snowy Mountain, and to suggest that the low-velocity body under Katmai Pass and Trident extends to depths shallower than 5 km and reaches toward Novarupta at its shallowest level. In summer 2008, they installed 11 temporary seismometers near Trident in an effort to acquire still better resolution of the velocity anomaly.

Principal Contributors to the Early Geologic Investigations at Katmai

George Martin

George Curtis Martin (1875–1943) was an intrepid field geologist with the Alaska division of the USGS for 20 years. A son of Massachusetts, he graduated from Cornell in 1898 and earned a Ph.D. in 1901 at the Johns Hopkins University, where he was a paleontology instructor 1901–1904. After studies of the geology and mineral resources of Maryland for the Maryland Geological Survey (1898–1904), he led three USGS Alaskan expeditions (1903–1905), investigating petroleum and coal resources of Controller Bay, Cold (Puale) Bay, and the west shore of Cook Inlet. In 1905 and 1910 Martin mapped the coal fields of the Matanuska Valley, in 1906 he returned to Controller Bay (Katalla oil field) and Bering River coal field, and in 1908 he reconnoitered the coal resources of Colorado east of the Front Range. His 1903 and 1904 trips to Cold Bay in an open boat made him uniquely familiar with the remote Katmai coast and thus the choice by the Research Committee of the National Geographic Society to investigate the Katmai eruption in the summer of 1912 (Martin, 1913). From 1909 to 1923, he spent 2–4 months in Alaska almost every summer, mapping the Lake Iliamna-Lake Clark region, Kodiak and Afognak Islands, the Kenai Peninsula, and the Matanuska and Nenana coal fields, as well as investigating gold-bearing placers near Cape Yakataga and in the Shumagin Islands. During the 1920s, he undertook expeditions to the Yukon and Koyukuk regions, the upper Kuskokwim drainage, and the Chignik district, as well as intensifing his study of the stratigraphy and petroleum resources of the Alaska Peninsula. Martin published numerous reports, notably a 1926 monograph on the Mesozoic stratigraphy of Alaska and a 1921 report on petroleum in Alaska. While Alfred Hulse Brooks served with the Army in Europe during World War I, Martin was acting Geologist-in-Charge of the USGS Alaska Division. While he was in charge, Martin examined the mines of the Chitina Valley and the Fairbanks district, and the USGS opened its first Alaskan office (in Anchorage). He retired from the USGS in 1924 and became a consulting geologist, investigating among other things the petroleum resources of Mexico. In 1935, while living in Corvallis, Oregon, Martin was appointed Executive Secretary of the Board of Geographic Names by the Secretary of the Interior. He had travelled to 47 states, Canada, Mexico, and Central America. Mount Martin Volcano and several Martin Creeks are named for one of Alaska's most adventurous and productive pioneering geologists, one who added much to both mineral wealth and pure science.

Robert Griggs

Robert Fiske Griggs (1881–1962) was a field-based botanist who graduated from Ohio State University in 1903 and earned a Harvard Ph.D. in 1911. He served on the botany faculties at Ohio State (1906–1921), George Washington University (1921–1947), and the University of Pittsburgh (1947–1952). He was a founding member of the Ecological Society of America (1915) and its President in 1943. His research included the cytology of fungi, the willows of Ohio, and the systematics of kelp. His expeditions sponsored by the National Geographic Society led to discovery and exploration of the Valley of Ten Thousand Smokes and to his leading role in establishment of Katmai National Monument. The Alaskan years turned him into an ecologist with interests principally in the colonization dynamics of marginal environments, including in his later career timberline studies at Mount Washington (New Hampshire) and in the Rocky Mountains. During World War II, he was Chair of the Division of Biology and Agriculture, National Research Council. In 1959, Knife Peak volcano adjacent to the VTTS was renamed Mount Griggs. In a sense, the present monograph is an update of Griggs' well-illustrated 1922 book, "The Valley of Ten Thousand Smokes," published by the National Geographic Society.

Clarence Fenner

Clarence Norman Fenner (1870–1949) was an analytical and field petrologist who graduated from Columbia as a mining engineer in 1892 and for 15 years studied ore deposits in Canada, Mexico, the western United States, and South America. Returning to Columbia in 1907, he completed his Ph.D. in 1910 and then joined the staff at the Geophysical Laboratory (1910–1938). His thesis on the Mesozoic basalts of the Watchung Mountains (New Jersey) included study of zeolites and secondary replacement minerals; these investigations and those on ore deposits fired a lifelong interest in hydrothermal and pneumatolytic processes involving "gaseous emanations" (Wright, 1952). This fascination predisposed him at Katmai to hypothesize exothermic reactions related to rhyolitic magma ascent and volatile escape, to which he attributed both large-scale assimilation and the explosive eruption. The Mesozoic sills of New Jersey may also have influenced his postulation of a shallow magmatic sill beneath the VTTS as the source of its tuff and fumaroles. In the lab, Fenner established the stability relations among the polymorphs of silica, and during World War I he pioneered the use of optical pyrometers in controlling glass furnaces when assigned to improve and extend U.S. capacity in producing optical glass (formerly imported from Germany). In the early years of U-Th/Pb geochronology, he made a study of U-Th minerals and devised chemical methods for separation and analysis. He also recognized the Fe-enrichment trend in differentiation of tholeiitic basalts, widely called the "Fenner Trend." At Yellowstone, he pioneered core drilling in the geyser basins, reached 180ºC at 124-m depth, and investigated alteration minerals, compositional gains and losses, and relative meteoric and magmatic contributions to hydrothermal fluids. For the Gardiner River mixed lava complex at Yellowstone, Fenner again invoked his "gaseous emanations" hypothesis, suggesting that solid basalt had been permeated and corroded by "tenuous rhyolitic solutions"—liquid and gaseous—causing "replacement analogous to granitization." He also advocated reconsideration of liquid immiscibility as an important magmatic process. At age 66, Fenner joined a research expedition to Antarctica (1936–1937). In retirement, Fenner returned to Peru (1938–1939) where he mapped the "incandescent tuff deposits" (ignimbrites) of the Arequipa region, likened them to the VTTS deposits, and adopted the local term *sillar* for nonwelded ignimbrites indurated by vapor-phase crystallization (and sintering). Fenner was one of America's most versatile, energetic, and imaginative geologists. He spent 2 months at Katmai in 1919 and nearly 3 months there with Yori in 1923 (fig. 8).

E.T. Allen

Eugene Thomas Allen (1864–1964) was a geochemist of exceptional breadth who was based at the Geophysical Laboratory for 37 years. Another son of Massachusetts and descended from its earliest colonists, he graduated from Amherst College in 1887 and completed a Ph.D. in chemistry at the Johns Hopkins University in 1892. Allen spent two years at Harvard (1893–95) as what would now be called a postdoctoral fellow, and he taught at the Women's College of Baltimore (now Goucher College), the University of Colorado, and Missouri School of Mines (Rolla) until 1901, when he joined the U.S. Geological Survey. In 1906, he transferred to the Geophysical Laboratory where he remained a research chemist (and repeatedly its acting director) until his retirement in 1933. Allen was a superior analytical chemist and a pioneer in high-temperature experimental methods and the synthesis and stability fields of silicate minerals. From 1910 to 1917 he shifted his emphasis to the stability of metallic sulfides and sulfates, clarifying processes of secondary enrichment of ore deposits. During World War I, along with Fenner and Zies, Allen researched the trace-element composition and production of optical glass, supply of which from Germany had been cut off. Participation in the 1919 expedition to Katmai confirmed another career shift, which had begun at Lassen Peak in 1915. The remainder of Allen's research career was devoted to field and chemical investigations of volcanic gases and thermal waters. His classic chemical study of fumaroles in the Valley of Ten Thousand Smokes (Allen and Zies, 1923) was followed by landmark papers on the fumaroles, geysers, and hot springs of Lassen, The Geysers (California), and Yellowstone. Allen was elected to the National Academy of Sciences in 1930. He died at home in Arlington, Massachusetts, 106 days after his 100th birthday.

E.G. Zies

Emanuel George Zies (1883–1981) was a geochemist, mineralogist, and pioneer investigator of volcanic gases and their sublimates. Born in Baltimore, he graduated from the Johns Hopkins University in 1906 and completed his Ph.D. in chemistry there in 1909 with a dissertation entitled "The osmotic pressure of cane sugar solutions at 0°." Zies joined the staff of the Geophysical Laboratory in 1913 as a geochemical volcanologist, retired formally in 1949, but remained an active emeritus researcher there until 1967. After participating in the 1919 expedition to the Valley of Ten Thousand Smokes (fig. 9), he conducted field studies at volcanoes in Java, Bali, Hawaii, South America, El Salvador, Guatemala (Santa María), Mexico (Parícutin), and Italy (Pantelleria). Ziesite is a copper vanadate deposited by fumaroles in the summit crater of Volcán Izalco (El Salvador), which was born in 1770. After Robert L. Smith published his classic 1960 review "Ash Flows," he and Zies met for lunch at the Cosmos Club, and Zies was highly congratulatory about Smith's understanding of pyroclastic flows and his insights about trace-metal transport by the VTTS fumaroles. Because Smith had written, however, that "Unfortunately, for students of pure magma chemistry, these deposits are formed from a mixed magma..." (Smith, 1960, p. 835), Zies genially jabbed him—"What the hell's a "pure magma"?

Garniss Curtis

Garniss Hearfield Curtis (1919–) is a field geologist, volcanologist, and geochronologist who graduated from the University of California at Berkeley in 1942 and worked 4 years as a mining engineer and geologist before returning to Berkeley in 1946, where he completed his Ph.D. and joined the Berkeley faculty in 1951. For his thesis he mapped a large area of the Sierra Nevada near Markleeville and Topaz Lake and investigated the origins of the extensive andesitic breccia deposits of the Mehrten Formation. As a member of the Katmai Project (1953–1954), he helped reinterpret the 1912 eruption in a modern volcanological framework and isopached the medial pumice-fall deposits, proving that Novarupta (not Mount Katmai) was their source vent. He was a pioneer in K-Ar dating, first at the University, later as founder of the Berkeley Geochronology Center. His ages calibrated the Tertiary land-mammal stages of North and South America and the fossil hominid stratigraphy of East Africa and of *Homo erectus* in Java. Curtis is also an authority on the geology of the Berkeley Hills, where he taught field methods to generations of geologists. He returned to Katmai with the authors (fig. 14) in 2003, the fiftieth anniversary of his initial summer in the VTTS.

Chapter 5

Products of the Eruption

The 60-hour-long explosive sequence at Novarupta consisted of three discrete episodes (fig. 2), separated by lulls of at most a few hours duration (Martin, 1913; Hildreth, 1983; Fierstein and Hildreth, 1992). **Episode I** began with plinian dispersal of purely rhyolitic fallout (Layer A) and synchronous emplacement of rhyolitic ignimbrite. Nowhere has any initial ash or phreatomagmatic deposit been found beneath Layer A, indicating that the plinian column first witnessed at 1300 h (Alaskan local time) on 6 June indeed marked the onset of eruption. Vent opening was accompanied, however, by a blast that spread sheets of high-energy proximal ignimbrite as far as 9 km radially. After ejection of ~3 km^3 of rhyolitic magma over the course of a few hours, small amounts of andesitic and dacitic magma began contributing to the sustained eruption column, marking (by definition) the onset of plinian Layer B (fig. 15). The juvenile ejecta in Layer B are zoned from >99 percent rhyolite at its base to only about 15 percent at its top, matching the progressive shifts in (andesite/dacite/rhyolite) pumice proportions in the main sequence of zoned ignimbrite (fig. 16) emplaced concurrently in the Valley of Ten Thousand Smokes (Fierstein and Hildreth, 1992; Fierstein and Wilson, 2005). Only a few thin andesite-dacite flow units capping the Episode I deposits lack a widespread fallout equivalent, suggesting that ignimbrite emplacement barely outlasted the plinian phase. Episode I lasted ~16 hours and produced almost all of the ignimbrite (~11 km^3) and roughly half of the fall deposits (8.8 km^3 out of 17 km^3).

Episode II began after an eruptive lull no longer than a few hours and consisted principally of plinian dispersal of phenocryst-rich dacite that deposited regional fall Layers C and D. Intercalated within these layers proximally are minor

Figure 15. Novarupta pumice-fall section on south flank of Mount Katmai, 10 km southeast of Novarupta and 4 km south of rim of Katmai caldera. Vaguely shower-bedded Layer A is 1.4 m thick, tripartite Layer B is 1.0 m thick, and bevelled Layer C is 1.6 m thick to its eroded surface. Deposit rests directly on glacial till.

intraplinian pyroclastic-flow and surge deposits made up predominantly of dacite and andesite (Fierstein and Hildreth, 1992; Houghton and others, 2004). Subordinate rhyolite (identical to that of Episode I) accompanied the dacite during deposition of the lower part of Layer C but was very sparse thereafter.

Episode III began after another lull as long as several hours, during which fines-rich Layer E settled from lingering ash clouds of composite origin (Fierstein and Hildreth, 1992). It was similar to Episode II in depositing plinian dacite Layers F and G regionally, along with thin, proximally intercalated, sheets of andesite-dacite ignimbrite and sector-confined lobes of fallout (Fierstein and others, 1997; Houghton and others, 2004). Layer H is another postplinian ashfall deposit, finer grained and more widespread than Layer E, that also settled from ash clouds of composite origin (Fierstein and Hildreth, 1992). The dacite pumice that dominates the products of plinian layers C, D, F, and G shows similar ranges of bulk and mineral compositions in all four layers (fig. 17). Eruptive styles were likewise similar, though Layers C and D were more strongly dispersed toward the southeast (Fierstein and Hildreth, 1992).

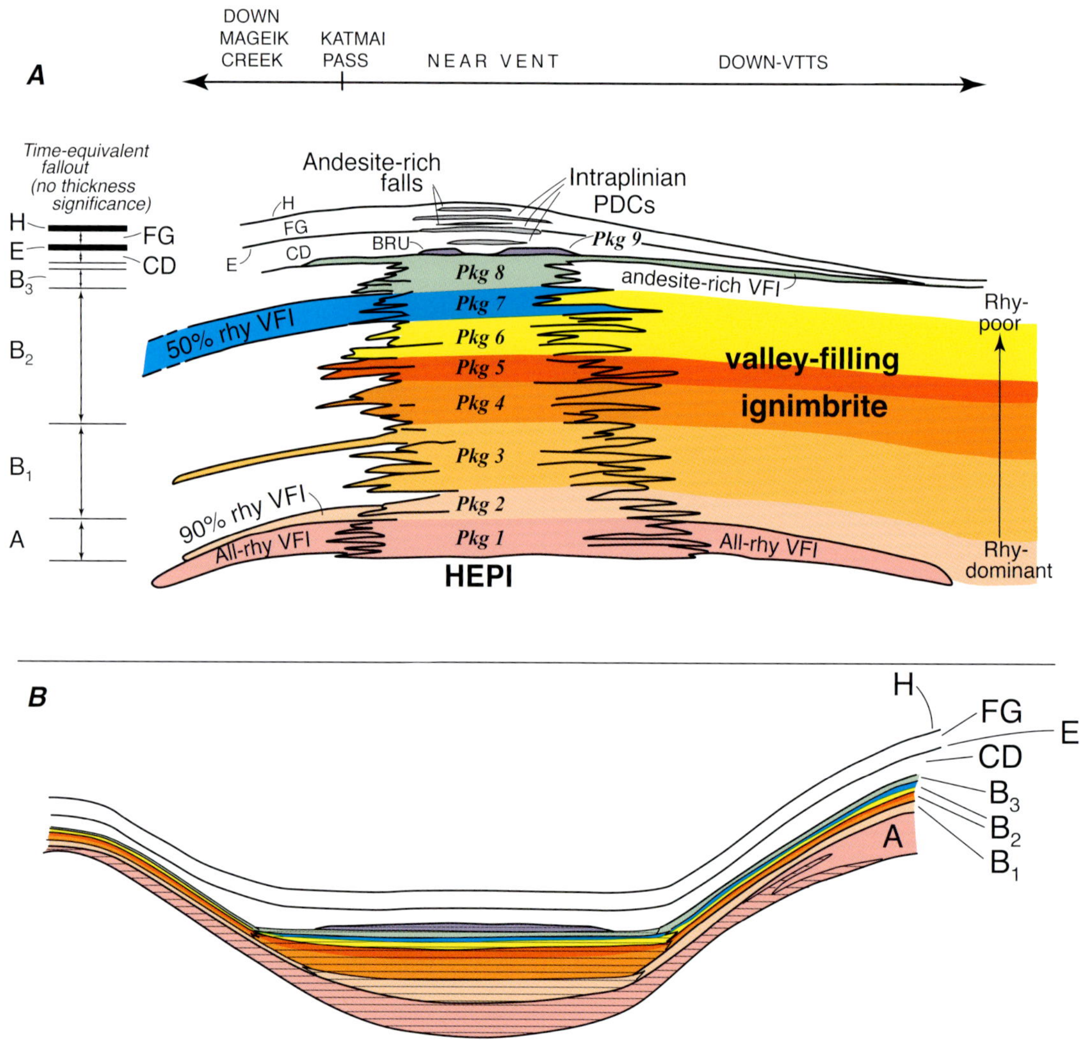

Figure 16. Schematic sketches illustrating relations among 1912 pyroclastic deposits. *A*, Cartoon summarizing ignimbrite facies and time-equivalent fall deposits (Layers A through H). Nine compositionally distinguishable ash-flow packages (Pkg) are represented in valley-filling ignimbrite (VFI) and eight of the nine in high-energy proximal ignimbrite (HEPI). Rhyolite is abbreviated, rhy. BRU, block-rich flow units that dominate small-volume Package 9, which was emplaced only on valley floor, not as HEPI. Thicknesses not to scale; vertical dimension scaled to depict time equivalence. *B*, Generalized sketch illustrating influence of topography on landscape-mantling fall deposits and contemporaneous ignimbrite packages (patterned) that are intercalated near Katmai Pass and along Mageik Creek. Episode I subunits are colored identically in both panels. Episode II and III regional fall deposits are uncolored, but intercalated ultraproximal falls and pyroclastic-density-current deposits (PDCs) are gray.

Figure 17. Pumice-fall deposit of Episodes II and III in upper Knife Creek, 4 km northeast of Novarupta. Geologist stands on fluvial pumice beds reworked during eruptive lull between Episodes I and II; they rest on sintered ignimbrite packages 8 and 9 at gulch floor. Coarse base of Layer C, just above geologist's head, contains subordinate rhyolitic pumice, but rest of 12-m-thick fallout sequence is 99 percent dacitic. Fines-rich layer, 2-3 cm thick, that crosses middle of image is "C–D parting," which separates 2.4 m of Layer C from 1.6 m of Layer D. Fines-rich Layer E, 10-15 cm thick, represents lull between Episodes II and III. Coarse Layer F grades up into Layer G, most of which is eroded here.

This threefold framework (fig. 2), worked out by stratigraphic analysis of proximal, medial, and distal deposits, fits well with the three timed intervals of ashfall at Kodiak (170 km downwind) and with the three downwind ash layers measured by Martin (1913) and correlated by microprobe analysis of shards by Fierstein and Hildreth (1992).

Fallout volumes calculated for Episodes II and III are 4.8 km^3 and 3.4 km^3 (Fierstein and Hildreth, 1992; Fierstein and Nathenson, 1992). Added to 8.8 km^3 for Episode I fallout, the total 17 km^3 of tephra is equivalent to ~6.5 km^3 of magma emplaced as fallout. Because the thickest ignimbrite deposits in the upper VTTS are welded and not deeply incised, their volume can only be approximated as 11 ± 2 km^3, equivalent to ~7 ± 2 km^3 of magma. For the three-episode magma total of ~13.5 km^3, we estimate that 55 percent was rhyolite, 35 percent dacite, and no more than 10 percent andesite.

The explosive episodes were followed by extrusion of three lava domes. No evidence is recognized that pins down the timing of these extrusions, but records of silicic eruptions elsewhere suggest that they could have taken place within days or as long a year or so after the plinian events. A small dacite lava dome that plugged the Episode III vent was subsequently destroyed by vulcanian explosions that scattered an 8-km^2 block bed atop Layer H (fig. 4 of Adams and others, 2006a). The same vent was then plugged by the rhyolitic Novarupta dome, which survives a century later (fig. 18). On the floor of Katmai caldera, which collapsed fitfully during the eruptive sequence at Novarupta (Hildreth, 1991; Hildreth and Fierstein, 2000), another dacite dome was extruded and then partially disrupted explosively. This was the crescentic or Horseshoe Island dome first seen by Griggs in 1916 and sampled by Fenner in 1923 before it was covered by the deepening caldera lake (fig. 5).

Eruptive Episode I

Reconstructing stratigraphic relationships between the main VTTS ignimbrite and plinian Layers A and B was surprisingly difficult. The opening plinian ejecta that produced Layer A were strongly dispersed east-southeastward in a 90° sector (fig. 19) across the snow-covered glaciers and ridges of Trident and Mount Katmai. Layer A is widely intact on the southern slopes of those peaks, but the ignimbrite did

Figure 18. Novarupta rhyolite dome, 380 m across and 65 m high, beyond which, viewed westward, Novarupta basin merges into the upper Lethe arm of the Valley of Ten Thousand Smokes. Buttress Range rises on skyline, pumice-cloaked Baked Mountain at right, and northern spur of Falling Mountain at left. Just right of that spur, in middle distance, where light-colored alluvial pumice fan (just left of center) abruptly enters narrow snow-filled gorge, compacted pyroclastic deposits dip gently back toward Novarupta, marking west margin of Episode I vent structure, which is backfilled with welded tuff. Unlike many lava domes, Novarupta is a simple, subcircular, axisymmetrical extrusion, not much wider than the orifice of its conduit.

not extend that far. On the northern slopes, serial pulses of high-energy proximal ignimbrite scoured fall deposits from highland ridges, ice movement disrupted most supraglacial pyroclastic deposits, and ignimbrite filling the upper VTTS is so thick that its base is nowhere exposed. Only in upper Mageik Creek south of Katmai Pass and at a few marginal sites where ash flows feathered out in the Trident saddles, on Mount Katmai, and in the Griggs fork of Knife Creek are intercalations of all-rhyolite fallout and all-rhyolite ignimbrite preserved and exposed. Because Fenner's 1923 ascent to the caldera rim was across the Knife Creek Glaciers (where 1912 deposits are disrupted) and all previous ascents had been from the south side of Mount Katmai, the early investigators appreciated neither that fallout and ignimbrite were interlayered nor that the pumice-fall deposits thinned and fined toward Mount Katmai, away from Novarupta.

Layer A

The all-rhyolite plinian fall deposit is as thick as 1.5 m near Katmai River (figs. 15, 20). It may have been thicker on the northern slopes of Trident, where it was intercalated with and scoured by ash flows or was subsequently eroded or sloughed over snow. Along its narrow east-southeasterly

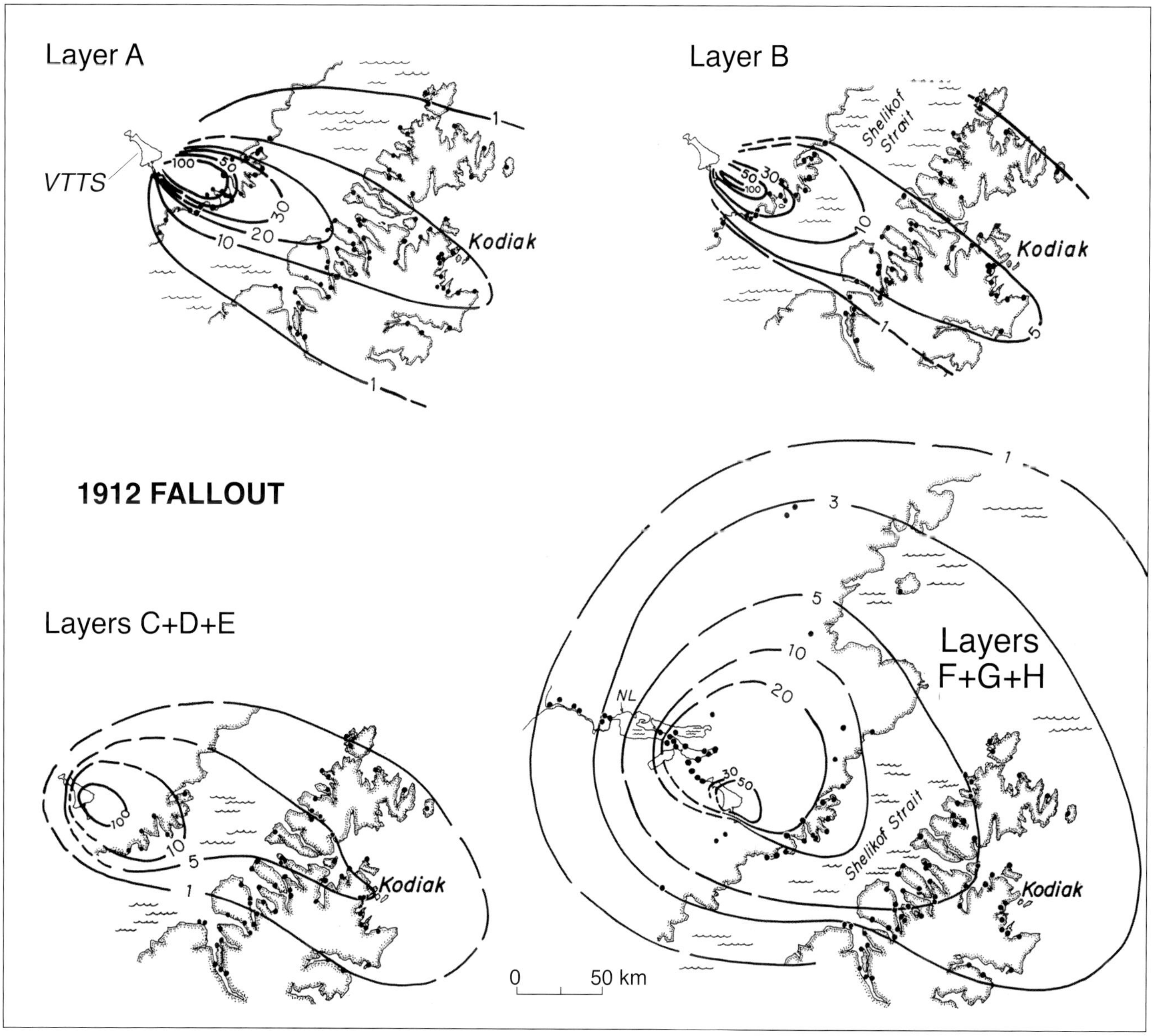

Figure 19. Isopach diagrams for 1912 tephra falls: Layers A and B of Episode I; combined Layers C, D, and E for Episode II; and combined Layers F, G, and H for Episode III. Thickness contours in cm. Outline of VTTS is shown for reference. Dots are sites of measured sections; for abundant proximal and medial sites, see Fierstein and Hildreth (1992). See figure 6 for wider geographic context of Shelikof Strait and Kodiak Island. NL, Naknek Lake.

dispersal axis, Layer A constitutes the basal 15 percent of intact fallout sections on Trident at distances of 6–8 km downwind, ~30 percent near Katmai River (10–15 km), and >60 percent on Kodiak Island (100–180 km). As far as ~13 km from vent, Layer A is a pumice fall that tends to be vaguely shower-bedded, exhibiting a few slightly coarser intervals within meter-thick sections. On the mainland the deposit is everywhere fines-poor, having ≤1 percent ash finer than 63 µm as far as 36 km downwind.

Medial deposits are dominated by white crystal-poor high-silica-rhyolite pumice, much of which has a lustrous silky sheen owing to finely elongate vesicles that impart a fibrous texture. Some pumice has more equant vesicles or has equant vesicles superimposed on the tubular or stretched generation. Maximum pumice size (average of 5 largest at each site) decreases from 17 cm on the south flank of Trident (7 km from vent), to 1.5 cm at Geographic Harbor (36 km), and to 0.2 cm on the west coast of Kodiak Island (110 km). Corresponding average maximum lithic sizes measured at the same locations are 9 cm, 0.4 cm, and <0.1 cm. Scattered pumice clasts as large as 25 cm and lithics up to 15 cm can be found near the dispersal axis on the south slopes of Trident and Mount Katmai. Representing the vent-opening outburst of ejecta, Layer A contains as much as 24 weight percent lithic fragments as far as 13 km downwind, but lithic content drops off to ~10 weight percent along the Shelikof coast (40–42 km)

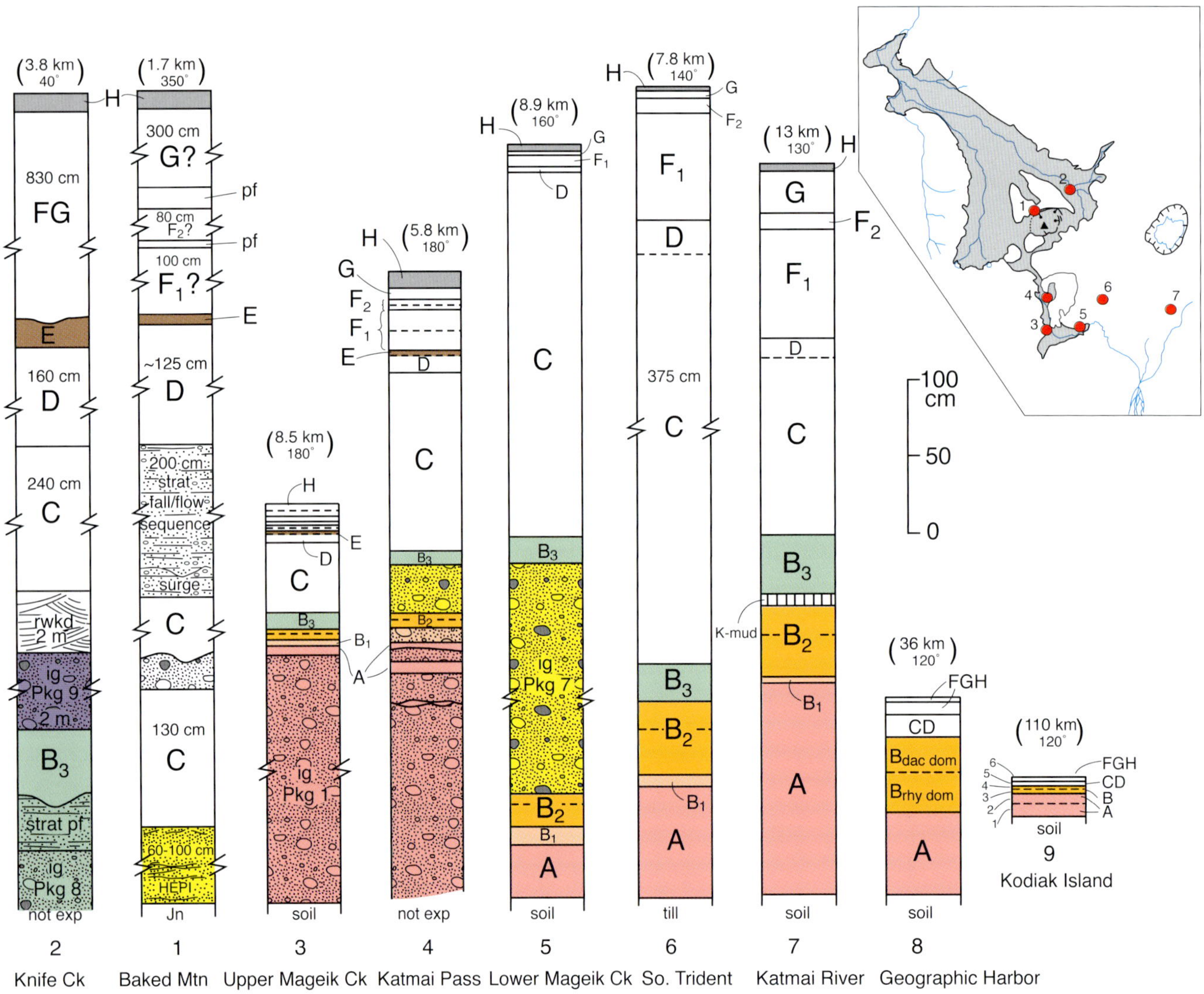

Figure 20. Representative fall-deposit sections measured at various distances and azimuths from Novarupta, as indicated at top of each column. Selected layers color coded for clarity of correlations. Inset shows numbered section locations on outline of VTTS ignimbrite and Katmai caldera (as in figure 5). For far-downwind section 9, color zones 1–6 in distal ash were correlated as indicated by microprobe and discussed by Fierstein and Hildreth (1992). Abbreviations: Jn, Jurassic Naknek Formation; rwkd, reworked; HEPI, high-energy proximal ignimbrite; ig Pkg, ignimbrite package; strat, stratified; K-mud, layer of hydrothermal mud ejected from Katmai caldera during its collapse.

and to only ~1 weight percent at Kodiak (170 km). The lithics are predominantly sedimentary rocks of the Naknek Formation, but they also include Trident andesites and Falling Mountain dacites, all three being lithologies that crop out adjacent to the vent at Novarupta.

The rhyolite pumice has only 0.5 to 3 weight percent phenocrysts, but liberation of crystals during eruptive fragmentation resulted in Layer A containing 2-3 weight percent free crystals on the south slopes of Trident and Katmai (5–13 km downwind), 4–9 percent at the Shelikof coast (40–42 km), and ~10 percent at Kodiak (170 km). Layer A is fairly well sorted (σ_{θ}=0.6–2.0), and sorting improves downwind (fig. 21). At a site 7 km from vent where median grainsize (Md) is 6.5 mm, σ_{θ} is 1.6, and at Kodiak where Md is 0.2 mm, σ_{θ} is 1.0. The quantitative data for this summary are tabulated in Fierstein and Hildreth (1992).

The only phenocrysts (Hildreth, 1983; Avery, 1992) common in the crystal-poor rhyolite pumice of Layer A are subequal amounts of plagioclase (≤1.2 mm; An_{25-29}) and dipyramidal quartz (≤0.8 mm), which contains abundant melt inclusions (Clocchiatti, 1972, 1975; Lowenstern, 1993). These are accompanied by sparse orthopyroxene (En_{49-50}) and titanomagnetite, much sparser ilmenite, apatite, and pyrrhotite, and rare zircon. There is little or no clinopyroxene in the rhyolite, although a trace is present within some Layer B rhyolite pumice, where contamination by coerupted dacite is a possibility. Although absent in most of the rhyolite pumice, trace amounts of unrimmed amphibole euhedra, as long as 0.4 mm, have been found in mineral separates from Novarupta lava-dome rhyolite, where dacite contamination is also a likelihood (Hildreth, 1983; Avery, 1992).

All-Rhyolite Ignimbrite

Accompanying fallout of Layer A during the first several hours of the eruption was emplacement radial to Novarupta of ash flows containing exclusively rhyolitic juvenile material. Because most of the all-rhyolite ignimbrite was soon covered by more voluminous ignimbrite that contains

Figure 21. Complete 1912 fallout section, 23 cm thick, including Layers A–H. Location is well south of dispersal axis, at site of Katmai village, 30 km southeast of Novarupta. Knife is 8.5 cm long. Section overlies organic-rich soil and is overlain by wind-reworked tephra.

a wide range of juvenile compositions, exposures of the former are limited to the periphery of the upper VTTS and to wind-stripped ridgecrests surrounding Novarupta basin. We described the all-rhyolite ash-flow deposits (Fierstein and Hildreth, 1992) as comprising three facies—high-energy proximal ignimbrite, valley-filling ignimbrite, and a thin marginal fringe facies—all emplaced concurrently with fallout of Layer A from the plinian column.

Rhyolitic High-Energy Proximal Ignimbrite (HEPI).—Within several kilometers of Novarupta, the only basal exposures of 1912 deposits are on windswept ridgecrests where the thick mantle of dacitic fall deposits (Layers C to H) have been stripped. At such locations, Episode I exposures are of stratified, lenticular, moderately to poorly sorted, pyroclastic-density-current deposits with wavy and lensoid bedding, scour and truncation of lower bedsets by overlying ones, and sparse ballistic ejecta. Many of the coarser layers and lenses are surgelike, reflecting high-velocity turbulent flow over proximal ridgecrests, and fairly well sorted (σ_ϕ=1–2). Most, however, represent more concentrated flows that left fines-rich, poorly sorted (σ_ϕ=2–4) deposits that drape the ridges and merge downslope into the thick valley-filling ignimbrite of the upper VTTS. Such high-energy strata were emplaced throughout Episode I, successive bedsets reflecting the compositional zonation of the overall sequence. The all-rhyolite bedsets are always basal, resting on basement rocks or on thin sections of Layer A fallout.

Good exposures of rhyolitic HEPI survive on Baked and Broken Mountains, Mount Cerberus (fig. 22), several northern spurs of Trident and Mount Katmai, atop Whiskey Ridge (fig. 23), and atop the Buttress Range (fig. 24). The all-rhyolite HEPI deposits range in thickness from 5–10 cm intervals interbedded with Layer A fallout in the saddles between the peaks of Trident, to 70 cm on the Buttress Range and as much as 4 m on Whiskey Ridge. Individual layers and lenses are typically 3–30 cm thick, lithic-rich, and commonly inversely graded; adjacent layers commonly exhibit contrasts in sorting, grain size (Md ~0.5 to 7 mm), and fine-ash contents (5 to 25 weight percent). As vent-opening proximal deposits,

Figure 22. All-rhyolite high-energy proximal ignimbrite (HEPI) deposits on northern ridge of Mount Cerberus, 3 km southwest of Novarupta and 250 m higher than margin of valley-filling massive ignimbrite at base of ridge. At top, darker heterogeneous HEPI deposit (containing andesite, dacite, and rhyolite pumice) overlies light-colored bedsets in which juvenile pumice is wholly rhyolitic. Both contain abundant lithic fragments. Ice-axe is 81 cm long. View is roughly westward.

the rhyolitic HEPI contain an abundance of lithic fragments, most layers having 10–15 weight percent, the extremes measured being 3 and 32 weight percent. As in Layer A, the lithics include sedimentary rocks of the Naknek Formation as well as lavas of Trident and Falling Mountain. In successive layers of various exposures, either Trident or Naknek lithics can predominate. Lithic fragments of Falling Mountain dacites are most abundantly exposed in basal HEPI sections on Baked Mountain.

Nearly all HEPI exposures lie along unobstructed direct lines from vent and are thought to have been deposited by moderately concentrated but turbulent upper parts of pyroclastic density currents that had developed gradients in particle concentration and were synchronously depositing valley-filling massive ignimbrite. Most HEPI exposures lie within 4 km of Novarupta, within lateral-blast range or close to the toe of a collapsing column. The all-rhyolite deposits on Whiskey Ridge (5 km from the Novarupta dome) and on the Buttress Range (8 km), however, were emplaced atop steep-walled ridges, respectively 400 m and 250 m above the adjacent VTTS floor, apparently demanding at least that thickness for the transporting, inferentially graded, blast-initiated currents. The thin all-rhyolite fringe deposit (described below) that extends across the slopes of Mount Griggs as far as 9 km from Novarupta is at least in part distal HEPI that was emplaced by a vent-opening lateral blast.

Rhyolitic Valley-Filling Ignimbrite (VFI).—Massive pink all-rhyolite ignimbrite is as thick as 13 m along the Griggs fork (fig. 25), ~6 m thick just south of Katmai Pass, and 4.5 m along Mageik Creek near Observation Mountain. In the Knife Creek and Lethe arms of the VTTS, such deposits are concealed by the main compositionally heterogeneous VFI, but the all-rhyolite VFI was certainly deposited earlier

Figure 23. Whiskey Ridge and the Knife Creek Glaciers, viewed east-northeastward from near Novarupta. Black, horizontal, marine Jurassic strata of Whiskey Ridge (right center) form steep-walled cleaver separating Glaciers 3 and 4, ~5 km east of Novarupta. Slender light-colored strip atop Whiskey Ridge is high-energy proximal ignimbrite (HEPI) deposit ~5 m thick (see fig. 42), which rests 400 m higher than margin of valley-filling ignimbrite on floor of Knife Creek valley in center of image. Two kilometers farther east, beyond tan ash-covered Glacier 4, another black cleaver separates Glaciers 4 and 5; along its narrow crest, HEPI deposit has thinned to ~1 m. In foreground is Glacier 1, overlain and surrounded by pumice-fall deposits of Episodes II and III, many meters thick. Glacier 2 is hidden from view between Glaciers 1 and 3.

throughout the upper VTTS because its all-rhyolite fringe facies flanks the Knife Creek margin of the main VFI as far down valley as the Juhle fork. In gorges incised to the base of the ignimbrite in the lower VTTS, however, no purely rhyolitic ignimbrite is exposed. Because the thin all-rhyolite VFI was emplaced more energetically than the later heterogeneous VFI, it extends laterally well beyond the upper-valley margins, but the heterogeneous VFI was far more voluminous and thus advanced incrementally twice as far down the valley. Lack of deep incision in the upper VTTS renders estimates fairly uncertain, but the volume of all-rhyolite ignimbrite is probably no more than 1 km^3 and that of the heterogeneous VFI roughly 10 km^3.

Rhyolitic VFI is poorer in pumice lapilli than most of the overlying heterogeneous VFI, pumice clasts larger than 1 cm rarely making up as much as 10 weight percent of it. Concentrations or trains of pumice that might mark flow-unit boundaries or depositional pulses are correspondingly uncommon in the massive rhyolitic ignimbrite, which contains 30–50 weight percent fines (≤63 µm). A few representative samples sieved yield Md~0.1–0.5 mm and σ_ϕ~2–3. Like Layer A and the rhyolitic HEPI, rhyolitic VFI is rich in lithic fragments, Naknek clasts generally exceeding lavas; measured lithic contents are in the range 13–15 weight percent.

At many places in the Mageik Creek drainage, the rhyolitic VFI is underlain, overlain, or intercalated with thin strata of plinian Layer A, but the fallout is commonly missing at the base, apparently having been synchronous with or scoured by the first flow. Along the Griggs fork, lenses of Layer A fallout are preserved locally within sections of all-rhyolite VFI. Such intercalations may be what led Curtis (1968) to interpret the pink rhyolitic VFI near Glacier 5 and the foot of Mount Griggs as Layer A fallout.

Figure 24. All-rhyolite high-energy proximal ignimbrite (HEPI) and dacite fallout section atop Buttress Range, in saddle above West Mageik Lake and 8 km west-southwest of Novarupta. Cross-stratified and lenticular HEPI deposit has alternating layers rich in fines or in rhyolite pumice granules. Resting on organic-rich soil, HEPI was emplaced ~200 m higher than nearby margin of valley-filling ignimbrite in Lethe arm of VTTS. Subsequent compositionally heterogeneous HEPI currents did not reach crest of Buttress Range. The all-rhyolite HEPI (R–HEPI), variously 20–70 cm thick (~30 cm here), is overlain by ~30 cm of Episodes II and III upwind dacite fall deposits (layers as labelled), which are here separated by prominent Layer E, which is fine-grained, cream-colored, and 3–4 cm thick. Shovel is 58 cm long.

Marginal Fringe Facies of All-Rhyolite Ignimbrite.—In contrast to the later-emplaced main heterogeneous ash flows (which outside the near-vent region were strictly valley-confined), the rhyolitic ash flows were more energetic and swept far up the surrounding slopes. The all-rhyolite HEPI grade into thin fringe deposits in and beyond the Trident saddles, and a fringe of rhyolitic ignimbrite 5–20 cm thick widely mantles the lava-flow benches at the eastern base of Mount Mageik (fig. 26). The latter are best exposed on lava-flow benches up to 250 m higher than the valley-confined ignimbrite along Mageik Creek, but they can locally be traced downslope to their merger with the 5-m-thick massive rhyolite ignimbrite along the valley axis.

A massive or vaguely stratified pink deposit on the lower flanks of Mount Griggs forms a run-up fringe continuous with the thick all-rhyolite VFI along the Griggs fork. It thins gradually from 13 m along the stream axis to only a few centimeters at distances 1–2 km upslope, and it extends ~7 km across the southwest slope of Mount Griggs as far as the Juhle fork at elevations up to 450 m above the VTTS floor. Across-slope transport of the feather edge of the ignimbrite is recorded by charred 1912 willows flattened in the down-valley direction. Snowmelt on the lower slopes of Mount Griggs remobilized parts of this fringe deposit, producing many small all-rhyolite gulch-confined debris flows that underlie and interfinger with the dacite pumice-fall deposits of Episodes II and III.

The rhyolitic fringe deposits, sampled and sieved to record their apparent diversity, have wide ranges in grain size (Md~0.5 to 3 mm), sorting (σ_ϕ~1.7 to 3), and structure, ranging from massive to crudely stratified. Samples sieved have ~5 weight percent lithics and 3–6 weight percent free crystals.

A few centimeters of fine rhyolitic ash beneath thin Layer A on Observation Mountain suggests that an ash flow

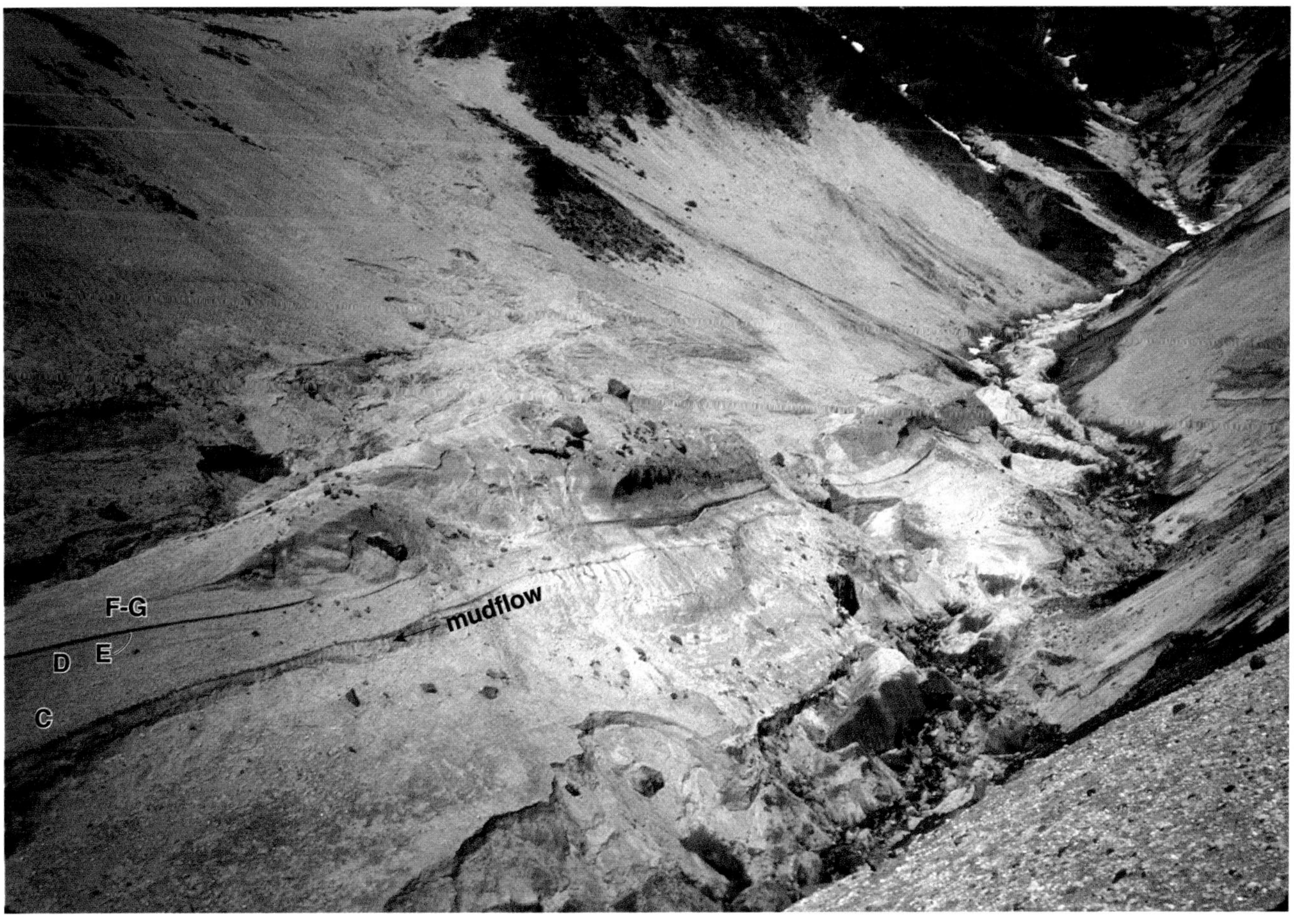

Figure 25. Pink all-rhyolite ignimbrite (Package 1) along axis of upper Griggs Fork, 9 km from Novarupta in an unobstructed northeasterly line-of-sight direction. It was far more mobile than all subsequent ignimbrite packages. Ignimbrite is ~8 m thick here but as thick as 13 m just downstream (to left). It is overlain successively by a thin brown ashy mudflow deposit, 2.4 m of Layers C–D dacite fallout, a prominent 4-cm-thick rib of Layer E, and remnants of Layers F–G, which are interrupted by mudflows and overlapped by pumice-rich scree and alluvium.

with a superjacent ash cloud went through Katmai Pass at the onset of eruption, at least minutes before fallout of plinian Layer A had expanded broadly enough to cover that area, at the southwest edge of its dispersal sector. Similar rhyolitic ash, probably likewise elutriated from the initial all-rhyolite ash flows, is locally the basal deposit atop the Buttress Range and on the cross-valley moraine near Three Forks (upwind sites where Layer A was not deposited). Thin ash at all three sites probably resulted from turbulent ash clouds that detached and extended beyond the termini of their associated valley-confined all-rhyolite flow units.

The evidence that the several kinds of purely rhyolitic deposits just described are depositional facies produced in common by an eruptive episode only a few hours long is summarized by Fierstein and Hildreth (1992, p. 662). Synchronous with a powerful sustained plinian column that produced regionally dispersed rhyolite fallout, voluminous density-stratified ash flows moved radially in all directions from Novarupta basin, distributing massive VFI along the valleys, wavy lenticular strata on ridgecrests, and fine rhyolitic ash dispersed laterally by billowing clouds.

Layer B

Use of this term has a confusing history because Curtis (1968) first applied it to fine-grained pink ignimbrite deposits along and near the Griggs fork of Knife Creek that contain both rhyolite and andesite-dacite components. These valley-fringing deposits are compositionally heterogeneous but still rhyolite-dominant; they thin rapidly away from the main heterogeneous VFI that fills the Knife Creek arm, and they nowhere overlie it. We now assign them to ignimbrite Packages 2 and 3 (Fierstein and Wilson, 2005). The more extensive fringe of all-rhyolite ignimbrite (fig. 25) that underlies these "Layer B" ash-flow deposits in that area was assigned by Curtis to Layer A. Sparse

Figure 26. View southwestward across upper VTTS from high on Mount Griggs. Knife Creek arm in foreground, River Lethe arm in right distance, and between them the intermediate valley separating siltstone ridges of Baked Mountain and Broken Mountain (nearer camera); all three valleys are floored by ignimbrite, margins of which are marked by abrupt color change. Baked and Broken Mountain basement rocks are draped by pale gray pumice-fall deposits. On center skyline, four-peaked Mount Mageik stands 19 km from the camera and rises 1,500 m above the valley floor. To its right are the black cone of Mount Martin and glacier-clad Alagogshak Volcano, both along the rangecrest. At left toe of Mount Mageik, a pair of dacite domes, Mount Cerberus (farther from camera) and Falling Mountain, frame the entrance to Katmai Pass, through which the VTTS ignimbrite extends. Overlying ignimbrite on valley floor in right distance, black fans are andesite-dacite alluvial gravels from Mount Mageik.

lenses of fallout did locally survive scour during emplacement of these marginal-facies ignimbrites along the Griggs fork, but only small fractions of Curtis' units A and B in this sector are true fallout layers. Hildreth (1983) referred to these deposits as the "feather edge" of all-rhyolite and compositionally heterogeneous ignimbrites emplaced successively. When we later studied in detail the 1912 deposits south of Trident and Mount Katmai, however, it became clear that true plinian fall deposits of heterogeneous composition accumulated atop Layer A in the downwind sector (figs. 15, 20). The fall deposits south of the volcanic axis thus provided a firm basis for redefining Layer B, independent of Curtis' original sites on the Griggs fork (Fierstein and Hildreth, 1992).

Layer B was accordingly restricted to the plinian fallout that accompanied emplacement of the main compositionally heterogeneous ignimbrite sequence that filled the VTTS and in part spilled through Katmai Pass into Mageik Creek (Hildreth, 1983; Fierstein and Hildreth, 1992; Fierstein and Wilson, 2005). Layer B was divided into three subunits (figs. 15, 16, 21) on the basis of grain size and compositional proportions (as elaborated in table 1 of Fierstein and Wilson, 2005), even though no depositional break is recognized within Layer B or between it and purely rhyolitic Layer A.

Layer B_1 is a conspicuous terra-cotta horizon 1–11 cm thick that caps the white pumice-fall deposit of Layer A. Its pumice lapilli show little difference in size from subjacent Layer A, but they include for the first time sparse dacitic and andesitic pumice as well as 35–45 weight percent oxidized co-ignimbrite ash, which coats lapilli and imparts the color. The layer is related to the first heterogeneous ash flows (Package 2) to spill into Katmai Pass. It thins laterally away from the valley-confined deposit of those ash flows (fig. 16), as well as downstream, where it is identifiable no farther than the lower slopes of Trident and Mount Katmai (presumably the limit of the co-ignimbrite ash cloud). Layer B_1 is very poorly sorted (σ_θ~4-5) owing to its co-ignimbrite ash contaminant, but it remains fairly coarse (Md~2 to 5 mm) owing to its plinian lapilli-fall component. As in Layer A, its lithic fragments are predominantly from the Naknek Formation but include subordinate Trident and Falling Mountain lavas. Pumice counts for Layer B_1 yield 82–99 percent rhyolite, similar to pumice proportions in ignimbrite Packages 2 and 3, which were the first to reach the lower VTTS near Three Forks.

Layer B_2 is a compositionally mixed fall deposit as thick as 73 cm on higher slopes of Trident and 30–50 cm thick at the south base of Trident and Mount Katmai. It typically shows slight inverse grading, and its top several centimeters is typically burnt orange in color owing to a small contribution of oxidized andesite-dacite co-ignimbrite ash. Numerous field and lab counts show pumice in its lowest part to be as much as 93 percent rhyolitic, the proportion declining upward to as little as 40 percent, as dacite and andesite pumice become progressively more abundant. Rhyolite dominance is maintained for half to two-thirds of Layer B_2 as dacite pumice increases from <5 percent near the base to as much as 60 percent in the upper third; the andesite proportion is small and varies erratically. The upward change in clast proportions in Layer B_2 is consistent and well documented from the proximal area all the way to Geographic Harbor and Kodiak Island (Fierstein and Hildreth, 1992). On the south slopes of Trident and Mount Katmai, average maximum pumice clasts are 9–12 cm and maximum lithic clasts ~5 cm. Lithic content is as great as 17 weight percent on the slopes of Trident but decreases to ~6 weight percent at Katmai River and Dakavak Bay. The lithic suite is still dominated by Naknek, Trident, and fewer Falling Mountain clasts, but fragments of welded-tuff vitrophyre recycled from vent-filling fallback ejecta (Hildreth, 1987) make their first appearance. Layer B_2 is moderately or well sorted (σ_θ~0.7–2.7), despite containing ~8 to 20 weight percent fine ash (≤63 μm), which is most abundant near its top. Its content of free crystals increases from ~5 weight percent on the slopes of Trident and Mount Katmai (7–13 km from vent) to 14 weight percent at Dakavak Bay (40 km downwind), a common pattern as the fraction of pumice lapilli declines and grain size diminishes with distance. A few centimeters of Layer B_2 that survived scour in upper Knife Creek are intercalated within ignimbrite Package 7 (Fierstein and Wilson, 2005).

Layer B_3 is a compositionally mixed fall deposit that extends the trend in pumice proportions represented by Layers A, B_1, and B_2. Numerous pumice counts give 20–38 percent rhyolite near its base and 11–25 percent near its top; the subunit contains as much as 6 percent andesite scoria and is the earliest fall deposit to be dacite-dominant (59–87 percent) throughout. Pumice proportions are like those of late (but not final) ash-flow packages in the VTTS. Layer B_3 is 30–50 cm thick on the southern slopes of Trident and Mount Katmai, where it is distinguished by a slightly finer-grained base, which separates it from B_2, and by a dusting of oxidized co-ignimbrite ash that coats lapilli and imparts a burnt-orange color throughout but most prominently at the top. Although it contains ~10 weight percent fine ash and ~25 weight percent free crystals, sorting is good to fair (σ_θ~0.9–2.5), even at the most proximal exposures. The subunit is coarsest in its middle, where extreme pumice clasts reach 20 cm, average maximum pumice reaches 16 cm, and average maximum lithic values are 4–7 cm on the southern slopes of Trident and Mount Katmai. The lithic suite still contains Naknek, Trident, and sparse Falling Mountain clasts, but for the first time in the fallout sequence, recycled welded fallback ejecta predominate. Thin remnants of Layer B_3 are locally preserved on the VTTS side of the volcanic axis, outside its main southeasterly dispersal sector; they occur directly beneath Layer C on Broken Mountain, Mount Griggs, the Buttress Range, and near Glacier 4, but wherever the main VTTS ignimbrite was deposited, compositionally equivalent Layer B_3 is absent.

Compositionally Heterogeneous (Main) Valley-Filling Ignimbrite (VFI)

The main sequence of ash flows that filled the VTTS was compositionally equivalent and synchronous with fallout of

Layer B. In the upper VTTS their marginal fringe directly overlies the all-rhyolite ignimbrite described earlier. The emplacement progression from rhyolite-dominant to rhyolite-poor ash flows has been characterized in progressively increasing detail by Curtis (1968), Hildreth (1983), Fierstein and Hildreth (1992), and Fierstein and Wilson (2005), who mapped the limits of nine successive compositionally distinguishable packages of ignimbrite (fig. 27). Of the nine exposed in the VTTS, four also passed southward into Katmai Pass (figs. 16, 28), and of those four, two feathered out shortly south of the pass, and two extended as far as 10 km down the Pacific slope to the foot of Observation Mountain. Of the latter pair, the first was the energetic all-rhyolite package, which is 3–6 m thick along the valley of Mageik Creek, and the other was

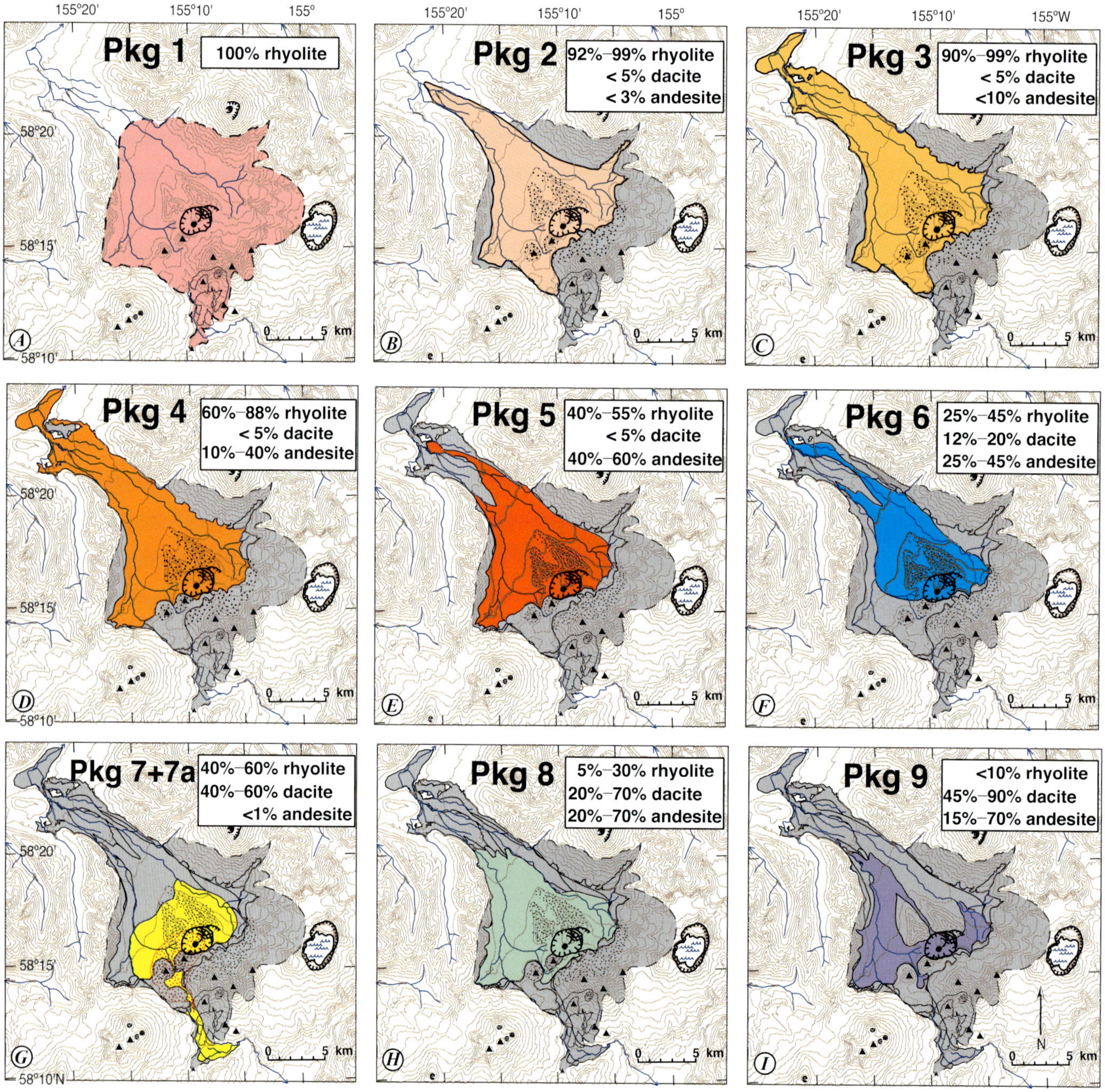

Figure 27. Distribution of nine successive packages of 1912 ignimbrite mapped by Fierstein and Wilson (2005) on basis of changes in pumice proportions, as given in insets. For each panel, gray shows extent of all previously emplaced packages. Black stipple indicates distribution of high-energy proximal facies (HEPI) that accompanied valley-filling deposits. Red stipple in panel G shows limits of Package 7a, deposit of a short-lived blast during eruption of Package 7. Emplacement sequence of nine packages is estimated to have taken 16 hours.

the seventh package, which is as thick as 20 m along Mageik Creek and is compositionally unique in having roughly equal proportions of dacite and rhyolite pumice but only a trace of andesite. In the VTTS, two of the packages (the third and fourth) extended at least 23 km down the glaciated valley of the Ukak River (fig. 29), and three more packages (the second, fifth, and sixth) extended as far as Three Forks (fig. 30), 17 km from Novarupta.

The combined valley-filling ignimbrite probably ranges between 100 m and 200 m thick in the upper VTTS (Curtis, 1968; Hildreth, 1983); it is ~35 m thick in the lower VTTS where incised to its base and 10–20 m thick at eroded distal termini (fig. 31). Because the sheet is not deeply incised in either the River Lethe arm or the Knife Creek arm of the upper VTTS, its total volume is poorly established but estimated to be 11±2 km^3. Of this total, no more than 10 percent is attributable to the opening all-rhyolite package and only ~1 percent to the later packages along Mageik Creek, so the compositionally heterogeneous ignimbrite in the VTTS has a volume of roughly 10 km^3.

Where exposed, much of the medial to distal sheet is nonwelded, but steep-walled gorges in the middle and upper VTTS are cut in indurated ignimbrite (fig. 32), both sintered vitric tuff and tuff indurated by vapor-phase crystallization. Welded tuff with flattened fiamme (fig. 33) is exposed in situ at only a few places, all in the upper VTTS, but blocks of such welded ignimbrite ejected at phreatic explosion craters provide evidence for widespread welding below the level of stream incision in both arms of the valley.

Following the initial all-rhyolite package, all eight subsequent packages contained andesite, dacite, and rhyolite pumice in varied proportions (fig. 34). Chemically analyzed rhyolite pumice from the main VTTS ignimbrite ranges in SiO_2 from 77.1 to 77.8 percent, and the accompanying andesite and dacite pumice form a compositional continuum extending from 57.9 to 68.6 percent SiO_2. Packages 2 through 4 are rhyolite-dominant with subordinate dacite and progressively increasing andesite; Packages 5 and 6 carried subequal amounts of rhyolite and andesite with subordinate dacite; Package 7 had subequal dacite and rhyolite with only sparse andesite; and, finally, Packages 8 and 9 (figs. 35, 36) carried abundant andesite and dacite but only minor rhyolite. Banded pumice is present throughout Packages 2 to 9, constituting 3–5 percent of pumice clasts overall but ranging locally between 0 and 37 percent.

Correlations of fallout layers and ignimbrite packages that erupted concurrently (Fierstein and Wilson, 2005) are based on field counts of pumice proportions (rhyolite, dacite, andesite,

Figure 28. Katmai Pass viewed southward from summit of Falling Mountain. Mount Cerberus is just out of view to right. Fumaroles from thick ignimbrite at lower right lured Griggs and Folsom into VTTS in 1916 and remained active when Fenner returned in 1923. Benches along pass are lava flows from Mageik (right) and Trident (left). All benches are veneered by thin ignimbrite and draped by dacite fall deposits, with exception of black 1958 and 1959 Trident lavas on left rear side of defile. 1912 ignimbrite Package 7 penetrated 7 km beyond the pass and is 10–20 m thick along axis of Mageik Creek at foot of Observation Mountain (left distance).

Figure 29. Terminus of VTTS ignimbrite viewed westward from basaltic plateau ("Sayre Mesa" of Hildreth and others, 2004). Far side of valley-filling ignimbrite (light colored) is 10 km from the camera and 1,000 m lower. Deposit is 10–15 m thick where Ukak River is eroding its terminus at right. In middle distance, ash flows were blocked by moraines and largely diverted to far west margin, where narrow Ukak River corridor fed terminal Ukak lobe. A few ash-flow lobes passed though saddles in moraines and feathered out into brush-covered hills of glacial debris. Margot Creek in distance; Mount Katolinat at upper right. Basalt on Sayre Mesa was studied and dated by Hildreth and others (2004).

Figure 30. Aerial view of lower VTTS, Three Forks, and Ukak River corridor. Windy Creek at left, River Lethe in center, and Knife Creek at right conjoin at Three Forks to form Ukak River. Ignimbrite here is nonwelded and 10–25 m thick. Base is exposed atop till along Knife Creek at right center. Beyond Knife Creek, ash flows banked against brown moraine and were diverted into narrow (400-m-wide) Ukak River corridor at left. In distance, white deposits are aeolian sand winnowed from 1912 deposits, and pink strips are tongues of ignimbrite that bypassed moraines and feathered out into brush.

and banded) as well as on physical intercalations (fig. 37). Flow packages, however, generally have higher andesite fractions than time-correlative fallout, reflecting the fact that little andesite pumice was distributed by the high plinian plume but most was, instead, preferentially partitioned into lower fountains ejected around the periphery of the plinian column (Hildreth, 1983). The same annular pattern recurred during Episodes II and III, when andesite-enriched flows, surges, and small proximal falls were released sectorially around the base of plinian columns that carried aloft little other than dacitic ejecta (Fierstein and others, 1997; Houghton and others, 2004).

The issue of what constitutes an ignimbrite "flow unit" has remained under discussion for decades, and nowhere is it more ambiguous than in the VTTS. Hildreth (1983, p. 52–3) called attention to several late flow units or packages thereof (figs. 31, 38) but thought that much of the ignimbrite exposed in the VTTS was emplaced by a continuously zoned outflow (fig. 39) that partitioned only distally into numerous secondary flows as it fitfully ground to a halt. Fierstein and Hildreth (1992, p. 664–5) illustrated and discussed further the problems of distinguishing among vent-derived flow units and secondary units produced by

Figure 31. Flow units and packages in massive nonwelded ignimbrite, southwest wall of Knife Creek, <1 km upstream from Three Forks. Its base resting on dark brown till, multipackage ignimbrite is here ~12 m thick and overlain by a thin white layer of dacite fallout, which is capped at surface by brown mudflows ~1–2 m thick. Dark brown ignimbrite Package 5 fills swale eroded into white Package 4 beneath it.

Figure 32. Knife Creek gorge ~3 km upstream from Three Forks, cut largely in ignimbrite Packages 2 through 4. Gorge here is ~30 m deep; stream cut down through nonwelded tuff into sintered or partially welded tuff, hackly jointed, that walls inner gorge. Flow units and variously sorted layers and lenses are conspicuous on walls. Left-bank rim is capped by thin white layer of dacite fallout, overlain in turn by thin brown mudflow deposit. Mount Katolinat on skyline.

budding, internal shear, overlapping flow paths, and discontinuous deposition from the base of moving ash flows. Because persistent interflow partings are seldom seen on gorge walls in the middle VTTS, they thought that "as much as 90 percent of the exposed sheet was emplaced quasi-continuously." Detailed mapping by Fierstein and Wilson (2005) of nine distinguishable packages of flows made an important advance in understanding the processes of outflow and deposition. Questions about distal partitioning of secondary units remain, but Fierstein and Wilson demonstrated contacts between successive packages, some sharp and some cryptic, as defined variously by changes in pumice proportions or ignimbrite matrix color, finer-grained bases of flow units, lobate accumulations of distinctive composition, compaction swales filled by later packages, shallow erosional channels filled by the flows that scoured them, slump surfaces left by remobilized ignimbrite and filled by a later package, and (largely south of the vent) fall deposits intercalated between flow units. Emplacement processes and flow units are discussed in more detail in chapter 6, below.

Compositionally Heterogeneous High-Energy Proximal Ignimbrite (HEPI)

Within ~3 km of Novarupta, the more widely distributed all-rhyolite HEPI is overlain by thicker deposits of compositionally heterogeneous HEPI. Like their earlier rhyolitic counterparts, these are stratified, lenticular, and highly varied in grain size and sorting. They are exposed on windswept proximal ridges wherever Episode II and III fallout has been eroded away—on Trident, Falling Mountain, Mount Cerberus, Whiskey Ridge, and all spurs of Baked and Broken Mountains. Deposited by high-velocity currents radial to Novarupta, the HEPI represent a dynamic facies that contrasts with time-equivalent VFI flows that were emplaced rather sluggishly by the time they reached the middle and lower VTTS (Hildreth, 1983). Nonetheless, the sequence of HEPI layers exhibits pumice proportions similar to most (but not all) of the sequence of VFI packages that filled the VTTS.

Most exposures are 1–5 m thick and consist of 5–25 discrete layers, each typically 1–15 cm thick but in places as

Figure 33. Eutaxitic welded ignimbrite of Package 8 along Glacier 1 fork of upper Knife Creek. Fiamme are both andesitic and dacitic. Ice axe is 81 cm long.

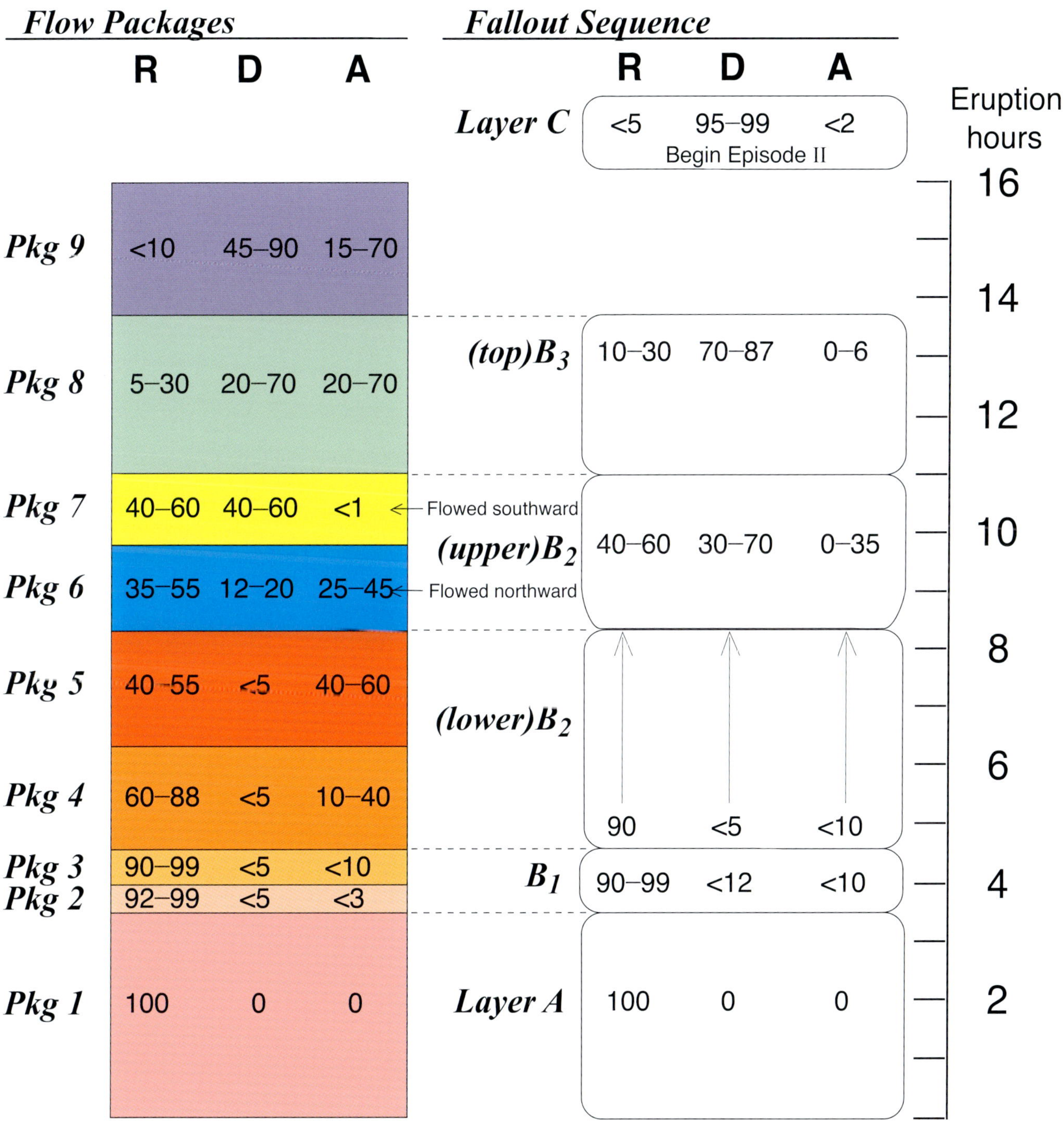

Figure 34. Schematic diagram illustrating eruption chronology for Episode I and correlations among compositionally equivalent ignimbrite packages and fall units. Ranges in proportions of pumice clasts—R, rhyolite, D, dacite, and A, andesite—based on field and laboratory counts, are indicated for all. Vertical scale is time; no thickness data intended. Inferred durations of plinian fall phases were elaborated by Hildreth and Fierstein (2000) and Fierstein and Wilson (2005), and they are summarized in figure 60, below. Durations of compositionally distinguishable plinian subunits provide constraints on time intervals available for emplacement of ignimbrite packages.

▲

Figure 35. Sintered to partially welded ignimbrite of Package 8, cut by upper Knife Creek ~400 m upstream from its confluence with Griggs fork. Welded platform is overlain by 1.5-m-thick ledge of sintered Package 9, rich in coarse dacite and andesite pumice. This in turn is overlain by ~2 m bedded sequence of fluvially remobilized pumice and ash (on which geologist is squatting) that marks eruptive lull between Episodes I and II. At upper left, dacitic Layer C and a remnant of Layer D are fumarolically oxidized. Terrace gravels at top were deposited before gorge incision into ignimbrite here.

◀ **Figure 36.** Stream-scoured surface of sintered ignimbrite Package 9, rich in coarse dacite pumice (along with subordinate andesite and sparse rhyolite). Underlying welded dark-gray Package 8 crops out behind geologist, along Knife Creek near confluence with Griggs fork.

thick as 35 cm. HEPI sections on Falling Mountain (fig. 40) and the long south spur of Baked Mountain (fig. 41) are as thick as 10–15 m and consist of as many as 110 layers. Atop Whiskey Ridge, 5 km from Novarupta, where the all-rhyolite HEPI is 4 m thick, the mixed HEPI is only ~1 m thick (fig. 42); and 2 km farther northeast, atop the cleaver separating Glaciers 4 and 5 (fig. 23), these packages thin to 104 cm and 9 cm, respectively. Successive pulses commonly scoured earlier layers, leaving numerous unconformities. Most layers are planar or gently undulating, but, because many pinch out laterally within 5–15 m, they are lenticular or lobate at outcrop scale. Layers are variously (1) fines-rich and matrix-supported, (2) dominantly coarse to medium ash with pumice granules, or (3) clast-supported beds rich in pumice lapilli.

Sieve data and componentry for the VTTS ignimbrite were given by Fierstein and Hildreth (1992). As in most ignimbrites, sorting in the VFI is generally poor ($\sigma_\varnothing$~3–4), ash finer than 63 µm makes up 30-50 weight percent of the deposits, and ash finer than 1 mm 60 to 80 weight percent, although pumice blocks and lapilli are ubiquitous. In HEPI, however, the variance in fine ash content is great, ranging from <10 to ~50 weight percent, even among adjacent layers. Median grain size ranges widely between 0.1 and 1 mm in most of the VFI but is as coarse as 4 mm in some HEPI layers and probably also in a few coarse VFI intervals not sieved. Sorting in the many lensoid HEPI layers is quite variable ($\sigma_\varnothing$~1.5–4), as is their lithic content, which can be lower than 1 weight percent or as great as 5–16 weight percent; in most VFI, lithic content ranges from 1–6 weight percent. Free crystals are generally 2–7 weight percent of VFI but locally are as abundant as 20–30 weight percent; in the lensoid HEPI, they range even more widely, from 10 to 50 weight percent. Many HEPI layers show coarse-tail inverse grading of pumice, but only a minority exhibit

Figure 37. Intercalated ash-flow and fall deposits 2.5 km southeast of Katmai Pass and ~300 m south of terminus of 1959–60 Trident lava flow. All-rhyolite ignimbrite Package 1 is ~1.5 m thick and has layers and lenses of all-rhyolite (Layer A) fallout within and above it. Bright salmon Layer B_1 is 8 cm thick here and consists of mixed fall and flow material. Ignimbrite Package 7 is 30 cm thick here, near the pass, but it thickens to 10–20 m downstream along Mageik Creek. It is sandwiched here by 8 cm of rhyolite-dominant Layer B_2 below and 10 cm of heterogeneous Layer B_3 above, which is in turn capped by >1 m of dacitic layer C. Knife (near center) is 8.5 cm long.

cross-bedding. No accretionary lapilli or other evidence of involvement of external water were noted in the HEPI.

The compositionally heterogeneous HEPI were less mobile than their all-rhyolite predecessors. They failed to crest the Buttress Range or to run up the lower slopes of Griggs and Mageik, and on Whiskey Ridge and Mount Cerberus they are much thinner than the rhyolite HEPI beneath them. As with the all-rhyolite HEPI, we envisage deposition on ridges well above the valley floor from turbulent but fairly concentrated upper levels of density-graded currents that were depositing poorer sorted VFI just downslope (fig. 43). The turbulence that favored scour, wavy bedding, and abrupt changes in sorting may have been enhanced by topographic roughness, by hydraulic jumps off ridgecrests, and by location within or near the deflation zone of collapsing fountains, consistent with a small ballistic contribution in some layers.

Time Constraints for Episode I

Intercalations of plinian fall Layers A and B_{1-3} with pyroclastic flow deposits in Katmai Pass, along Mageik Creek, and locally in upper Knife Creek show clearly that these falls and flows were produced by the same source column and emplaced contemporaneously (Fierstein and Hildreth, 1992). Details of compositional fluctuations linking the fallout sequence with the sequence of ignimbrite packages are elaborated by Fierstein and Wilson (2005). They graphically summarized the fall-flow correlations (fig. 34) and placed the sequences in a temporal framework spanning 16-hour-long Episode I. General constraints on emplacement times are imposed by recorded intervals of ashfall at Kodiak (Hildreth, 1983), seismicity recording fitful collapse of Katmai caldera (Abe, 1992; Hildreth and Fierstein, 2000), and phreatic syncollapse Katmai

Figure 38. Contact between compositionally contrasting ignimbrite packages above west shore of West Mageik Lake. Nonwelded rhyolite-dominant Package 4 is overlain by sintered andesite-rich Package 5. Thin contact zones are locally vaguely stratified (as at right) and commonly scoured (as at left). Ice axe is 81 cm long.

Figure 39. Massive nonwelded ignimbrite showing change in compositional proportions without any obvious depositional break. Exceptional exposure, ~10 m high, is near northeast edge of VTTS where Juhle fork of Knife Creek turns abruptly from valley margin toward its interior.

Figure 40. Matrix-rich oxidized ignimbrite containing abundant coarse clasts capping summit of Falling Mountain dome, 1.2 km from Novarupta dome. Pyroclastic deposits are underlain by dacite dome lava, which was sheared off in 1912 eruption, forming steep scarp. Section is overlain by white dacite pumice falls of Episodes II and III. Ignimbrite is 4–5 m thick here but thickens downslope to >13 m where plastered against scarp nearby (see fig. 71); lower part of section consists of stratified HEPI with highly variable sorting and fines contents. At upper left is scarp marking southeast wall of vent structure, here consisting of Trident-derived andesite lava flows heavily mantled by 1912 ejecta.

Figure 41. Compositionally heterogeneous HEPI bedsets on crest of south ridge of Baked Mountain, 2.2 km northwest of Novarupta dome. Variably sorted, oxidized strata are 10–15 m thick here, consisting of ~110 layers and lenses, each 1–30 cm thick. Ridgecrest is 250 m above valley-filling massive ignimbrite on adjacent valley floor and lies directly athwart main western outlet from Novarupta basin. In background, section wedges out ~100 m higher against Jurassic bedrock; on summit pinnacle (180 m higher than camera) is preserved a remnant of earlier all-rhyolite HEPI but none of these heterogeneous packages. Off-white dacite fall deposits, once several meters thick here, have been widely stripped by windscour. Ice axe (left) is 81 cm long.

mud layers intercalated within the Layer B fallout sequence (Hildreth, 1991). Interpretive linkage of the Katmai seismicity and mud layers places the B_2-B_3 transition near midnight, ~11 hours into the plinian eruption (Hildreth and Fierstein, 2000). Because the distinctive seventh package is directly overlain by Layer B_3, it can thus be inferred that at least 80 percent of the VTTS ignimbrite was emplaced within the first 11 hours. If accumulation times for the $B_{1\text{-}3}$ intervals were approximately proportional to fallout thickness, then Layer A represents ~3.5 hours, and Layer B_1 would represent ~1 hour, Layer B_2 ~6.5 hours, and Layer B_3 and ignimbrite Packages 8 and 9 together ~5 hours (Fierstein and Wilson, 2005).

Episode I ended with a complete eruptive hiatus, marked by deposition of water-reworked pumice and ash atop sintered Packages 8 and 9 in upper Knife Creek (figs. 35, 44).

Eruptive Episodes II and III

A sequence of fall units, predominantly dacitic, rests directly upon the main VTTS ignimbrite (both VFI and HEPI), or upon Layer B_3, or locally upon water-reworked deposits below the snouts of the Knife Creek Glaciers. Curtis (1968) divided this fall sequence into six units (Layers C through H), largely on the basis of the relatively simple medial fall succession in the upper Knife Creek arm, 3–10 km northeast of Novarupta. Despite a few complexities, Fierstein and Hildreth (1992) were able to extend Curtis' framework northwestward beyond Three Forks and southeastward across Mount Katmai to Shelikof Strait and Kodiak Island (figs. 19, 20). They recognized that the C–D and F–G pairs were each deposited virtually continuously but that fine-grained Layer E, which separates the pairs in the VTTS, represented a time break of hours—probably longer than the break between Layers B and C. On this basis, the C–D and F–G time intervals were defined as Episodes II and III, respectively. Proximally to medially, all four of these pumice-fall layers (figs. 17, 45) have faint plane-parallel internal bedding, moderate to good sorting, and aerodynamic equivalence among pumice and lithic clasts, characteristics of material deposited from the umbrella cloud of a high sustained buoyant plinian plume. Layers E and H are ash-rich beds of composite origin, formed principally by slow-settling coplinian fines, as well as by ash that had been elutriated from small proximal pyroclastic density currents during C–D and F–G plinian episodes and that settled after termination of those episodes.

Using regional isopach data and microprobe analysis of coplinian ash partitioning, Fierstein and Hildreth (1992)

Figure 42. Stratified oxidized HEPI deposits atop Whiskey Ridge, 5 km east of Novarupta and 400 m above valley-filling ignimbrite on floor of Knife Creek arm at foot of ridge (see figs. 23, 73). Resting on Jurassic bedrock, section consists of ~4 m of all-rhyolite HEPI, ~1 m of heterogeneous HEPI, and coarse lag of windscoured dacite pumice-fall at surface. Mount Griggs at upper right.

calculated eruptive volumes of 4.76 km^3 and 3.44 km^3 for Episodes II and III, respectively; these reduce to magma volumes of 2.0 km^3 and 1.66 km^3. Duration of Episode II ashfall at Kodiak was 26 hours and of Episode III ~10 hours. There need be no simple correlation, however, between fallout intervals 170 km downwind and the duration of eruptive episodes at vent, owing to altitudinally varied transport velocities and azimuths, wind shear, progressively slower settling of ash that becomes finer with distance, and turbulent retention of fine coplinian ash aloft. Ashfall intervals recorded at Kodiak are certainly longer than plinian eruptive intervals, but they may not be far wrong if settling times for Layers E and H are included.

Unlike Episode I deposits, in which compositional equivalence permits confident correlation of ignimbrite packages with regional fall deposits, there is considerable compositional contrast between the dacitic plinian fall deposits (Layers C–D and F–G) and coeval proximal deposits intercalated with them near Novarupta. Within ~3 km of the Novarupta lava dome, proximal products of Episodes II and III include alternations and mixtures of locally and regionally dispersed fall ejecta and several thin deposits of

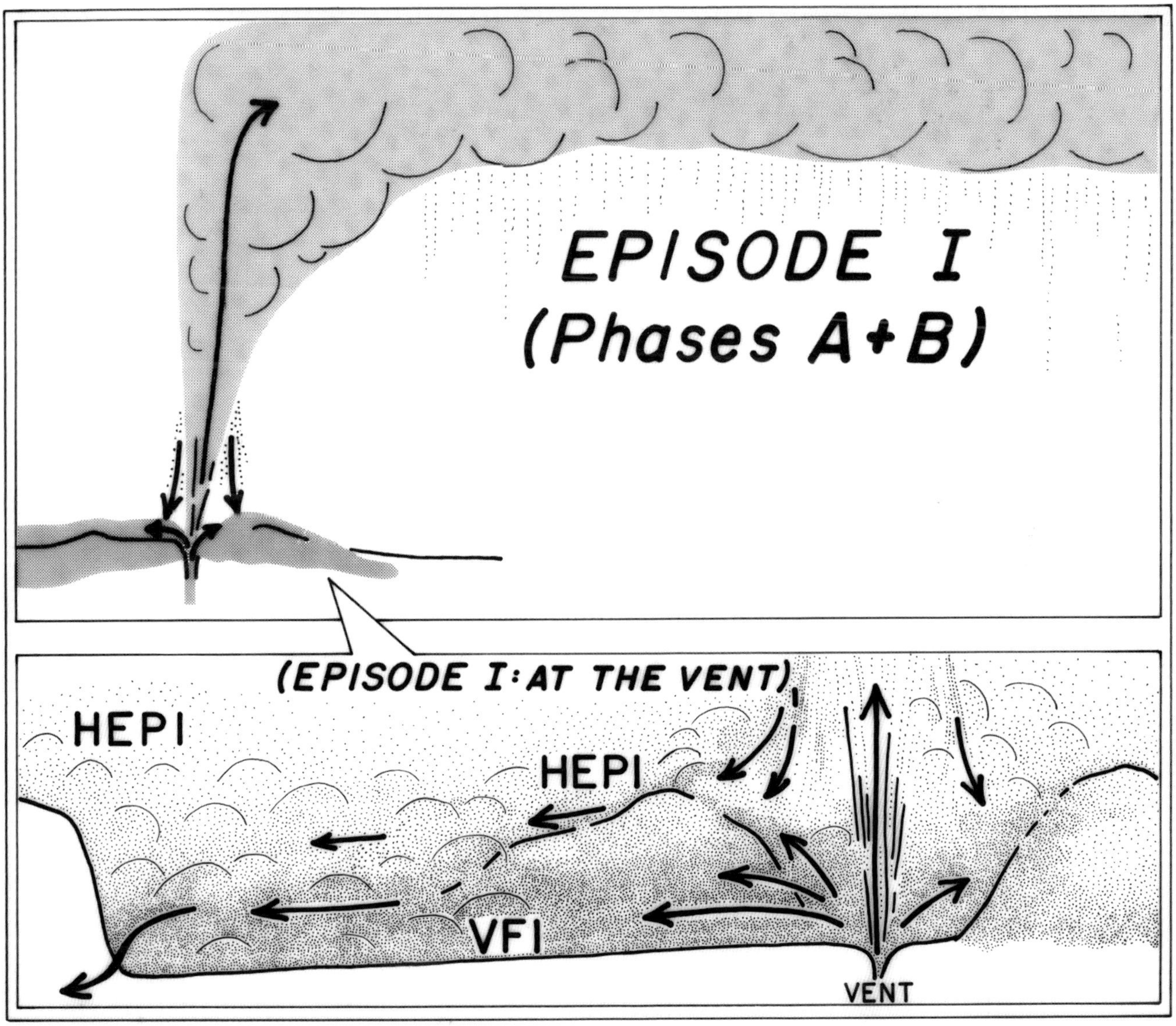

Figure 43. Conceptual sketch illustrating Episode I regional fallout from high plinian column and synchronous emplacement of massive valley-filling ignimbrite (VFI) and stratified high-energy proximal ignimbrite (HEPI) from radially collapsing currents that were graded in particle concentration. Modified from Fierstein and Hildreth (1992). Near-vent Baked and Falling Mountains (schematically drawn at right) were overwhelmed by such stratified currents. Represented at left, Whiskey Ridge (5 km from vent) blocked and diverted denser parts of currents (as VFI) while turbulent upper parts deposited 5 m of HEPI on ridgecrest 400 m higher.

pyroclastic density currents (PDCs—flows and surges) that have no regional equivalents (Fierstein and others, 1997; Houghton and others, 2004). The locally dispersed components of the fall deposits form sector-confined lobes and wedges for which thicknesses halve radially away from the vent within 100-300 m (in contrast to several kilometers for the medial-distal plinian falls). Both the locally dispersed fall material and most of the proximal PDCs are far richer in lithics and in andesitic and banded pumice than the coeval plinian falls (C–D and F–G). The near-vent deposits are interpreted by Houghton and others (2004) in terms of four transport regimes: (1) sustained buoyant plume, (2) column-margin collapse to produce PDCs, (3) fountaining, where coarser fractions decouple on parabolic trajectories to produce conical deposits, and (4) direct lateral (ballistic) ejection. Here, we first describe the plinian layers and then summarize the complex proximal deposits.

Layer C

The first of the postignimbrite plinian fall deposits, Layer C, is the thickest and coarsest among them, and, like the Episode I falls, its dispersal was strongly east-southeastward. Layer C is ~3 m thick (figs. 17, 46) in upper Knife Creek (4–6 km from vent) and still ~2 m thick on the southern slopes of Trident and Mount Katmai (fig. 20; 7–10 km from vent). Close to Novarupta, it may be as thick as 6 m but is interrupted by one or more PDCs. Downwind distally, Layer C merges indistinguishably with Layer D (Fierstein and Hildreth, 1992), and they are together ~15 cm thick near the Shelikof coast (50 km) and ~5 cm at Kodiak (170 km).

In the upper VTTS, Layer C is coarsest in its lower half, where dacite pumice as large as 50 cm and welded-tuff lithics as big as 20 cm are present, along with sparse andesitic and banded pumice and 5–20 percent rhyolite pumice. The

Figure 44. Geologists examining high-energy fluvial beds of pumice and ash, reworked in part from Knife Creek Glaciers, during hours-long eruptive lull between Episodes I and II. Geologists stand on main Episode I ignimbrite, which floors gulch cut along Glacier 1 fork of upper Knife Creek. Fluvial beds are overlain by coarse base of dacitic Layer C. On far wall, two fine-grained dark ribs are C–D parting and Layer E, which is overlain in sequence by coarse Layer F, finer pumice-fall Layer G, and thick section of surge-bedded strata emplaced explosively from post-eruption phreatic craters.

upper half is a lapilli fall weakly graded to a thin crystal-rich coarse-ash top, and its pumice is almost entirely dacitic. The lower part of Layer C marks the last eruption of an appreciable amount of rhyolite until extrusion of the Novarupta lava dome. South of Trident and Mount Katmai, Layer C is a single lapilli fall, poorly graded except in its top ~10 percent, which is likewise crystal-rich coarse ash. Upwind to the northwest, Layer C splits into two normally graded subunits, and it fines and thins rapidly to ~1 m in the central VTTS (fig. 45) and to only a few centimeters near Three Forks.

Dacite pumice of Episodes II and III is unusually crystal-rich (27–42 wt %), and the fraction of free crystals in Layer C along upper Knife Creek (4–7 km from vent) is also within that range. There is, however, for each of the plinian dacite Layers C, D, F, and G, an annular crystal-enrichment zone in which the free crystal content of the deposit exceeds that in pumice clasts, generally at distances where Md is close to 1 mm, the size typical of the abundant feldspar phenocrysts. In the lower VTTS and at Katmai River, Layer C thus contains 50–70 weight percent crystals. The fraction drops back to 4550 weight percent near Shelikof Strait (40 km downwind) and to 22 weight percent on Kodiak Island (110 km), as glass shards become dominant distally. Median grain size of Layer C is typically 4–16 mm in upper Knife Creek, as great as 50–80 mm close to Novarupta, ~1.5 mm at Katmai River, ~0.5 mm at Geographic Harbor, and <0.1 mm on Kodiak Island. Sorting for Layer C shows a remarkably wide range (σ_ϕ~0.5–3.3), owing in part to the high crystal content and the unusually proximal distribution of some sites from which samples were sieved. Lithic content is 2–7 weight percent, predominantly welded-tuff vitrophyre recycled from Episode I vent fill. The 5-cm lithic isopleth for Layer C extends ~13 km southeast and 7 km north of the vent (Fierstein and Hildreth, 1992).

Although the top 10 cm or so of Layer C commonly grades normally to coarse ash and small lapilli, it does not grade to fines that might indicate a lull in plinian activity. Instead, it is capped by a thin bed of fines-bearing crystal-rich ash, 1–5 cm thick, that separates plinian Layers C and D in the upper VTTS (fig. 17). This bed appears to incorporate the fine ash associated with emplacement of a poorly sorted ignimbrite (<1 m thick) that formed by partial column collapse and is exposed only within 2.5 km south and west of Novarupta (PDC 5 of Houghton and others, 2004). This finer grained "C-D parting" is present only in the upper VTTS.

Layer D

Layer D represents a continuation and resurgence of the dacite plinian episode that began with eruption of Layer C. In the upper Knife Creek arm (3–6 km from vent), Layer D is 1–1.7 m thick. It may exceed 2 m closer to Novarupta, but it is interrupted there and shaved by PDCs. Still ~1 m thick just west of Katmai caldera, it thins downwind from ~60 cm on the upper south slopes of Trident to ~20 cm at Katmai River, and, still farther southeast, it merges with the graded top of Layer C. Like all plinian layers that preceded it, Layer D was strongly dispersed east-southeastward. Off-axis, it thins to only 10–15 cm along Mageik Creek, and upwind it thins rapidly to ~40 cm in the central VTTS and to only 1–2 cm at Three Forks and the Ukak River.

Much of Layer D is a massive lapilli fall rich in free crystals and normally graded to coarse and medium ash only in the top ~10 percent of its thickness. Unlike Layer C, it was followed by a temporary cessation of plinian activity, as it grades up normally to coarse ash and is overlain throughout the VTTS and on adjacent peaks by fines-rich Layer E (fig. 17). In the upper VTTS, Layer D rests on the fines-bearing C–D parting, and on the southern slopes of Trident and Mount Katmai its lapilli-rich base rests on the normally graded crystal-rich top of Layer C. Sorting is only fair in Layer D, again owing to the high crystal content of the dacite pumice; the matrix of medium-to-coarse crystal-rich ash fluctuates in abundance vertically, contributing to vague internal stratification. In upper Knife Creek (3–6 km from vent), Layer D contains 40–45 weight percent free crystals and 2–4 weight percent lithics. Its lithic content is smaller than in Layer C but is similar in consisting predominantly of recycled welded vent-filling ejecta. The largest pumice clasts are 15–30 cm in the upper VTTS and 8–10 cm at Katmai River; average maximum lithic sizes at corresponding locations are 4–8 cm and 2–4 cm. The pumice is predominantly dacitic, but ~1 percent of small pumice lapilli are crystal-poor rhyolite proximally, and this fraction increases to ~10 percent at Katmai River (where most lapilli in Layer D are small). The scarcity of rhyolite pumice proximally, the near-absence of coarse lapilli and blocks of rhyolite pumice, and the rarity of banded rhyolite-dacite pumice suggest that much or all of the rhyolite in Layers D, F, and G was entrained by the column from near-vent surfaces and that little or none of it coerupted with the dacite.

Analysis of vesicularity and bubble-wall textures of dacite pumice from Layers C and D (Adams and others, 2006b) shows little evidence for slowing of magma ascent or a trend toward open-system outgassing as a precursor to the break in plinian activity at the end of Episode II. More likely is an interruption by conduit blockage caused by vent-wall collapse, which was soon cleared by an explosive blast that radially emplaced a lithic-rich PDC deposit and permitted resumption of plinian activity as Episode III.

Layer E

Layer E is an ash bed, variously 3–9 cm thick throughout the VTTS, that accumulated during temporary shutdown of the stable plinian plume between Episodes II and III. It is recognizable between coarser layers D and F (figs. 17, 24, 45) for at least 23 km northwest of Novarupta, as well as in the Trident saddles and along Mageik Creek. Farther downwind, from Katmai River to Kodiak, it is not separable from the normally graded ashy top of the Episode II (C–D) plinian deposit. In the VTTS, it stands out in relief as a fines-rich rib that separates the enclosing pumice-fall deposits (Curtis, 1968).

In and near the VTTS, Layer E maintains fairly constant thickness (fig. 6 of Hildreth, 1983), whereas the enclosing plinian dacite layers thin rapidly away from Novarupta. Except within ~3 km of the vent, 50 weight percent or more of Layer E consists of ash finer than 63 µm, sorting is good (σ_ϕ=1–2), lithic content is <1 weight percent, and pumice as big as 2–16 mm is rare. Componentry shows that Layer E consists of 81–96 weight percent vitric ash, and microprobe analysis of the glass shards shows 86–95 percent of them to be derived from dacite magma (Fierstein and Hildreth, 1992). Rhyolite shards contribute 3–11 percent and andesite shards 1–5 percent of the vitric ash fraction. Crystal content ranges from 9 to 19 weight percent, declining with distance from vent, and most crystals are smaller than 0.25 mm. By comparison, the enclosing plinian layers contain 35–70 weight percent free crystals (many of which are ≥1 mm), and their dacite pumice has 27–42 weight percent phenocrysts.

The ash constituting Layer E had at least three separate sources (Fierstein and Hildreth, 1992): (1) The plinian plumes of Episode II dacite left slow-settling fines aloft. (2) Lesser fractions of rhyolitic and andesitic shards suggest that coignimbrite and coplinian ash contributions were left over from Episode I. As Episode I products were about two-thirds rhyolitic, however, they could have contributed no more than ~10 percent of the shards in dacite-dominant Layer E. (3) A multiflow package of ignimbrite, 1–3 m thick, is intercalated proximally at the level of Layer E,

Figure 45. Complete section of dacite fall deposits, 11 km upwind northwest of Novarupta, near Lethe Hills (fig. 6) at west margin of VTTS. Fine-grained pale-gray Layer E, here 9 cm thick, separates deposits of Episodes II and III. Layers C and D are together only 13 cm thick, while Layers F and G are together ~70 cm thick. Southeast of vent, C–D versus F–G thickness proportions are opposite, with Episode II deposits (C–D) always thicker than those of Episode III (F–G). At bottom of image, Layer C rests directly on oxidized ignimbrite (ig) of Episode I with no intercalation of coignimbrite ash. At top, fine-grained dacite-dominant Layer H (4 cm thick) is overlain conformably by a pumice-rich debris-flow deposit (df). Ruler is 16 cm long.

extending 1–3 km radially away from Novarupta. This represents the most laterally extensive PDC deposit of Episodes II and III (PDC 4 of Houghton and others, 2004), a thin landscape-mantling radially distributed veneer, which may have been produced by the outburst that reopened the vent, initiating Episode III. Although dominantly dacitic, this poorly sorted, lithic-rich ignimbrite also contains 10–20 weight percent andesitic and banded pumice and smaller fractions of rhyolite pumice. Ash clouds generated by these small proximal PDCs certainly contributed to Layer E, especially in the upper VTTS, but the fairly constant thickness of Layer E from Mageik Creek to the Ukak River suggests that its regional deposition was in large part from lingering clouds of coplinian dacite ash from Episode II.

Layer F

After the few-hours-long lull recorded by accumulation of Layer E, Episode III began with plinian dispersal of crystal-rich dacite compositionally identical to that of Episode II. Just as Layer C is coarser and more widely dispersed than D, Episode III was most vigorous in its opening phase, causing Layer F to be thicker and coarser than subsequent Layer G. In the upper Knife Creek area, Layer F is a lapilli fall 3–6 m thick, rich in crystal ash and consistently marked by a coarser lower half (fig. 17). Closer still to Novarupta, Layer F maintains its integrity (Fall 3 of fig. 9 of Houghton and others, 2004), although it is sandwiched there by proximal PDC deposits. Its coarser and better sorted lower half is weakly graded normally, and its moderately sorted upper half is richer in crystal-ash matrix, which fluctuates on a decimeter scale, imparting a weak plane-parallel internal stratification. Dacite pumice coarser than 30 cm is common in the lower half, within 4 km of Novarupta. Farther than ~5 km from vent, this twofold division transforms into a pair of normally graded subunits (F_1 and F_2 of fig. 20; Fierstein and Hildreth, 1992) that remain distinguishable as far as Three Forks and Katmai River. Altogether, Layer F

Figure 46. Flow unit of ignimbrite Package 9 (lower left), its lateral margin exposed in cross section, near upper Knife Creek just downstream from Glacier 4. Flow unit is rich in coarse dacite (and subordinate andesite) pumice blocks (see figure 36), some of which were here concentrated toward edge of flow lobe. Lobe is overlain by pale-gray ash-cloud deposit 14 cm thick, which settled during the lull between Episodes I and II. Ash is overlain by coarse base of dacite-dominant Layer C, which opened Episode II and is as thick as 3.5 m where intact nearby. Hammer is 28 cm long.

is thicker than 1 m on Trident and 50–80 cm along Katmai River; as it fines downwind, it merges with Layers G and H to yield a combined thickness of ~10 cm along Shelikof Strait and 2-3 cm on Kodiak Island. Northwest of Novarupta, Layer F thins to ~60 cm in the central VTTS and to ~20 cm near Three Forks. Because it thins northwestward less rapidly than do Layers C and D, Curtis (1968) inferred a change in wind direction between Episodes II and III. More regional isopach data (Fierstein and Hildreth, 1992), however, show that Layers F and G were still dispersed preferentially toward the east-southeast (as during the Episodes before them) but not so strongly. That the proximal isopachs for Episode III are more equant may reflect lower column height as well as slackening of wind velocity (Fierstein and Hildreth, 1992).

Field counts show the pumice clasts in Layer F to be 97–99 percent crystal-rich dacite in the VTTS and 84–94 percent dacite south of Trident and Mount Katmai. As in Layers C, D, and G, downwind dilution by several percent of small lapilli and granules of rhyolite pumice (inconspicuous in the coarser proximal deposits) is probably caused by reentrainment of nonwelded Episode I material from near-vent surfaces.

In the upper VTTS the largest pumice clasts in Layer F commonly exceed 30 cm and are as coarse as 50–90 cm proximally. Average maximum pumice size is still ~25 cm in the Trident saddles and 8–12 cm along Katmai River, where corresponding average maximum lithic sizes are ~12 cm and 3–4 cm, respectively. Lithics make up only ~1 weight percent of the deposit in upper Knife Creek and merely a trace at the Shelikof coast 40 km downwind; they are predominantly recycled welded-tuff vitrophyre. Free crystal content of Layer F exceeds 50 weight percent in the VTTS and remains 48–56 weight percent at the Shelikof coast. Owing to the crystal-rich nature of the dacite, median grain size (Md~1 to 32 mm) and sorting (σ_θ~1.8–3.0) range widely (by sublayer, distance, and azimuth) in the upper VTTS, but Md declines to 0.6 mm in the central VTTS and, downwind, to 1.2 mm at Katmai River, ~0.25 mm at Geographic Harbor, and 0.06 mm on Kodiak Island (110 km). Within 15 km of vent, the deposit is fines-poor throughout and grades only to crystal ash at the top where overlain by Layer G.

Layer G

The final plinian deposit of the Episode II-III sequence is also the most irregular and complex (figs. 45, 47). Curtis (1968) observed two or more ashy partings within his Layer G lapilli-fall deposit along upper Knife Creek, where rapid thickness variations (<0.5 m to 3.5 m) forced him to draw peculiar isopachs for that limited area, only 3–6 km from vent. Our work on more proximal deposits equivalent to Layer G shows that it consists of two parts separated by a set of thin ignimbrites that extend only ~2 km radially from Novarupta but supplied ash clouds for partings within the medial lapilli falls. The lower half proximally is a well-sorted lapilli fall (Fall Unit 2 of Houghton and others, 2004), coarser than the falls above and below it and ~2–3 m thick, that contains dacite pumice as coarse as 50 cm and only subordinate crystal ash. This lower half is itself partitioned into two parts, each with a weakly normally graded top, by thin ignimbrites and derivative ash laminae; each part is typically 1–2 m thick but exhibits rapid sectorial changes in thickness and coarseness. The upper half (Fall Unit 1 of Houghton and others, 2004) is not quite as coarse, though it locally contains dacite pumice as big as 40 cm; it is subdivided into three parts by very thin ignimbrites and derivative ash laminae, and its thickness varies sectorially from typically 1–1.5 m to as much as 4 m. Much of the ejecta ring around the Novarupta dome, including the sector-confined andesite-rich "S-bed" (fig. 48) of Fierstein and others (1997), is time-equivalent to this uppermost part of Layer G. The complex stratigraphy of this closing stage of the plinian sequence is illustrated and described in detail by Houghton and others (2004).

The proximal intricacy grades laterally into medial lapilli-fall deposits as thick as 3.5 m along upper Knife Creek, where grain-size fluctuations suggest three pulses. The three intervals, however, are defined by variation in proportions of pumice lapilli and coarse crystal-rich ash, not by grading to fines that might indicate time breaks in plinian emission. Likewise, south of Trident and Mount Katmai, Layer G consists of two (and locally three) weakly graded subunits, each 5–35 cm thick. Only the top interval grades up into fines-rich Layer H, which accumulated after final extinction of the plinian column.

In the VTTS, pumice clasts in Layer G are >99 percent dacite, but south of Trident and Mount Katmai (where most rhyolitic pumice is 0.5–4 mm) as many as 12 percent are rhyolite, probably reentrained from Episode I material. Average maximum pumice size exceeds 20 cm in the upper VTTS and declines to 7 cm on Trident and 3–4 cm at Katmai River; corresponding average maximum lithic sizes are 7 cm, 2 cm, and only 1–2 mm. As throughout Episodes II and III, most lithics are welded tuff recycled from Episode I vent fill. The fraction of free crystals in Layer G is 40–60 weight percent in the VTTS, along Mageik Creek, and all the way to Shelikof Strait (40 km downwind). Median grain size ranges widely in the proximal falls (Md~1–64 mm), as does sorting (σ_θ~1.7–3.3), depending on proportions of crystal ash and pumice lapilli in successive sublayers (Houghton and others, 2004). Sorting improves downwind, and Md declines from ~1 mm in upper VTTS medial deposits to ~0.2 mm in distal deposits at Three Forks and ~0.1 mm at the Shelikof coast. As grain size diminishes between Katmai River and Shelikof Strait, dacite-dominant Layers F, G, and H merge indistinguishably and become contaminated by fallout of coplinian fines left over from Episode I (table 3a of Fierstein and Hildreth, 1992). Although principal downwind dispersal of Episode III ash was still east-southeastward, part of the cloud was driven northeastward toward Cook Inlet (fig. 14 of Fierstein and Hildreth, 1992). The vitric shards that dominate the 4-cm-thick layer of 1912 ash on Augustine Island (160 km northeast of Novarupta)

Figure 47. Photograph and annotated sketch of section in gulch ~1.1 km southeast of Novarupta dome (seen at top), illustrating proximal intercalation of Episode II–III fall units with several thin pyroclastic density current (PDC) deposits. For scale, total Fall 2 is 2.8 m thick. Several such gulches radial to Novarupta expose such sequences, in which coarseness and thickness of fall and flow units change sectorially, as documented in detail by Houghton and others (2004). Falls 1 and 2 here are equivalent to regionally dispersed Layer G; Fall 3 to Layer F; PDC 4 to Layer E; Fall 4 to Layer D; PDC 5 to the C–D parting. Layer C and Episode I ignimbrite are below level of exposure in this gulch.

are >88 percent dacitic, ~10 percent rhyolitic, and ~1 percent andesitic (table 3a of Fierstein and Hildreth, 1992).

The fluctuations of Layer G were followed by termination of plinian activity and transition to effusion of the lava domes of Episodes IV and V. Analysis of vesicularity and bubble-wall textures of dacite pumice from Layer G (Adams and others, 2006b) shows a progressive shift toward smaller bubbles, thicker vesicle walls, and teardrop shapes, indicating onset of bubble-wall collapse with open-system degassing. Unlike the temporary plinian stoppage after Episode II, which was probably caused by conduit blockage, this final cessation of plinian activity reflects slowing of magma ascent and maturation of vesicular melt, permitting freer escape of gas, transition to open-system outgassing, and accumulation of degassed magma that ultimately blocked the conduit, precursory to effusive dome growth.

Layer H

Throughout and near the VTTS, on Trident and Mount Katmai, and from Mageik Creek to Katmai Canyon, Layer H is widely preserved as a bed of muddy vitric ash that caps the 1912 primary deposits (figs. 21, 45). The blue-gray or terra cotta (oxidized) mud is typically 5–50 cm thick, but its thickness appears not to vary systematically with distance from Novarupta. Its primary thickness, however, is typically hard to assess, owing to disturbance by scree, mudflows, windscour, and other surface processes. Sections thicker than 40 cm have been measured in upper Knife Creek, high on the flanks of Mount Katmai, and at Katmai Canyon. Farther from vent than 20 km northwest or 40 km southeast, Layer H is not separable from the fine-grained top of Layer G. Accretionary lapilli are locally conspicuous in Layer H, as

Figure 48. The "S bed" atop the Turtle, ~600 m northeast of Novarupta dome. Rich in andesite scoria, the layer is enclosed within dacitic Layer G of Episode III and extends as a 150-m-wide lobe for only 1.1 km northeast of vent (Fierstein and others, 1997). The bed thins with distance, and its andesite proportion is progressively diluted by dacite pumice, ultimately fading into a scattering of sparse clasts among the dacite. Distance over which its thickness halves is ~70 m, in contrast to many kilometers for dacite-dominant fall layers. Containing 50–90 percent andesitic clasts, S bed is andesite-richest unit of whole 1912 eruption. It may reflect viscous segregation of low-viscosity andesite along conduit margin, resulting in a sector-confined collar of low-fountaining ejecta peripheral to a high-velocity core that simultaneously fed dacitic plinian column. Best exposed here on wall of radial graben atop the Turtle. View westward toward spurs of Baked Mountain. Geologist is 1.63 m tall.

are vesicles that probably reflect syndepositional incorporation of atmospheric moisture.

Even proximally, Layer H includes >50 weight percent ash finer than 63 μm and contains ≤10 weight percent crystals and a negligible fraction of lithics. Microprobe analysis shows that, in and near the VTTS, 85–100 percent of the glass shards are dacite derived. Relative to the phenocryst-rich dacite pumice of Episode II and III, therefore, Layer H is strongly depleted in crystals. Downwind near Shelikof Strait, however, microprobe data indicate that about a third of the glass shards in the top few centimeters of the ashfall are rhyolitic, evidently Episode I fines that were held aloft since the first day of the eruption (Fierstein and Hildreth, 1992). Upwind to the northwest, where the expanding ash cloud did not bring ashfall at Naknek until 12 June (Dumond, 2005), the rhyolitic shard fraction falls off from 19 percent at Brooks Camp (fig. 49; 50 km from vent) to only ~1 percent at King Salmon (100 km).

Clearly, then, most of Layer H is coplinian dacitic ash and dust that settled slowly after the dacite plinian eruptions of Episodes II and III. Mass balance indicates that no more than 10 percent of Layer H in and near the VTTS can be coignimbrite ash, although far downwind roughly a third of Layer H fallout is coplinian ash left over from Episode I (Fierstein and Hildreth, 1992). Detailed investigation of the proximal deposits (Houghton and others, 2004) shows that no pyroclastic density currents were emplaced subsequent to Layer G that might have contributed ash to Layer H.

Sieving of Layer H indicates median grain size between 0.05 and 0.1 mm and moderate to good sorting (σ_θ~1–2). Locally, however, as many as three medium-to-coarse crystal-ash laminae are intercalated within dominantly vitric Layer H. On Mount Katmai, some of these are related to the phreatic mud layers expelled from the fitfully collapsing caldera (Hildreth, 1991). In upper Knife Creek and the uppermost Lethe arm of the VTTS, some are probably related to the numerous phreatic explosion craters in the ignimbrite there (Hildreth, 1983; Hildreth and Fierstein, 2003). On and adjacent to many slopes, uncommon sandy layers thicker than several millimeters (some containing sparse rounded pumice granules) may in many cases have been wind deposited. No ultraproximal evidence is recognized for postplinian explosions at the Novarupta vent that might have deposited coarse ash laminae during settling of Layer H, but there is little doubt that small outbursts could have occurred.

Proximal Pyroclastic Density Currents

Within 3 km of the Novarupta dome, which plugs the Episode III vent, the sequence of dacite fallout deposits equivalent to medial-distal Layers C through G contains thin packages of pyroclastic density current deposits (figs. 47, 50) intercalated at five different horizons (Houghton and others, 2004). Most are sector-confined multiflow packages that show rapid radial and sectorial changes in thickness and grainsize. All are dacite-dominant, but nearly all also contain more andesitic and banded pumice (and lithics) than the medial fall deposits. The various packages range in thickness from 0.1 to >3.5 m and pinch out within 1.5 to 2.5 km of Novarupta, although most also generated ash clouds that deposited thin ashfall layers extending a kilometer or two farther. They include (1) matrix-rich channel-confined ignimbrites, typically 1–2 m thick and generally consisting of a few thin massive flow units defined by pumice-concentration zones; some are rich in dacite pumice blocks, but all contain more andesite pumice than the enclosing fall deposits; (2) radially distributed, thin bedded ash-rich packages that show pinch-and-swell lamination, interbedding with thin lapilli falls, and dacite-andesite-rhyolite pumice proportions similar to enclosing thin fall deposits; (3) radially distributed massive landscape-mantling ash-rich veneers deposited by concentrated but energetic currents; and (4) channel-confined matrix-poor grain-flow avalanche deposits formed by local slumping of coarse pumice-fall deposits. These complex proximal deposits are described, illustrated, and interpreted by Houghton and others (2004), and their grain size characteristics are illustrated in figure 51. Their total volume is small (<0.05 km^3), but their importance for revealing the processes of proximal sedimentation from the margins of the gas-thrust jet and the base of the convecting column is profound.

The variety of thin sector-confined flow deposits radial to Novarupta (fig. 52) and intercalated within deposits of sustained plinian pumice fall suggests column-margin instabilities on very short spatial and time scales. It indicates a variety of transporting currents initiated from a range of heights and on a range of trajectories, reflecting sporadic heterogeneities—in particle density and in lithic and andesite proportions—within an annular collar around the emerging column (fig. 53).

Proximal Fall Deposits: Contrasting Facies

Within 3 km of Novarupta, in addition to the proximal flow deposits just discussed, there are also several coarse-grained locally dispersed fall deposits interbedded with the regionally dispersed dacite fall deposits of Layers C, D, F, and G. Even ultraproximally, the latter remain moderately sorted lapilli-to-block falls that are radially distributed and pass gradationally in grain size and thickness into their medial equivalents; thicknesses halve over distances of several kilometers. Subtle fluctuations in grain size and in fraction of crystal-rich coarse-ash matrix define faint shower bedding. Like the medial-to-distal plinian fall deposits, the material in these "normal" proximal fall units was derived from high in convective plumes and their ensuing umbrella clouds.

The locally dispersed proximal fallout facies is identified as six narrow wedge-shaped lobes that halve in thickness within only 100–300 m, terminate 1.5–2 km from vent, and pass radially and laterally into the normal fall deposits. These lobes are generally coarser (Md~8–64 mm) than the normal facies (Md~2–6 mm), apart from crystal-enriched intervals that are similar (Md~1–3 mm) in both facies. Neither proximal facies, however, is well sorted (σ_θ~1.7–3.2), owing to the crystal-rich nature of their dacite and andesite pumice and to concurrent sedimentation from diverse parts of the eruptive jet and plume. The lobate facies is much richer than the normal fallout in andesitic pumice and lithic fragments, both of which decrease rapidly away from vent. The andesite-rich "S-bed" (fig. 48) of Fierstein and others (1997) is a striking example that extends narrowly northeastward for ~900 m from Novarupta, across the Turtle sector of the ejecta ring. Like the "S-bed," the several larger lobes mapped by Houghton and others (2004) were emplaced by intermittent processes limited to local sectors of the periphery of the eruptive jet and lower plume. Rapid lateral and radial changes in proportions of andesite pumice and lithic fragments within such fallout lobes further emphasize the role of small-scale transient heterogeneities around the margin of the emerging jet.

Figure 49. Structureless white fine ash, 20 cm thick, at Brooks Camp, 49 km upwind northwest of Novarupta (fig. 6). Coin is 2.4 cm across. Point counts by microprobe showed dacite shards to make up 87 percent of this deposit at base and 79 percent at top (Fierstein and Hildreth, 1992). Sparse andesite shards are 1–4 percent, and rhyolite shards 9–19 percent. Rhyolite-rich plinian ash was driven strongly southeastward from vent, and rhyolite-rich ignimbrite produced rather little coignimbrite ash. Upwind ashfall continued for several days after eruption ended and consisted predominantly of fine fractions of plinian dacite of Epiodes II and III. Lush herbaceous plant life rapidly recolonized ashfall blanket.

The excessive proportion of lithics and especially of andesite in the ultraproximal fall and flow lobes may be explicable in terms of inhomogeneous two-phase flow in dikes and conduits, both before and after fragmentation of the magma, as explored by Carrigan and others (1992) and Houghton and others (2004 and references therein). Viscous segregation of lower-viscosity andesite magma into lateral zones during joint sheet flow with dacite or rhyolite may have been succeeded, following fragmentation, by lateral zones of higher-density high-concentration particulate material along the sheet or conduit walls. Changes in direction or from sill to dike or from dike to pipe-like conduit could also promote inhomogeneities during transport. At the vent, such a dense annular collar would emerge at lower velocity than the turbulent core and thus be more likely to contribute to marginal fountains than to the central plinian plume (fig. 53). Some such processes certainly influenced distribution of near-vent ejecta during Episodes II and III, and they probably operated during Phase B of Episode I as well (producing the greater fractions of andesite in ignimbrite packages than in coeval fall deposits).

Eruptive Episode IV: Dacitic Phantom Dome and Block Bed

Sometime after cessation of the pyroclastic eruptive sequence, following the days or weeks required for complete accumulation of fines-rich Layer H, a small dacite dome that plugged the Episode III vent was explosively destroyed. The resulting apron of decimeter-to-meter-sized fragments covers an elliptical 4-km by 2-km dispersal area that surrounds Novarupta and is elongate north-south (fig. 6 of Adams and others, 2006a). The approximately 700 blocks studied represent four main lithologies—pumiceous, dense, flow-foliated, and variably welded breccia. All lithologies are crystal-rich dacite compositionally like that erupted as pumice during the preceding Episodes II and III (62.2–66.7% SiO_2; n=8). Many blocks of all lithologies became breadcrusted after fragmentation and dispersal, indicating that the plug material was still hot and capable of exsolving residual gas after explosive disruption (fig. 54). The distribution of blocks of different size and density suggests complex or serial ejections, and fan-shaped dispersal of blocks that shattered on impact indicates ballistic trajectories for many.

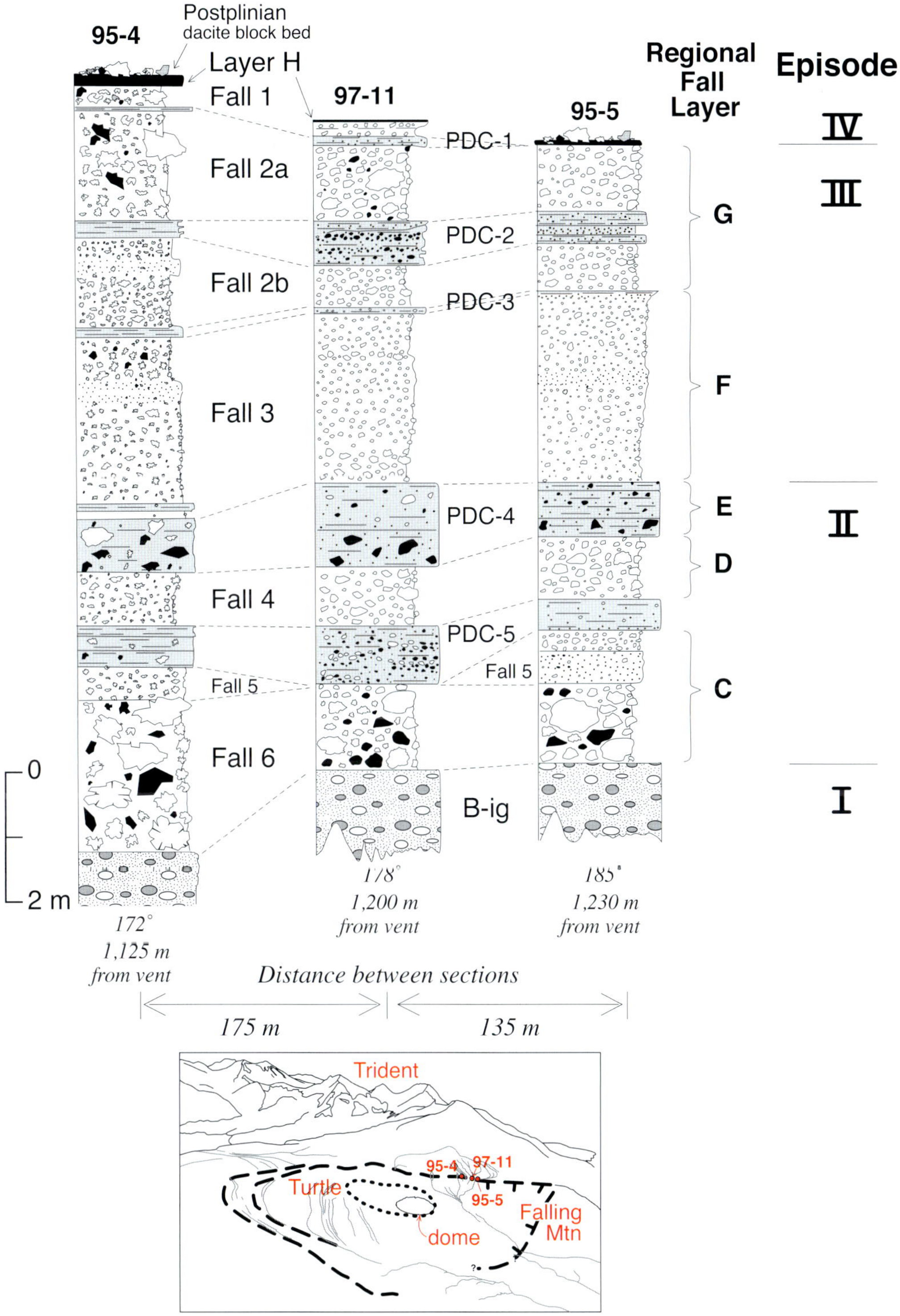

Figure 50. Stratigraphic sections in three confluent forks of gulch system adjacent to east side of Falling Mountain, ~1.2 km south of Novarupta, illustrating complex ultraproximal deposits investigated in detail by Houghton and others (2004). At least five pyroclastic density current (PDC) deposits are intercalated (only proximally) in the fallout sequence, for which regionally dispersed equivalents are indicated at right, along with eruptive episode sequence. Distances between sections and azimuths and distances from Novarupta are given at bottom. Inset sketch shows locations of sections relative to Novarupta dome, ejecta ring (dotted), rim of vent structure (dashed), and Falling Mountain scarp (see photo figure 70). Compare sequence with that of Section 94-F, which lies 1.1 km east, as illustrated in figure 47.

Vesicle population studies (Adams and others, 2006a, b) indicate for Episode IV dacite a continuation of the trends toward bubble coalescence and collapse already underway late in Episode III, marking the transition from closed- to open-system degassing of conduit magma. A short interval of passive extrusion culminated in explosive disruption of a low-permeability plug as pressurized gas was suddenly released after accumulating from subjacent dacite that was incompletely degassed and still somewhat permeable. Modelling ballistic ejection of Episode IV blocks by transient Vulcanian explosions suggests a role for excess trapped gas or for external water, augmenting the gas provided by the vesicular facies of the ejected material alone. Explosion could have been triggered by gravitational carapace collapse, by build-up of internal pressure of trapped gas, or by infiltration of rainwater through cracks toward the hot interior.

Episode IV material dispersed across the strewnfield adds up to a hard-to-estimate volume that is certainly less than a tenth that of the subsequent Novarupta dome (~5 million m^3). What is remarkable is that this small destructive event cleared the vent and upper conduit of mobile dacite magma, permitting egress of the rhyolite magma that then formed the Episode V Novarupta dome without further ejection of pyroclastic material.

Eruptive Episode V: Novarupta Rhyolite Dome

The Phantom dacite dome of Episode IV was never observed, and the subsequent rhyolitic dome (Novarupta) was not observed until July 1916, when Folsom and Griggs first entered the VTTS. Although it has cooled substantially and virtually ceased steaming, the morphology of the dome is essentially unchanged since it was first seen. Nestled within an extraordinarily asymmetrical ejecta ring (fig. 55), the

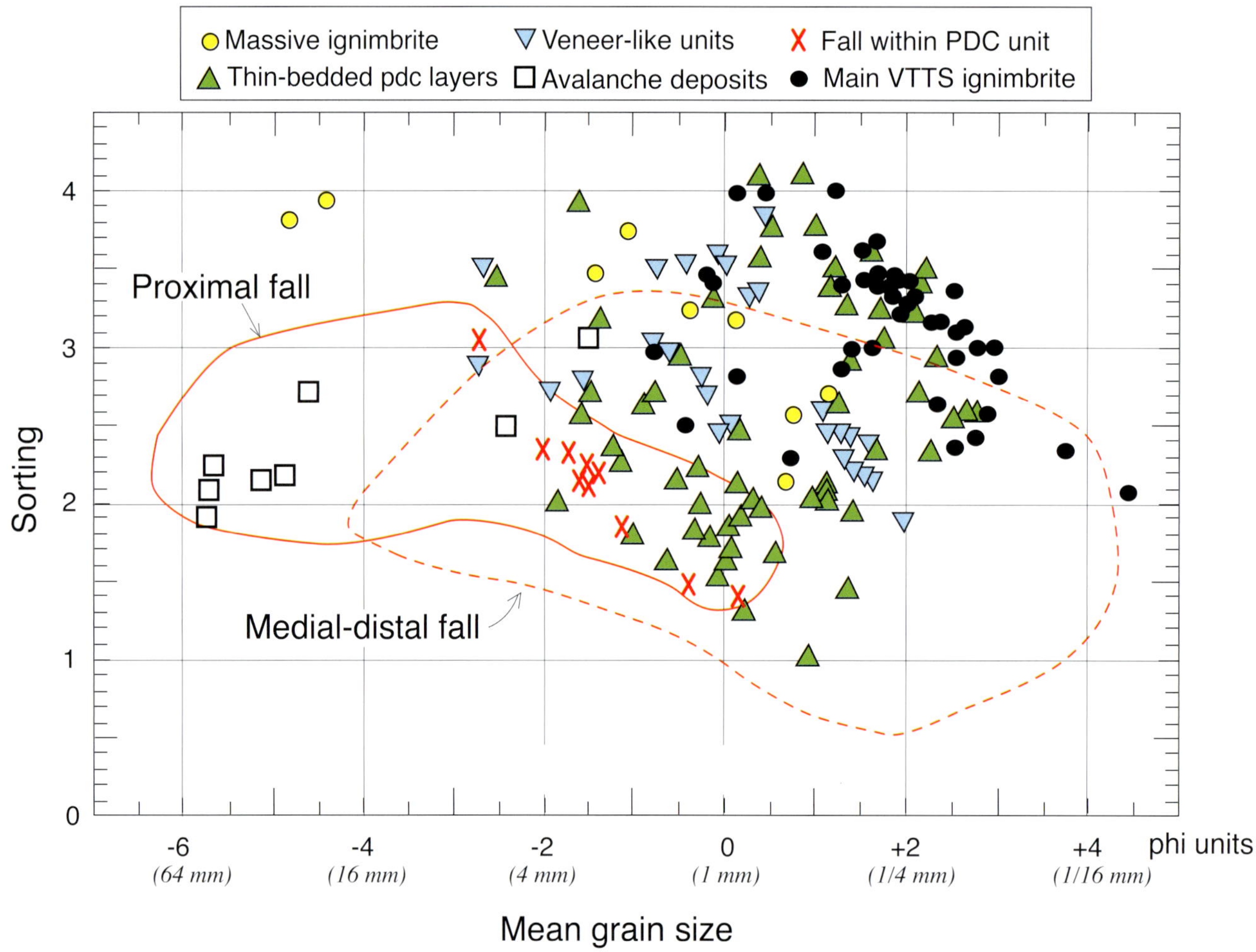

Figure 51. Grain size characteristics of the pyroclastic density current deposits (PDCs) intercalated ultraproximally with Episode II and III fall deposits within 1-3 km from Novarupta dome. For comparison, red lines enclose fields of abundant data for proximally and regionally dispersed dacite fall deposits (Layers C, D, F, G). For elaboration of PDC facies distinguished in key, see Houghton and others (2004). Sorting is conventional sedimentological parameter $\sigma\phi$.

Novarupta dome is ~380 m in diameter, circular in plan, and 65–70 m high, and it plugs the vent of Episodes III and IV (and less certainly II). Its coarsely blocky surface is almost entirely glassy and flow-foliated, steep-sided and rugged on top, where there is no sign of any tephra (either uplifted by the dome or deposited subsequently). Spreading of the dome opened numerous fissures on its upper surface, some with 5–20 m of relief. Principal among these radiate 80°, 235°, and 295° from a small 20-m-deep pit just south of the dome's center (fig. 56); a few more trend northeast across the northeast rim, and a concave-outward arcuate set marks the northwest slope of the dome. Exposed deep on the wall of one fissure is an abrupt transition from glassy carapace into a denser devitrified interior facies (Cowee and others, 1999). The dome has a peripheral apron of coarse angular talus blocks, most of which were shed during extrusion and have remained undisturbed since first observed. Many early photographs published by Griggs (1922) and Fenner (1923, 1950) show the dome and its talus much as they are today.

The exposed part of the dome has a volume of roughly 5,000,000 m^3. The dome surface is ~95 percent rhyolitic, and most of its ubiquitous flow foliation reflects textural variation in the rhyolite. Much of the exposed rhyolite is pale gray and finely vesicular to microvesicular, but some layers and lenses are coarsely vesicular and oxidized brown, while fewer are variously pink, massive, incipiently devitrified, or finely spherulitic. The rhyolite lava carries the same assemblage of sparse phenocrysts present in the rhyolite pumice erupted earlier (quartz, oligoclase, orthopyroxene, Fe-Ti oxides), as well as traces of augite and amphibole that are probably contaminants from interlaminated andesite and dacite. As pointed out by Fenner (1950, p. 720) and Curtis (1968), crystals from the intermediate components are sparsely but widely disseminated within the deceptively clean-looking dome rhyolite. Such contamination explains why most (normalized) chemical analyses of dome rhyolite (sampled for apparent purity) yield 74.5–76.9 percent SiO_2, whereas virtually all samples of 1912 rhyolite pumice have more than 77.1 percent.

Conspicuous layers, lenses, and rounded to angular chunks of dark-brown to black intermediate lava are included within the dome rhyolite (fig. 57). They make up ~5 percent of exposures around the dome margins and appear to be much sparser in the dome core, but the rugged top of the dome has not been as thoroughly examined. Many of the dark layers are 1–15 cm

Figure 52. Interbedded deposits of pyroclastic density currents (PDC) and pale-gray plinian dacite fallout on southeast spur of Baked Mountain, 1.4 km northwest of Novarupta dome and ~100 m above floor of Novarupta basin. All layers are products of Episode III. Note that topmost PDC crosscuts all below it. Layer G pumice-fall deposit tops section here as well as on Baked Mountain ridge in background. For scale, backpacks are ~1 m high.

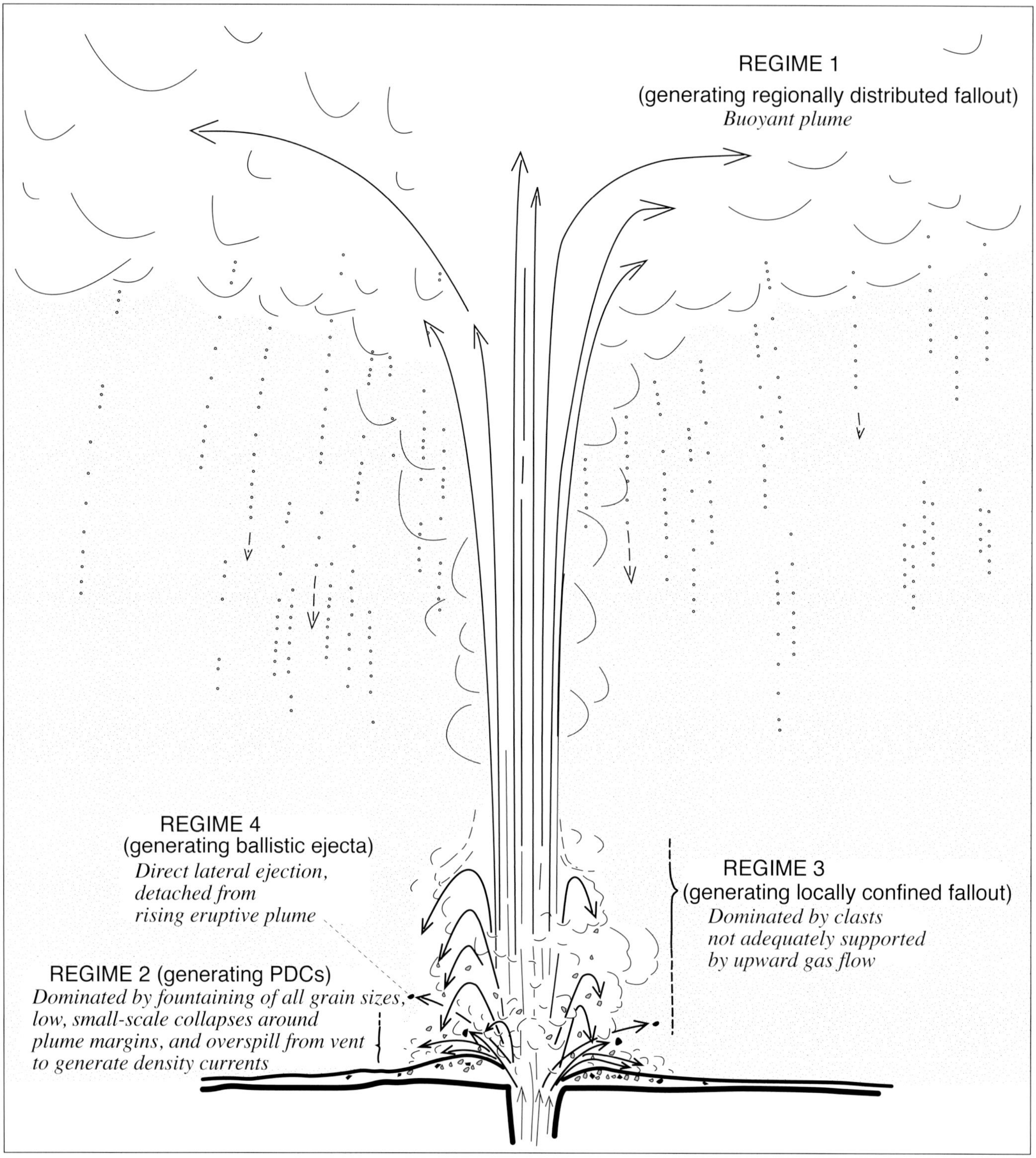

Figure 53. Cartoon illustrating four eruptive regimes capable of simultaneous operation from a sustained plinian column. Model was elaborated in detail by Houghton and others (2004) to account for deposition of ultraproximal sector-confined falls and thin pyroclastic density currents (PDCs) concurrent with regional plinian fallout.

thick and meters long, and some are bundles of thin parallel laminae that intimately alternate with rhyolite. All dark layers are at least finely vesicular, and some are coarsely scoriaceous or vuggy and contain vapor-phase minerals. Discrete pieces of dark intermediate lava are mostly 0.5–10 cm across and rarely as long as 15–25 cm. Most are within or associated with the dark layers and lenses, but such clasts occur in rhyolite layers as well. They include clasts that are angular and solitary, broken and slightly separated, rounded, lenticular, or viscously distended. Like the dark layers and lenses, most clasts are glassy, phenocryst-rich, and variably vesicular. Much of the clast population may form a deformation continuum with the dark layering, their brittle through viscous behavior depending on local cooling rate and strain rate. Rare clasts of dark lava are crystal-poor (unlike any 1912 andesite-dacite ejecta) and may be true xenoliths. Xenoliths of Naknek sandstone and siltstone, undeformed and usually oxidized, make up fewer than 5 percent of the non-rhyolite clasts in the dome.

Nearly all of the dark layers, lenses, and clasts are crystal-rich pyroxene-plagioclase lava, compositionally indistinguishable from the andesite-dacite juvenile continuum previously erupted during Episodes I through IV. Layers and clasts chemically analyzed range in SiO_2 content from 55.2 to 65.1 percent.

Bulk densities measured for numerous rhyolite samples from Novarupta range from 1,300 to 2,170 kg/m^3 (Adams and others, 2006a). These values compare with ranges of 310 to 1,250 kg/m^3 for the pumiceous facies of the Episode IV block bed and 600 to 1,300 kg/m^3 for pumice in Episode III fallout (Layers F and G). A few Novarupta rhyolite samples representative of the density range measured were selected for textural analysis, yielding 18–38 volume percent vesicles and 3–6 volume percent crystals (Adams and others, 2006a), significantly more crystals than held by rhyolite pumice. As in the Episode IV dacites, it was found that bulk density increases as vesicle size decreases and glass walls thicken. Novarupta vesicles, however, are usually foliated, with vesicles of all sizes elongate in the same direction but generally not segregated into layers. The low-density dome sample (1,490 kg/m^3) studied in detail has vesicle diameters ranging in size from 8 μm to 630 μm, with a median diameter near 100 μm (0.1 mm). Absence in Novarupta dome of the very-low-density pumiceous facies present in Episode IV dacite probably reflects days to months of open-system degassing of rhyolite stored shallowly in the conduit system, possibly since the close of Episode I. After the Episode IV dacite was flushed, such rhyolite was thus able to extrude passively without major build-up of overpressure by gas still exsolving beneath the plug.

Figure 54. Block of massive breadcrusted dacite on slope just northwest of rhyolitic Novarupta dome. Hundreds of such dacite blocks, variously dense, pumiceous, flow-foliated, or brecciated, are strewn across an area of 8 km^2, overlying Layer H and representing explosive destruction of a small plug-dome (Episode IV) that preceded extrusion of Novarupta (Adams and others, 2006a). Knife is 8.5 cm long.

A

B

Figure 55. Novarupta and its ejecta ring. Views northeastward of the 380-m-wide lava dome and its asymmetrical ring of Episode III ejecta: *A*, 1919 and *B*, 1982. Above the dome-encircling moat, the ejecta ring ranges in height from a low of ~7 m in the southwest to 225 m in the northeast, where it broadens into the 1-km-wide northeast-elongate mound called the Turtle. Configuration of the dome, ring, faults, and fissures has changed little since 1919. The dome steamed vigorously for decades, as did several fumarolic areas along the ring that are now conspicuously discolored. Measured temperatures of acid fumaroles in 1918–19 were as high as 290°C on the west rim of the ring and 166°C–264°C along faults atop the Turtle (Sayre and Hagelbarger, 1919; Allen and Zies, 1923). In the same areas, wispy emissions were visible only close at hand and gave temperatures of only 25°C–90°C in the 1980s (Keith, 1991a). At lower right of both images, a fallen block of shattered dacite of Falling Mountain crumbled in place to form a 12-m-high hummock that has sometimes been mistaken for a domelet or "volcanette." 1919 photo courtesy of National Geographic Society, taken from toe of Falling Mountain; 1982 photo by authors, taken from summit of Falling Mountain.

Figure 56. Vertical view (from tightly banking helicopter) of Novarupta dome, 65 m high, nearly circular in plan view, and 380 m across. Note fumarolic alteration outside and along crest of fissured ejecta ring. Dome surface is almost entirely glassy, flow-foliated, coarsely blocky, and tephra-free. Spreading of dome on extrusion opened variously oriented fissures on its upper surface, some with 5–20 m local relief. Apron of coarse talus fringes steep sides of entire dome. Orientation indicated at margins of image.

Figure 57. Flow foliation in Novarupta lava dome. Medium- and light-gray layers are rhyolite, dense to finely vesicular. Darker brownish layers and laminae are both dacitic and andesitic. Dark enclaves are variously crenulate, deformed, or broken, and they include some of the most mafic andesite released in whole 1912 eruption. Only a few percent of dome surface is visibly contaminated with such nonrhyolitic material, but even apparently clean rhyolite has ubiquitous xenocrysts. Knife is 8.5 cm long.

Graffito inscribed in oxidized sintered ignimbrite along a hinge-line fault on the rim of the "high sand line" bench near the northwest base of Mount Cerberus. The "Glacier Priest," Bernard Hubbard, a faculty member at the University of Santa Clara, California, led small teams of rugged male students on several summer expeditions to Alaska in the 1920s and 1930s. The first of his three trips into the Valley of Ten Thousand Smokes (VTTS) was in July 1929. Fault fissures like this one have provided many VTTS expeditions with shelter from the severe sandblast attending storm winds intensified by funneling through adjacent Katmai Pass. July 1982 photo of one of the present authors by the other.

Chapter 6

Volcanological Aspects of the Primary Eruption Products

In this section we summarize the volumes of material that erupted during the five episodes of the 1912 eruption. Following that, we discuss interesting physical aspects of the fall and flow deposits, their emplacement, distribution, and consolidation.

Volumes Erupted

Griggs (1922) estimated 4.75 cubic miles (19.8 km^3) of regional fallout, which he attributed to Mount Katmai, plus ~0.5 mi^3 (2.08 km^3) of proximal fallout attributed to Novarupta, and ~1 mi^3 (4.17 km^3) for the VTTS ignimbrite, yielding a total ejecta volume of ~26 km^3. Curtis (1968) estimated ~20 km^3 of fallout and, modelling much thicker tuff concealed in the upper valley, ~11 km^3 of ignimbrite, giving a total ejecta volume of ~31 km^3.

By logging 382 fallout sections from Bristol Bay to Katmai River (on foot) and on both sides of Shelikof Strait (by boat), Fierstein refined the volume estimates and subdivided them by episode (Fierstein and Hildreth, 1992). Estimates included integration to vanishing thickness (Pyle, 1989; Fierstein and Nathenson, 1992), showing that 9–14 percent of each fallout layer lies beyond its 1-cm isopach. As determined by microprobe, proportions of rhyolite shards in the fine-ash fraction of each layer varied considerably by azimuth and with distance from vent. Owing to atmospheric turbulence and intrinsically slow settling velocities for low-density glass shards of irregular shape, considerable volumes of Episode I rhyolitic ash ultimately accumulated within dacitic Layers C through H, requiring complex reapportionments to obtain true eruptive volumes for each episode (Fierstein and Hildreth, 1992). As so adjusted, the fallout volumes are 8.8 km^3 for Episode I (6.1 km^3 for A; 2.7 km^3 for B), 4.8 km^3 for Episode II, and 3.4 km^3 for Episode III, yielding a total of 17 km^3.

The 11-km^3 volume estimate by Curtis (1968) for the 120-km^2 VTTS ignimbrite is as good as the uncertainties warrant; addition of the Mageik Creek lobe and the eroded Ukak terminus would raise the estimate by only 0.1 to 0.2 km^3. The base of the ignimbrite is not exposed in the middle or upper parts of either the Knife Creek or River Lethe arms of the VTTS. Weaker development of marginal benches (attributable to differential compaction; see section titled "High Sand Mark," below) along the Knife Creek arm and an abundance of phreatic craters that burst through the ignimbrite sheet in upper Knife Creek suggest that the sheet might be thinner there than in the Lethe arm. If so, the ignimbrite volume might be as much as 20 percent smaller than estimated by Curtis.

Equivalent magma volumes are about 6.5 km^3 emplaced as fallout and 7±2 km^3 emplaced as ignimbrite, yielding a total of ~13.5 km^3 of magma erupted as ejecta in 1912. The subsequent lava domes added trivially to the total, and neither was accompanied by significant additions of tephra. The Episode V Novarupta Dome has a volume of ~0.005 km^3 (5 million m^3), and the Episode IV Phantom dome (the plug destroyed to produce the block bed) was an order of magnitude smaller still.

The observation that roughly equal proportions of explosively fragmented magma were emplaced as fall and flow deposits in 1912 contrasts strongly with some other 20th century plinian eruptions. In both the Santa María (Guatemala) eruption of 1902 and the Quizapu (Chile) eruption of 1932, pyroclastic density currents provided only trivial fractions of voluminous plinian outbursts that deposited widespread tephra falls. During Episode I, fall Layers A and B accumulated continuously, nowhere showing any evidence for a significant interruption in deposition from the plinian column. Nonetheless, the transition from Phase A to B represented not just a compositional change but a fundamental shift in eruptive style, in that ~90 percent of Phase A was emplaced as fallout and ~80 percent of Phase B as ignimbrite. Episodes II and III were again predominantly fallout, as the several accompanying flow deposits amount to no more than 1 percent of their respective volumes.

Coignimbrite Ash and Coplinian Ash

Compositional proportions (rhyolite/dacite/andesite) of glass shards, determined by microprobe for numerous distal ash samples (40–155 km from vent; table 3a of Fierstein and Hildreth, 1992), show that downwind fallout (on both sides of Shelikof Strait) equivalent to dacite-dominant Layers C-D and F-G-H incorporate 23–40 percent rhyolitic ash that had stayed aloft after Episode I. The 1912 dacitic glass shards are far richer in Ti, Fe, Mg, and Ca than the rhyolitic glass shards and are thus readily distinguishable. From Katmai River to Geographic Harbor (13–36 km from vent, where coarse ash and pumice granules still permit visual clast counts), the rhyolite fraction added to these dacite fall layers is 7–16 percent (table 2 of Fierstein and Hildreth, 1992). Although a limited fraction of rhyolitic ash and lapilli was no doubt entrained from near-vent surficial material by the Episode II and III columns, this was probably a modest contribution, because most of the near-vent Episode I rhyolite was by then welded (as represented by the lithic suite predominant in the Episode II and III deposits). Phase B of Episode I (Layer B and the main VFI) was the only interval during which voluminous rhyolite and dacite (and andesite) erupted concurrently.

The abundance of rhyolite shards in downwind ashfall deposits of Episodes II and III could thus be attributable to (coignimbrite) ash elutriated from rhyolite-rich packages of the main Episode I ignimbrite or to slow-settling (coplinian) ash that remained aloft after deposition of Layers A and B. The former is unlikely to have provided a large fraction of the rhyolite shards that settled with the Episode II and III downwind dacite fallout because (1) Layer C lies directly on the main ignimbrite (fig. 45) with no coignimbrite intercalation, and (2) in proximal and medial sections from the VTTS to Katmai River, fine-grained Layer H (which caps the Episode III pumice falls) contains little rhyolite ash (0–15 percent; table 3 of Fierstein and Hildreth, 1992). Scarcity of coignimbrite ash near the VTTS and downwind may reflect (1) persistence of the Phase B column during generation of the main ignimbrite, favoring updraft of circum-vent ash clouds into the plinian column itself; (2) relatively low-energy emplacement of VTTS ignimbrites, promoting rather limited elutriation of fines; (3) flushing of lingering Episode I fines by the dacite pumice falls, and (4) muffling of any coignimbrite plume from the ignimbrite sheet itself by the blanket of Episode II-III fallout.

There is no discrete coignimbrite ashfall layer (rhyolite-rich or otherwise) between lapilli-fall Layers B and C (representing the lull between Episodes I and II) proximally, medially, or downwind. Upwind (northwest) of the VTTS, where most of the rhyolite shards might be expected to be of coignimbrite origin, they make up only 3–19 percent of the unitary layer of fine ash (17–30 cm thick) at distances of 20–50 km from vent (fig. 49) and only ~1 percent of shards in the 2-cm-thick distal layer at King Salmon (100 km northwest of vent). Moreover, medially and distally, downwind Layers A, B_1, and rhyolite-dominant B_2 are actually enriched in crystals relative to the crystal content of rhyolite pumice, indicating that they do not contain a significant contribution of (typically crystal-depleted) coignimbrite ash. Likewise, as far downwind as Shelikof Strait (~40 km from vent) Layers C-D and F-G themselves remain enriched in free crystals (relative to the crystal content of their dacite pumice), despite their downwind-increasing contamination by rhyolite shards. It can thus be concluded that most of the rhyolitic fines that mixed with the dacite-dominant downwind ashfall of Episodes II and III were leftovers from the Episode I plinian phase, which reached a greater elevation, was more voluminous, and had a greater fraction of fines than the subsequent dacitic plinian outbursts.

Shards that settle together distally thus need not, as a generality, be all from the same eruptive episode. As the 1912 rhyolitic downwind shards (mostly finer than 125 µm) were predominantly plinian in origin, their settling was delayed and decoupled from fallout of coarser coerupted ejecta. By analogy with the term coignimbrite ash, such suspended fines have been called coplinian ash. Atmospheric turbulence and intrinsically slow settling velocities of raggedly fractured tubular and bubble-wall shards inevitably decouple some fraction of the fines from the bulk of coerupted ejecta that settles systematically to produce a deposit that thins exponentially with distance. Figure 58, reproduced from Fierstein and Hildreth (1992), illustrates isopach data for the 1912 fall deposits and for three others that all yield two-segment best-fit lines (Pyle, 1989; Fierstein and Nathenson, 1992). The breaks in slope between medial and distal deposits occur at distances where slow-settling fines become an important or dominant fraction; the flatter segments represent distal tephra deposits where the exponential fallout is augmented by fines that remain aloft longer and are thus transported farther downwind than would be predicted by terminal fall velocities (Fierstein and Hildreth, 1992). Because many plinian eruptions, especially rhyolitic ones, produce ejecta consisting of ≥50 weight percent fine ash (finer than 63 µm), such a two-segment medial/distal fallout pattern is common. And, in the 1912 eruption and elsewhere, such a pattern may be intrinsic to the plinian plume and need not always reflect dispersal of coignimbrite ash elutriated from ash flows or rising as a secondary buoyant plume above hot ignimbrite sheets. The voluminous 1932 rhyodacite eruption of Quizapu (Chile) produced a continent-wide plinian deposit with ~50 weight percent of ejecta finer than 63 µm (Hildreth and Drake, 1992) and a similarly inflected thickness-area relationship (fig. 58), even though ash-flow generation was negligible.

A small fraction of the finest ejecta remained aloft longer still. Volz (1975a, b) reviewed 1912–1914 atmospheric records and solar-radiation data for Europe and North America, concluding that (1) tropospheric dust was largely washed out within a few months; (2) stratospheric aerosol was concentrated at 15–20 km altitude and, though diminishing gradually after September 1912, persisted measurably through late 1914; and (3) mean atmospheric residence time for the aerosol was 9–12 months. Although the records could not distinguish proportions of silicate dust and acid aerosol, Volz inferred from atmospheric turbidity measurements that 12.7 M metric tons of material remained aloft in September 1912 and 2.8 M metric tons still in September 1913. Taking an average magma density of 2,500 kg/m^3, such estimates would mean that in September 1912, ~100 days after the eruption, roughly 0.08 weight percent of the 6.5 km^3 of the magma erupted as fallout remained suspended in the atmosphere. If, after three months, the dust had settled and most of the stratospheric aerosol remaining had converted to sulfuric acid droplets (density 1,788 kg/m^3), the mass fraction aloft would still be ~0.06 weight percent.

Atmospheric and Climatic Effects

Kimball (1913) summarized advance of the ash and gas cloud driven eastward by stratospheric winds and reported over Wisconsin on 8 June, Virginia 10 June, Algeria 19 June, and at various European cities 20–27 June. Precise measurements demonstrated for the first time that the volcanic haze, which had spread widely across the Northern Hemisphere, attenuated the solar radiation reaching the surface (Abbot, 1913). Abbot proposed insightfully that the haze affected the radiation budget by reflection, absorption, and scattering; the haze was at that time generally assumed to consist mainly of silicate dust rather than sulfuric-acid aerosol, the importance of which was not widely appreciated until the 1960s.

Satellite measurements of the mass of aerosol, its particle-size distribution, and its latitudinal, altitudinal, and temporal distribution were unavailable in 1912, and estimates of atmospheric optical density, radiation budget, and average surface cooling were still fairly primitive. It is instructive to ponder the 1912 aerosol in light of multifaceted analyses of the globe-encircling aerosol from the 1991 eruption of Pinatubo, based on satellite data and summarized by Sparks and others (1997). Despite important differences in latitude and sulfur content, masses of volcanic aerosol that entered the stratosphere in 1912 and 1991 were similar.

It is now well established that the role of volcanic silicate dust in affecting the radiation budget is complex but small and short-lived, because most of it falls out within weeks to months (Rampino, 1988; Lacis and others, 1992). Longer lived stratospheric aerosols are predominantly liquid droplets (<0.1 to 1 µm) that nucleate homogeneously as sulfuric acid formed by complex oxidation of SO_2 gas, principally involving reaction with OH radicals produced by photo-dissociation of water vapor. Conversion of SO_2 to H_2SO_4 proceeds slowly over many months, probably maximizing optical depth 3-6 months after an eruption. Optical depth, which measures depletion of incoming solar radiation and is closely related to aerosol mass, is the most important factor in aerosol climatic forcing.

Volz (1975a, b) compiled a large body of written observations and solar radiation data for 1912–14, principally from Europe and North America, suggesting that atmospheric dust had completely disappeared by October 1912 but that colorful twilights persisted at least until December 1913. From atmospheric turbidity records, Volz inferred that the persistent

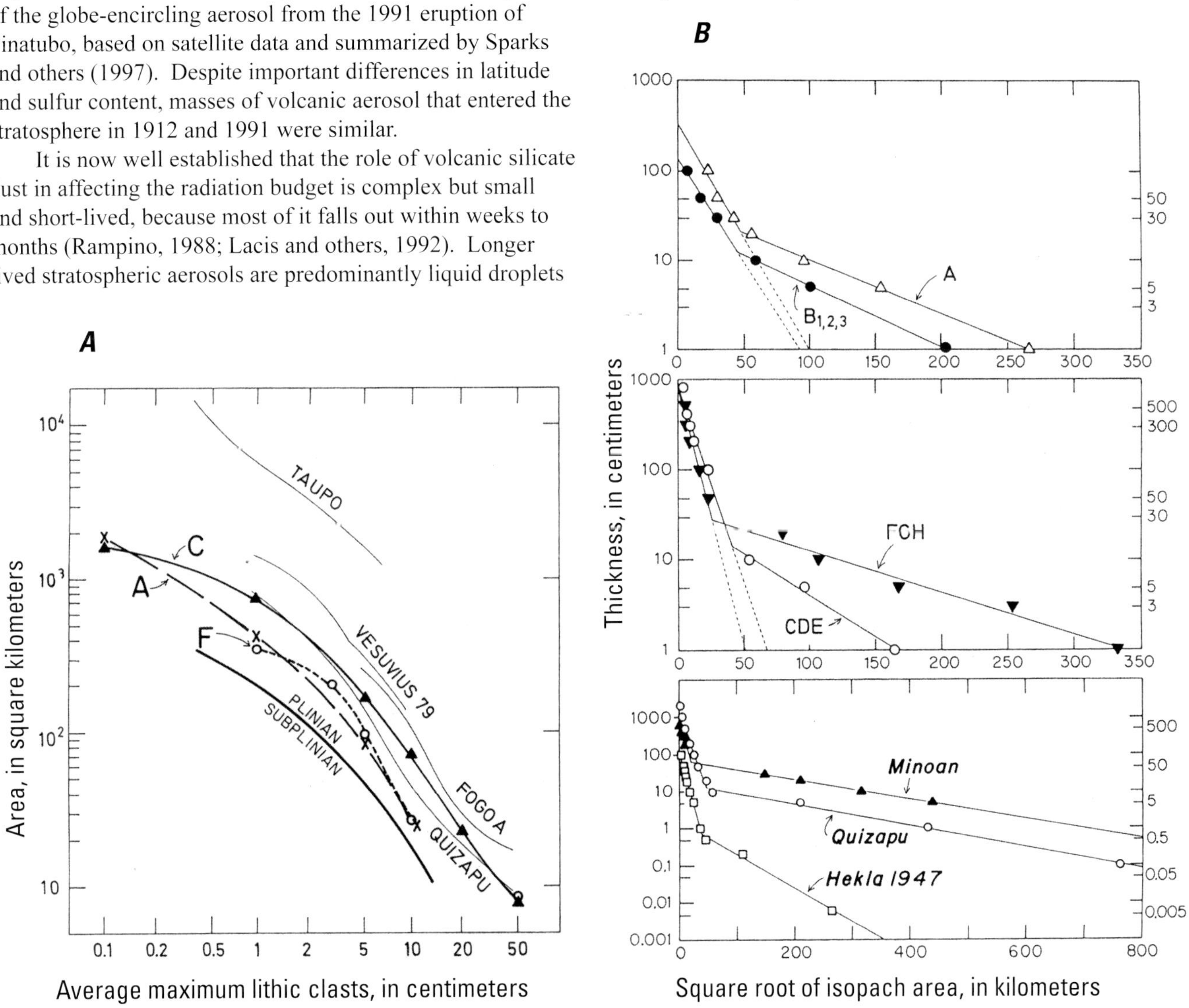

Figure 58. Dispersal and thickness of 1912 fallout layers compared to those of a few other well-studied eruptions. *A*, Area in km² enclosed by isopleths of average maximum lithic clasts (ML) for Layers A, C, and F. *B*, Plots of thickness (log scale) versus square root of isopach area ($A^{1/2}$) for Layers A, B, CDE, and FGH. Straight-line segments define exponential thinning (Pyle, 1989; Fierstein and Nathenson, 1992). Breaks in slope represent changes in settling regime with diminishing grain size (Fierstein and Hildreth, 1992). For ultraproximal data, also see figure 24 of Houghton and others (2004).

fraction of Katmai aerosol was largely at altitudes of 15–20 km and predominantly north of 44°N latitude, that mean atmospheric residence time was 9–12 months, and that stratospheric aerosol remained significant at least as late as December 1914. Settling rates for spherical droplets of sulfuric acid from a stratospheric injection altitude of 25–30 km (Fierstein and Hildreth, 1992) suggest cumulative settling times to the tropopause (~10 km) of about 1 month for radii of 1 µm, 1 year for radii of 0.1 µm, and more than a decade for radii of 0.01 µm (Sparks and others, 1997). Progressive aggregation of aerosol droplets increases settling velocities sufficiently to induce thorough removal to below the tropopause within 2 years. In the troposphere, aerosols are flushed quickly with normal precipitation. Scrutiny of ice-core records in Greenland suggests that acid fallout from the Katmai eruption lasted ~1.5 years there (Hammer, 1977).

From the mean acidity of annual layers in Greenland ice cores, global fallout of acid aerosol from the 1912 eruption was estimated at 20 million metric tons (Hammer, 1977) or 30 million metric tons (Hammer and others, 1980), whereas calculations based on global-average peak optical-depth measurements suggest 13–20 million metric tons of stratospheric sulfate aerosol (Rampino and Self, 1984; Palais and Sigurdsson, 1989). Palais and Sigurdsson (1989) applied the microbeam-based "petrologic method" of calculating the mass of acid aerosol emitted by subtracting the amounts of S, Cl, and F left in the matrix glass of ash or pumice from those contained by pristine melt inclusions sealed in phenocrysts, and then multiplying by the total mass of magma erupted. Unfortunately, all of their 1912 samples (supplied by J. R. Riehle) were dacitic fallout from 17 km northeast of Novarupta and did not include the volumetrically important Episode I rhyolite, which was dispersed southeastward. Applying the sulfur-yield data of Palais and Sigurdsson (1989) to the ~4 km^3 of dacite magma erupted as plinian fallout during Episodes I, II, and III reduces their estimate from ~7.9 to ~3.5 million metric tons of H_2SO_4 released to the atmosphere. The 2.5 km^3 of rhyolite magma erupted as fallout was neglected, but similar microbeam work by Westrich and others (1991) found sulfur contents of melt inclusions in 1912 rhyolite phenocrysts to be below their detection limit (<65 ppm). Accordingly, their petrologic estimate of H_2SO_4 released in 1912 is ~3.5 million metric tons for the entire eruption, similar to the mass recalculated from the data of Palais and Sigurdsson for dacite fallout alone.

On the other hand, because Westrich and others (1991) analyzed the rhyolite, which is halogen-rich (though sulfur-poor), their estimate of HCl released to the atmosphere is 2.5 times that of Palais and Sigurdsson. As for HF, analytical imprecision (at 400–600 ppm F) and rather small differences between F contents of melt inclusions and matrix glass weakens application of the "petrologic method" for estimating the HF yield to the atmosphere in 1912. Nonetheless, the 1912 eruption is one of few historic events to have produced a strong HF signal in ice cores on the Greenland icecap (Lyons and others, 1990); others include Tambora (1815) and several Icelandic eruptions. If HF yield to the atmosphere was 25 percent that of HCl, then the petrologic data of Palais and Sigurdsson (1989) and Westrich and others (1991), recalculated for the compositional proportions and total fallout volume (6.5 km^3 DRE) estimated by Fierstein and Hildreth (1992), yield 4.3 M metric tons of HCl, 1.1 M metric tons of HF, and 1.6 M metric tons of H_2SO_4, for a total of 7 million metric tons of acid gases.

All the estimates discussed above deal with the acid gases released by the 6.5 km^3 of magma erupted as plinian fallout. If the ~7 km^3 of 1912 magma emplaced as ignimbrite, which has a much greater ratio of rhyolite to dacite than the total fallout, released comparable fractions of its acid volatiles to the atmosphere, then the total acid gases emitted could be more than double that estimated above, thus at least 15 metric tons. Some unknown fraction of the gases emitted from collapsing ash flows would have ascended to the stratosphere with the column, and another unknown fraction would have been evolved during outflow of successive valley-filling packages; the latter may or may not have reached the stratosphere if a large coignimbrite plume failed to develop from the sheet (as argued by Fierstein and Hildreth, 1992). For the whole eruption, fallout plus ignimbrite, we recalculate ~9 million tons of HCl, ~4 million tons of H_2SO_4, and 2.3 million tons of HF, for a total of ~15 million tons of acid gases. The dominance of halogen acids is striking. Cl species are known to react with ozone in the stratosphere, potentially thinning the ozone layer (Sigurdsson, 1990; Symonds and others, 1988), but the impact (if any) of halogen gases in augmenting the role of sulfate aerosol in depleting transmissivity of solar radiation remains poorly known.

The microbeam data of Westrich and others (1991), augmented by that of Lowenstern (1993) and Wallace (2005), recalculated for a total erupted magma volume of 13.5 km^3, indicate that at least 700 million tons ($7x10^{14}$ g) of H_2O was released to the atmosphere by the 1912 eruption. Lowenstern found <50 ppm CO_2 in melt inclusions in both rhyolite and dacite, and Wallace measured 20–50 ppm CO_2 in quartz-hosted melt inclusions in the rhyolite. Owing to cracking or crystallization of melt inclusions in 1912 andesite, their pre-eruptive CO_2 content has not been established. Nonetheless, the small fraction of andesite among the 1912 ejecta and the low CO_2 contents of rhyolite and dacite indicate that the CO_2 released was trivial compared to the vast atmospheric CO_2 reservoir and thus of no climatic consequence.

If the 20–30 million metric tons of total 1912 acid aerosol estimated from Greenland ice cores (Hammer and others, 1980) and the 13–20 million tons of stratospheric aerosol (presumed to be largely H_2SO_4) estimated by optical density (Rampino and Self, 1984) are not greatly exaggerated, then estimates by the petrologic method are too low. The latter method assumes that all gases released were derived from the magma erupted and that there were negligible contributions to volatile emissions from (1) unerupted magma left in the reservoir, (2) gas bubbles dispersed in the magma before eruption, (3) the preeruptive hydrothermal system that collapsed with Mount Katmai, or (4) ignimbrite-derived fumaroles. Failure of any such assumptions would cause underestimation by the petrologic method. Because sulfur is strongly partitioned into any vapor phase coexisting with low-temperature silicic melt, and because 1912 rhyolite probably contained dispersed gas bubbles (Lowenstern, 1993; Wallace, 2005), there may well have been a significant preeruptive mass of excess sulfur not accounted for by the petrologic method. On the other hand, estimates from ice-core acidity and optical density require assumptions about latitudinal distribution of the globe-encircling aerosol (Hammer, 1977) that might lead to overestimation. Aerosol dispersal from the 1912 vent at 58°N appears to have been strongly latitude-parallel, tending at high elevations to migrate poleward, probably accounting for observation of strong turbidity peaks in 1912–13 only north of 40°N (Volz, 1975a), and perhaps enhancing the aerosol fraction deposited at 65°–77°N in Greenland. For high-latitude eruptions, upper tropospheric winds could also favor acid deposition in Greenland (Hammer and others, 1980), leading to overestimation of total global aerosol from the ice-core record. Because the tropopause is lower than ~10 km at high latitudes, such winds might also have transported fumarolic gases to Greenland, helping further to reconcile the differences between emissions estimated by petrologic and ice-core methods.

The acid fumaroles from the ignimbrite sheet were observed to have continued issuing halogen-rich gas for a decade after the eruption, though much of this is likely to have suffered tropospheric rainout. Zies (1929) estimated that in 1919 the VTTS fumaroles were annually emitting 1.25 million tons of HCl, 0.2 million tons of HF, and 0.3 million tons of H_2S; if such outputs approximated the average for 10 years (1912–21), they would represent 17.5 million tons of acid gases from fumaroles alone. Indeed, the microbeam data of Westrich and others (1991) indicate that potential yields of halogens remaining in ignimbrite glass (after decompressive losses during eruption) were on the order of 7 million tons of HF and 20 million tons of HCl, available for release to the vapor phase during devitrification and leaching of the sheet. How much of this gas might have reached the stratosphere (above a 10-km tropopause) is unknown.

Crystallization of glass (devitrification) in the ignimbrite was recognized by Fenner as one source of the halogen gases. Smith (1960) discussed Fenner's inferences, pointed out that halogens in the glass greatly exceeded the mass of free acid gases trapped interstitially in the tuff, and reasoned that gas liberation from crystallizing glass should long outlast the welding process, as field evidence has verified.

Acid aerosol in the stratosphere certainly reduces solar transmissivity, thus lowering surface temperature, and evidence is good that sulfuric-acid droplets are more effective in doing so than halogen gases. Surface cooling was recorded widely in the Northern Hemisphere during summers of 1912 and 1913, reaching a maximum measured deficiency of –0.9°C for September 1912 (Kimball, 1913; Humphreys, 1913; Griggs, 1922, p. 39). Average daily deficiency for the whole Northern Hemisphere for the 16-month interval following the eruption was reported to be –0.16°C. Modern analysis of monthly mean surface-temperature data for stations throughout the Northern Hemisphere (Jones and others, 1982) found striking anomalies (relative to a 1946–60 reference period) of –0.37°C to –1.18°C for nearly every month from July 1912 through October 1913. Rampino and Self (1984), however, attributed to the eruption a Northern Hemisphere mean temperature decrease of only –0.2°C.

Yields of acid gases to the atmosphere have been estimated by the petrologic method for numerous other large eruptions (Devine and others, 1984; Palais and Sigurdsson, 1989). A log-log plot of sulfur yield versus Northern Hemisphere temperature decrease produces a good linear relation embracing the data for most of the eruptions studied (Palais and Sigurdsson, 1989; fig. 18.1 of Sparks and others, 1997). For example, a sulfur yield of 30 million tons by the Laki eruption of 1783 was followed by a cooling of 1°C; a yield of 17 million tons by the Tambora eruption of 1815 produced a cooling of 0.7°C; and a yield of 1 million tons by the Krakatau eruption of 1883 cooled Northern Hemisphere surface temperature by 0.3°C. On those plots, however, data for the 1912 eruption fall far below the trend defined by most eruptions studied, plotting at either too small a cooling or too great a sulfur yield. The anomalous position for Katmai can be eliminated either by doubling the cooling to 0.4°C (as the 1912–13 data suggest) or lowering the sulfur yield to half a million tons (1.6 million tons of plinian H_2SO_4)—as we recalculated above, based on the data of Westrich and others (1991) and our own compositional proportions and eruptive volumes (Fierstein and Hildreth, 1992).

Changes in Compositional Proportions During the Eruptive Sequence

Few eruptions worldwide have produced a compositional range of ejecta as wide as that of 1912—57.9 to 77.8 percent SiO_2, plus sparse blocks of mafic scoria down to 50.4 percent SiO_2. The tight andesite-dacite compositional array is continuous (fig. 59), though pumice less silicic than 62 percent is typically black while pumice having 63–68 percent SiO_2 is generally pale gray (fig. 4). None of the 250 or so (nonmixed) 1912 pumice clasts analyzed contains between 68.6 and 77 percent SiO_2, manifesting a broad compositional gap that separates 1912 dacite and rhyolite. Numerous field counts of pumice clast proportions in all ignimbrite packages and all fallout layers (Hildreth, 1983; Fierstein and Wilson, 2005), combined with shard proportions in distal ash layers counted by microprobe (Fierstein and Hildreth, 1992), allow us to estimate that at least 55 percent of the 1912 magma released was high-silica rhyolite, 35 percent dacite, and no more than 10 percent andesite.

Although the first ~3 km^3 of magma erupted was strictly rhyolite, the coerupting proportions of andesite, dacite, and rhyolite fluctuated greatly from the middle of Episode I onward (fig. 60). Neither the rhyolite magma nor the

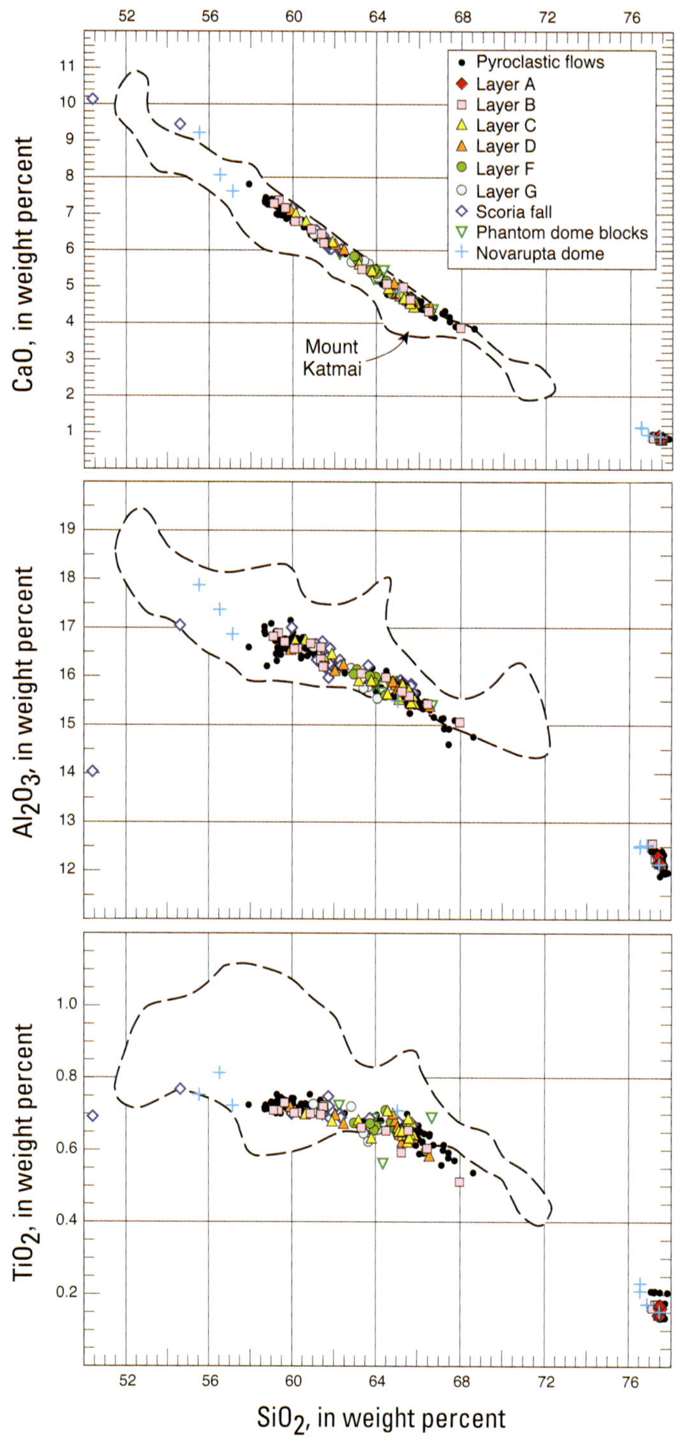

Figure 59. Variation of CaO, Al_2O_3, and TiO_2 versus SiO_2 content (all in weight percent) for ~265 samples of 1912 Novarupta pumice and dome lava, subdivided as indicated in inset key. Principal features are continuity of andesite-dacite array, wide compositional gap from dacite to rhyolite, slight contamination of dome samples, and five (rare) mafic scoria outliers. Outlined enclosure is field for ~150 samples from Mount Katmai, illustrating its compositional affinity with Novarupta. Data in appendix A. For more chemical data and discussion, see chapter 13.

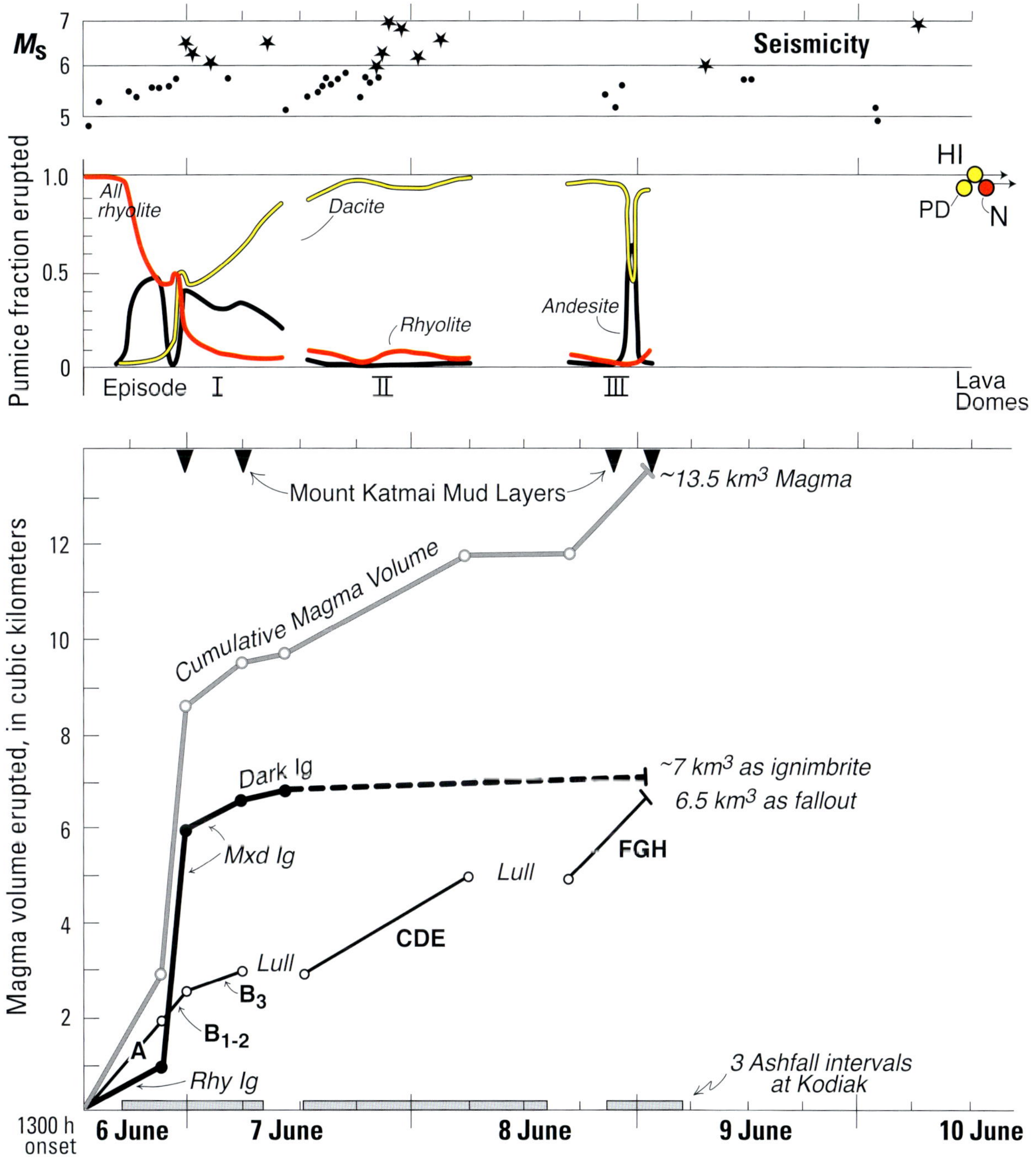

Figure 60. Eruptive sequence of June 1912, showing estimated timing and magma volumes of eruptive units, magnitudes (M_s) of earthquakes accompanying caldera collapse, stratigraphic positions of main mud layers ejected from collapsing caldera, intervals of ashfall at Kodiak, and fluctuating proportions of rhyolite, dacite, and andesite pumice that erupted concurrently. Compare with figures 2 and 34. Rhy Ig, all-rhyolite ignimbrite Package 1; Mxd Ig, compositionally heterogeneous main ash-flow sequence; Dark Ig, late Packages 8 and 9, both rich in andesite and dacite pumice. Fall units A and B represent Episode I. Segment CDE is Episode II; segment FGH is Episode III. Mount Katmai mud layers are sheets of lithic ejecta expelled by hydrothermal explosions. Three lava domes were emplaced at unknown times after Episode III: PD, Phantom dome, dacite plug explosively distributed to produce block bed that overlies Layer H; HI, Horseshoe Island dacite dome on caldera floor; N, Novarupta dome, ~95 percent rhyolite but streaked with andesite and dacite. Stars emphasize earthquakes of magnitude 6.0 or greater.

andesite-dacite magmatic continuum was exhausted during the series of explosive eruptions, as shown by subsequent extrusion of three lava domes. The Horseshoe Island dome on the Katmai caldera floor was dacite (66% SiO_2; Fenner, 1930, 1950). The Phantom dome that transiently plugged the Episode III vent was also dacitic but was explosively destroyed and never observed; texturally varied samples from the derivative block bed (Adams and others, 2006a), scattered as far as 2 km from Novarupta, range from 62.2 to 66.7 percent SiO_2 (n=8). The succeeding Novarupta dome, which is predominantly rhyolite (76-77% SiO_2) but streaked with ~5 percent andesite and dacite (55-65% SiO_2), plugged the same vent and was first observed in 1916.

The compositionally varied nature of the deposits and the rapid fluctuations in proportions of coerupting pumice types facilitated (1) extraordinarily detailed correlations between fall units and contemporaneous ash flows (fig. 34); (2) subdivision of the main Episode I ash-flow sequence into nine successive, compositionally distinct, packages of flows (fig. 27; Fierstein and Wilson, 2005), illustrating stepwise assembly of a 120-km^2 ignimbrite sheet over an interval of ~16 hours; (3) linkage of compositional packages in the HEPI facies on near-vent ridges to particular packages in the voluminous valley-confined facies (Fierstein and Hildreth, 1992; Fierstein and Wilson, 2005); (4) quantitative discrimination of mixed contributions from different eruptive episodes to fine-grained ash layers, upwind, far downwind, and in VTTS (Layers E and H, as elaborated above; Fierstein and Hildreth, 1992); and (5) recognition of narrow lobes of proximal fallout and sector-confined proximal PDC deposits intercalated with regional fall deposits of Episodes II and III (figs. 47, 48, 52), and inference therefrom of column-margin instabilities and processes of ultraproximal sedimentation from lower margins of the emerging jet and convective eruption column (Fierstein and others, 1997; Houghton and others, 2004).

For the first ~3.5 hours of the eruption, the magma released was exclusively rhyolitic. For the next ~5 hours, andesite gradually increased its proportion from trace amounts until it rivaled rhyolite as the principal component, while dacite consistently contributed only a small fraction of the ejecta (<5%). After the 8th hour, the fraction of dacite began to increase, and after the tenth hour, dacite became the dominant component, fluctuating in its fraction of the eruptive mixture until the end of Episode 1 (figs. 34, 60). During Episodes II and III, from shortly after the 16th hour until the end of the explosive sequence on the third day, dacite was consistently predominant, with small bursts of andesite largely limited to ultraproximal falls and flows shed from the periphery of the main dacite column.

Banded Pumice

Syneruptive mingling of the 1912 rhyolite, dacite, and andesite magmas produced an abundance of banded pumice (fig. 61) as well as conspicuous dark-colored lenses and layers in the light-colored rhyolitic Novarupta lava dome. Banded pumice is most abundant in Layer B and in ash-flow Packages 4 through 9 that make up the VTTS ignimbrite sheet. Field counts of pumice clasts at hundreds of sites, including all subunits, yield wide ranges of banded pumice content—0–25 (count) percent in VFI, 0–18 percent in HEPI, and 0–24 percent in Layer B (Hildreth, 1983; Fierstein and Hildreth, 1992; Fierstein and Wilson, 2005). Banded pumice is absent in Layer A and concurrent all-rhyolite flow deposits, and it is rare in regionally dispersed dacitic fall Layers C–D and F–G. Intercalated proximally with the dacite pumice falls, however, most of the thin sector-confined fall and flow deposits of Episodes II and III (Houghton and others, 2004) are rich in banded pumice, including as much as 27 percent in the andesite-rich S bed (Fierstein and others, 1997). We estimate that, for the entire 1912 eruptive sequence, 3–5 percent of pumice clasts are banded.

The banded pumice was provocative during the early Katmai investigations because products of magma mixing had by then seldom been described. Fenner advanced a radical interpretation that dark components of the banded pumice and lava dome had been solid andesite of the Katmai edifice and of valley-floor moraines that were melted and assimilated into juvenile rhyolite magma by virtue of exothermic reactions that he imagined were linked to delayed liberation of magmatic gases. He attributed sieve-textured plagioclase in andesite bands to attack by such gases, and he interpreted xenoliths in the dome lava as fragments of the rocks being melted. Xenolith-bearing clasts of variably welded vent-filling tuff recycled as lithic ejecta during Episodes II and III were interpreted by Fenner as "obsidian" (fig. 62) and "scum" (fig. 63) from the surface of a lava lake he imagined to occupy a crater on Mount Katmai while part of the edifice was being dissolved. Fenner seems also to have thought that the dacite pumice was

Figure 61. Banded pumice clasts from VTTS ignimbrite. Knife is 8.5 cm long.

Figure 62. Block of recycled welded ejecta near Novarupta dome. One of many textural varieties of Episode I vent-filling tuff expelled from nested inner vent during Episodes II and III, this type was called "obsidian" by Fenner. Welded block contains abundant lithic fragments of Jurassic siltstone, a discrete pale-gray ejected block of which lies at upper right. Pen is 13.5 cm long.

Figure 63. Block of recycled welded ejecta ~1 km from Novarupta dome. Such agglutinated pumice breccia, rich in rhyolite, dacite, and andesite pumice clasts, is a common variety of recycled vent-filling tuff that was expelled as lithic ejecta during Episodes II and III. Called "scum" by Fenner, lapilli of such material are the principal lithic component of fallout Layers C, D, F, and G and are identifiable as far as 25 km downwind. Geologist is 1.63 m tall.

a thoroughly hybridized result of such assimilation. How he rationalized complete hybridization while dismissing simple syneruptive mixing of different liquids is not clear. Fenner nonetheless persuaded Griggs (1922) of his hypothesis and defended assimilation of solid mafic lavas by rhyolitic magma for the rest of his life (Fenner, 1923, 1926, 1934, 1937, 1938a, 1938b, 1944, 1950).

By 1953, petrologic and thermal relations of rhyolite and andesite were better appreciated, Fenner's mechanism had been discounted, and Williams and others (1953) reinterpreted the banded pumice in terms of mingling of confluent magmas. Curtis (1968) went on to point out that most contacts between bands were sharp (with gradations only microns thick), that glassy andesite was chilled against rhyolite, and that sieved and nonsieved variably zoned plagioclase had coexisted in the andesite before encountering the rhyolite, which has its own nonsieved unzoned, easily distinguishable, more sodic plagioclase.

The "bands" in the banded pumice are extremely varied, ranging from hairline streaks to layers many centimeters thick and including isolated blebs and blobs, multistrand stringers, and sparser convolutions. Individual banded pumice blocks and lapilli consist predominantly of either rhyolite, dacite, or andesite, or they can have subequal proportions of any two of them. Pumice clasts consisting of all three are rare. Rhyolite-andesite and dacite-andesite combinations are abundant, but dacite-rhyolite banded pumice is uncommon except in ignimbrite Package 7 (Fierstein and Wilson, 2005). Proportions of the three compositional combinations generally reflect those of the nonbanded pumice clasts in each emplacement unit.

The sharpness of compositional contacts within clasts is consistent with syneruptive confluence in the magma feeding system and with contiguity for only minutes or hours prior to fragmentation. Wide fluctuations in proportions of andesite, dacite, rhyolite, and varieties of banded pumice in successive eruptive packages indicates that feed rates of the three magmas from reservoir(s) to their ultimate mutual conduit were unequal and not steady. Although joint conduit transport was able to produce extensive mingling of the viscous magmas in a short time interval, the interval was long enough for andesite to be partitioned preferentially into a peripheral collar around the emerging jet and thence into fountains and flows collapsing around the margins of plinian columns, the interior cores of which carried far smaller proportions of andesite into the regional fallout (fig. 53).

Syneruptive mingling of the compositionally different magmas is not known to have produced any homogeneously hybridized pumice, but limited mutual contamination by phenocryst transfer is not uncommon. Sparse quartz grains have been identified in a few andesite and dacite pumice clasts, where they evidently lacked time to be resorbed or rimmed before eruptive quenching. Sparse hornblende prisms in a small fraction of the dacite pumices and in the Novarupta rhyolite dome are compositionally identical, and in neither host have they developed reaction rims, even though they were in equilibrium with neither (see section below titled "Experimental Petrology"). Rare clinopyroxene crystals, mostly fragments, have been found in a few rhyolite pumices and in the rhyolite dome but are absent in most rhyolite samples. Subordinate populations of Fe-Ti-oxide minerals that give much higher equilibration temperatures than most of the accompanying titanomagnetite-ilmenite pairs are present sparsely in some dacite pumices and in the dome rhyolite. Some pairs in an andesite host were found to have developed Ti gradients at their contacts (Hammer and others, 2002), but most such transferred oxide crystals remained homogeneous and unmodified. Such contaminants, widely scattered and easily overlooked, require cautious sample selection and scrutiny of mineral microbeam data. The general absence of reaction between such transferred crystals and their new host melts supports the interpretation that the mingling that produced both the subtle contamination and the conspicuous banding was syneruptive and thus occurred over only hours or days.

Contemporaneity of Falls and Flows

Plinian fallout, high-energy proximal ignimbrite (HEPI), and voluminous valley-filling ignimbrite (VFI) were deposited contemporaneously throughout most of Episode I. This is demonstrated by (1) comparably progressive changes in rhyolite/dacite/andesite pumice proportions in the sequences of fall layers and flow packages (fig. 34); (2) intercalation of layers and lenses of Layer A with all-rhyolite HEPI and VFI deposits; (3) intercalation of compositionally heterogeneous HEPI and VFI packages with compositionally similar parts of Layer B (fig. 37); and (4) absence of the Episode I fall deposits in the VTTS wherever equivalent ignimbrite is present. Irrespective of whether the A–B fallout near Novarupta and on the VTTS floor was incorporated by concurrently emplaced flows or scoured by them, the flow/fall intercalations that crop out nearby, atop peripheral ridges, in the Trident saddles, and along the Mageik Creek valley from Katmai Pass to the toe of Observation Mountain, prove their contemporaneity.

During Episodes II and III, on a lesser scale, intercalation of several small proximal PDC deposits within the stack of regionally dispersed plinian pumice falls (Layers C–D and F–G) likewise demonstrates generation of laterally emplaced flows from various heights in a sustained column (figs. 47, 52). These were mostly sector confined, just a few meters thick, and extended only a few kilometers, but they nicely illustrate radial generation of pyroclastic currents by transient instabilities in an annular collar around the emerging jet and sustained column.

During Episode I, such lateral currents were proximally turbulent, far more voluminous, and represented greater mass fractions of the ongoing eruption column than during Episodes II and III; some were blast like and others soon densified in the valleys, where some extended >20 km from vent. Despite the complexly changing compositional proportions in the 16-hour sequence of discrete packages of flows that built up the VTTS ignimbrite (fig. 34), and despite the 4:1 volume dominance of

Phase B ignimbrite over Layer B fallout, no hiatus is recognized within Layer B (fig. 15) that might suggest discontinuous accumulation of plinian fallout during assembly of the main VTTS ignimbrite.

Characteristics of the Eruption Columns

Initial rise of the plinian column was seen in clear weather from the S.S. *Dora* in Shelikof Strait at 1300 Alaskan time on 6 June. No observers were able to see any column after a few hours, but the three intervals of ashfall recorded at Kodiak (170 km downwind) correlate well with the three layers discernible there on the ground (Martin, 1913; Griggs, 1922; Hildreth, 1983; Fierstein and Hildreth, 1992). The three fallout intervals lasted ~20 h, ~26 h, and ~10 h at Kodiak, but there need be no simple correlation between fallout intervals downwind and duration of eruptive episodes at vent, owing to altitudinally variable transport velocity, wind shear, progressively slower settling of ash that systematically fines with distance, and turbulent retention of fine ash aloft. Ashfall at Kodiak first ceased between 0910 and noon on 7 June, but it was reported to have diminished gradually after 0300. If so, and if a 4-hour transport time (based on first sighting of the column at 1300 and onset of fallout at Kodiak at 1700) can be generalized, then the A and B plinian phases of eruptive Episode I may have together lasted only 10–14 hours. This possibility is supported by an additional time constraint—the first of several layers of nonjuvenile lithic debris ejected during caldera collapse at Mount Katmai is intercalated between Novarupta Layers B_2 and B_3 and is thought to coincide with the first major teleseism (M_s 6.5) at 2356 on 6 June, apparently the onset of major collapse (Hildreth, 1991; Abe, 1992; Hildreth and Fierstein, 2000). By that hour, all of Phase A, two-thirds of Layer B, and as much as 85 percent of the main Phase B ignimbrite had erupted, equivalent to a magma volume of ~8.5 km^3 (figs. 34, 60). For an average magma density of 2,400 kg/m^3 and a duration of 11 hours, these data suggest an average mass eruption rate of 5x10^8 kg/s, our best constrained estimate for most of Episode I. For the plinian fraction alone (~2.6 km^3 of magma dispersed as fallout), the mass eruption rate would be ~1.6x10^8 kg/s.

No similar narrative or geophysical constraints can be applied with any confidence in estimating durations of eruptive Episodes II and III. Local flooding and reworking of pumice atop the ignimbrite along upper Knife Creek (fig. 44) indicates a time break of unknown duration before opening of Episode II. One interpretive possibility is that the interval of extraordinary seismicity from the evening of 7 June through 0300 on 8 June (including six earthquakes of magnitude 6.0 to 7.0) represented fitful caldera collapse, during which Episode II dacite was withdrawn from under Mount Katmai. Such a scenario could mean that the main eruptive interval that produced Layers C and D lasted only 7–9 hours and that the 26-hour-long ashfall interval at Kodiak included slow settling of fines both before and after eruptive Episode II. If the ~2 km^3 of Episode II dacite magma were released in 8 hours, then its average mass eruption rate was 1.7x10^8 kg/s.

The time break between eruptive Episodes II and III permitted accumulation of Layer E, dominantly fine dacitic vitric ash, which is 3–9 cm thick proximally and medially out to at least 23 km from vent. Layer D grades up normally from lapilli fall to a fine-ash top, showing that fallout of pumice lapilli ceased entirely, the only clear lull in the Episode II–III sequence. The eruptive hiatus is inferred to have ended with radial blast like emplacement of lithic-rich PDC 4 (Houghton and others, 2004), which reopened the vent for resumption of dacite magma release as plinian Layers F and G. Vent blockage by wall collapse was probably responsible for the hiatus, rather than the slowing of magma ascent and the shift to open-system degassing (inferred from vesicle textures) thought to have terminated Episode III (Adams and others, 2006b).

At Kodiak, the second ashfall hiatus lasted less than 6 hours and was followed by the final period of fallout ~10 hours long, roughly equivalent to Episode III. During this interval, seismicity resumed after a 20-hour lull, and two final phreatic mud layers were ejected from the enlarging caldera of Mount Katmai and intercalated within the pumice falls of Layers F and G (fig. 60). There are no tighter constraints on the duration of eruptive Episode III. If it truly lasted 10 hours, the average mass eruption rate for its 1.63 km^3 of dacite magma was 1.1x10^8 kg/s (or twice that for a duration of 5 hours).

The mass eruption rates estimated above (1.6, 1.7, and 1.1x10^8 kg/s for Episodes I, II, and III, respectively) are based on volumes of the plinian fall deposits and their approximate eruption durations. The model of Wilson and Walker (1987) provides a relation between mass eruption rate and column height: $H = 0.236\,(M_0)^{1/4}$ where H is column height in kilometers and M_0 is mass eruption rate in kilograms per second. This yields column heights of 27, 27, and 24 km, respectively, for plinian Episodes I, II, and III.

An alternative method developed by Carey and Sparks (1986) reconstructs characteristics of the source plume from measurements of clast dispersal in the fall deposits. Dispersal pattern is controlled mainly by column height, local winds at various altitudes, and grain-size distribution of the ejecta. Apart from ultraproximal deposits augmented by lower column-margin fallout (Houghton and others, 2004), regional fall deposits typically display exponential decay of thickness and grain size away from vent, reflecting sedimentation from the umbrella region of the spreading plume. As reviewed by Sparks and others (1997), dispersal of clasts of a given size is quite sensitive to column height, especially lithic clasts that are less varied in density than pumice and less susceptible to breakage on impact. Plotting the lithic-isopleth data (fig. 58) of Fierstein and Hildreth (1992) on the nomograms of Carey and Sparks (1986) yields column heights of 28, 28, and 24 km, respectively, for plinian Episodes I, II, and III. The data also yield rough estimates of maximum wind velocities (at the tropopause), respectively >30 m/s, 30 m/s, and <10 m/s for the three successive episodes. The more symmetrical distribution, less vigorous plume, and weaker southeastward dispersal of Episode III ejecta, relative to those of I and II, were noted by Curtis (1968) and Fierstein and Hildreth (1992).

Column Collapse and Emplacement of Pyroclastic Density Currents

Buoyant plinian columns commonly collapse, partially or extensively, into fountains that generate pyroclastic flows. Wilson and others (1980) analyzed eruptive conditions that favor collapse—decreasing exsolved magmatic gas content and increasing vent radius, or, closely linked respectively, decreasing vent-exit velocity and increasing mass eruption rate. Many subsequent theoretical and field studies of explosive eruptions and their deposits were reviewed by Sparks and others (1997). During Episode I in 1912, a sequence of pyroclastic flows filled the VTTS and was more voluminous than the plinian fallout that preceded and accompanied it. On the other hand, during Episodes II and III, several small pyroclastic density currents (flows, surges, and blasts) were largely sector confined, extended only 1–3 km from vent, and amounted in total volume to less than 1 percent of concurrently deposited plinian dacite Layers C to H.

Episode I Eruptive Sequence

In the VTTS, valley-filling ignimbrite (VFI) extends ~23 km northwest from Novarupta and has been shown to consist of nine successive, compositionally distinguishable, packages of ash flows (figs. 16, 27). Of the nine, several also entered Katmai Pass, and two of them (1 and 7) thence extended southward as far as 10 km down Mageik Creek. In Package 1, the pumice is entirely rhyolitic but generally small (few clasts larger than 2 cm), and lithics are abundant owing to initial vent-reaming and vent-wall collapse. Package 1 is distributed radially, extending 9 km from Novarupta, in part as a blast like vent-opening HEPI. In Packages 2 through 5, the fraction of andesite pumice progressively increased from ~3 to ~50 percent, while dacite pumice maintained a fairly constant fraction (<5 percent); each of these four packages filled the upper VTTS and reached at least as far as Three Forks. Taken together, Packages 3 and 4 volumetrically dominate the VTTS sheet; they were partly diverted around the Three Forks moraine (figs. 30, 64), becoming the only packages that extend all the way to the Ukak terminus (fig. 29). Where it banked against the moraine, a thick section of Package 4 developed diffuse laminae (fig. 65), most layers being 1–10 cm thick (fig. 66). The distal tongues of dark-colored andesite-rich Packages 5 and 6 were confined to swales that were produced on a time scale of hours by axial compaction of earlier packages and commonly enhanced by active basal scour (figs. 31, 67). Package 6 was the first to contain abundant dacite pumice (12–28 percent), along with subequal proportions of rhyolite and andesite. Although not as widespread as earlier packages around the upper VTTS (presumably because less energetically emplaced), it nonetheless issued swale-confined dark-colored andesite-rich tongues, which, like those of Package 5, reached Three Forks (figs. 27, 30).

Package 7 is compositionally unique in consisting of equal proportions of dacite and rhyolite pumice with only a trace of andesite. It marks the arrival of dacite as a major contributor to the eruptive mixture, an importance that expanded in Packages 8 and 9 and became predominant in Episodes II and III. In the VTTS, Package 7 is distributed only within 6 km of Novarupta, but it flowed at least 10 km down the valley of Mageik Creek. Intercalated within it proximally is a thin blast like deposit, confined to the southwest sector and containing abundant andesite scoria and blocks of welded vitrophyric lithics (Fierstein and Wilson, 2005). Package 8, though rather variable in composition, contains grossly equal proportions of dacite and andesite pumice and is the first ignimbrite unit in which rhyolite is clearly subordinate (5–30 percent). It is limited to and caps much of the upper and middle VTTS, where it is dark gray (or locally brightly oxidized), 5–10 m thick, widely sintered or vapor-phase indurated, and locally welded and eutaxitic (fig. 35). Package 9, which terminated Episode I, is a set of small-volume ignimbrite lobes and tongues rich in dacite pumice blocks but typically only 0.5–3 m thick (fig. 46). Dacite is dominant, but andesite scoria is abundant and the rhyolite fraction is less than 10 percent. Variously nonindurated or sintered, scoured exposures of Package 9 locally form blocky pavements (fig. 36) on the floors of the upper Lethe and Knife Creek arms. Although no HEPI equivalent is recognized, currents of Package 9 also crossed the low saddle between Baked and Broken Mountains, leaving deposits on the floor of the narrow valley that trends northwest from that saddle (fig. 26).

With the exception of block-rich Package 9 and local pumice-concentration zones in other packages, most of the VTTS contains familiar massive fines-rich ignimbrite, with poor sorting (σ_ϕ~2.5–4.5), median grain size generally 0.1–0.5 mm, modest lithic fractions (1–6 weight percent), and varied fractions of scattered pumice lapilli and blocks. Much of the sheet is nonwelded (fig. 68), but many exposures in the middle and upper valley are sintered, vapor-phase indurated (fig. 69), or rarely eutaxitic (fig. 33). Ignimbrite blocks ejected from postemplacement phreatic explosion craters are commonly dense and some have flattened pumice, indicating widespread welding at levels deeper than that of present-day stream incision. Exposed total thickness ranges from 10 to 20 m distally to as much as ~33 m in some lower valley gorges (fig. 32) and on marginal benches. In most of the central and upper VTTS, however, where the sheet is indurated and certainly thickest, streams have cut only a few meters deep (fig. 35). Along the Lethe arm, differential welding (Hildreth, 1983) and comparative stream profiles (Curtis, 1968) suggest an axial total thickness of at least 150 m, perhaps locally >200 m.

The near-vent ignimbrite facies called high-energy proximal ignimbrite (HEPI) by Fierstein and Hildreth (1992)

Figure 64. Main VTTS ignimbrite banked against cross-valley moraine ~2 km northeast of Three Forks, 16 km from vent at Novarupta. Ignimbrite, ~20 m thick just to right of moraine, failed to override it, but thin ash-flow tongues penetrated a few low saddles like that in foreground. Backside of moraine at left has 15–25 m relief and is veneered by thin ash-cloud, plinian, and aeolian ash deposits.

Figure 65. Extreme development of diffuse lamination in distal VTTS ignimbrite ~1 km upvalley from cross-valley moraine and 15 km from Novarupta. Layers are 1–10 cm thick, and exposure is ~8 m high. Mount Katolinat on skyline.

Figure 66. Diffuse layering in distal VTTS ignimbrite where it banks against moraine 16 km from Novarupta. Layers are 5–10 cm thick. Ice axe is 81 cm long. Layering continues to top of deposit, which does not extend beyond moraine.

Figure 67. Boundary between ignimbrite Packages 6 and 3 just east of Three Forks. Package 6 (above) has diffuse internal layering and its own thin fine-grained basal layer, which crosscuts subhorizontal layering in rhyolite-richer Package 3 (below). Broad swale is not compactional; earlier partial drainaway of upper Package 3 is not excluded, but swale was at least partly scoured by currents emplacing Package 6. A modest time break is implicit. Ignimbrite is overlain by ~0.5 m of stratified white dacite fallout of Episodes II and III, then by several meters of brown ashy mudflow deposits. Shovel just left of center is 58 cm long.

Figure 68. Classic flow-unit contact in nonwelded ignimbrite on left bank of Knife Creek just above Three Forks. Fine-grained basal layer of ignimbrite Package 4 (upper third of cliff) conformably overlies Package 3. Exposure ~12 m high; base rests on glacial and fluvial deposits. At top, thin strata of white dacite fallout of Episodes II and III are capped by tan ashy mudflow deposits.

was deposited on all ridgecrests within 6 km of Novarupta. Typically 1–5 m thick but as thick as 12–15 m on Baked and Falling Mountains, these stratified veneers were deposited by high-velocity turbulent currents in all directions from Novarupta basin. The wide ranges in median and maximum grain size, fines fraction (5–50 wt percent), lithic fraction (1–16 wt percent), sorting (σ_ϕ~1–4), and thickness (1–35 cm) of individual layers, which number from 5 to >110 in various sections, were described in chapter 5. The compositional sequence of pumice proportions in HEPI sections roughly matches that of the sequence of packages that make up the main VFI, although the final VFI package (Package 9) apparently lacks a HEPI equivalent. If HEPI deposits mantle a liberally estimated area of ~50 km^2 in the upper VTTS and if their average thickness is 5 m, then their total volume would be ~0.25 km^3 or only 2–3 percent that of the VFI emplaced during Episode I (~11 km^3).

Concurrent Deposition of HEPI and VFI

Concurrent production of HEPI and VFI at Novarupta exemplifies some of the issues that underlie various models of ignimbrite deposition, particularly those concerning solids concentration in the depositing currents, the degree to which such currents are turbulent, the mechanisms of particle support in such currents, whether currents are stratified (graded in particle concentration) or have sharp upper surfaces, and whether currents deposit incrementally at the base or slide to a halt (more or less) en masse (Fisher, 1966; Sparks, 1976; Wilson and Walker, 1982; Druitt, 1998; Freundt and Bursik, 1998; Branney and Kokelaar, 2002).

Properties of gas-charged particulate masses as they collapse from an eruption column and initiate the currents that deposit ignimbrite remain inadequately known. At Novarupta it can be inferred that collapse was not complete, because intercalation of plinian Layer B with packages of the main ignimbrite and their mutually equivalent compositional sequences indicate sustained coexistence of a (reduced but substantial) buoyant plume with the more voluminous collapsing fountains. It can also be inferred that the Phase B fountains were not very high because most currents had insufficient momentum to flow more than ~2 km laterally (and 300 m in elevation) up the adjacent slopes of Trident. Phase A currents, however, though far less voluminous, passed through the Trident saddles (~800 m higher than Novarupta) and swept across the slopes of Mount Griggs (as much as 450 m higher than Phase B VFI in the Knife Creek arm). The radial distribution

Figure 69. Jointed ignimbrite, sintered and vapor-phase indurated along Juhle fork, at northeast edge of VTTS ~600 m downstream from mouth of side valley where stream enters valley-filling tuff. Waterfall is ~25 m high. Fumarolic alteration of ignimbrite in foreground. Jointed zone grades down into massive ignimbrite that ranges locally from nonindurated to sintered.

of the Phase A all-rhyolite ignimbrite, its greater momentum, greater proportion of stratified HEPI, and sparsity of coarse pumice probably reflects a radial blast like component to its initial momentum rather than simply a greater collapse height relative to the currents of Phase B. This may in turn reflect the greater gas content of the rhyolite and perhaps, too, a component of lateral momentum owed to decompression of the abruptly unconfined mass emerging from an immature vent at the eruptive onset.

Absence of Phase B HEPI on the upper vent-facing slopes of Trident and on the summit of Baked Mountain (only ~1.5 km laterally and ~250 m higher than the vent-funnel rim) suggests that vertical momentum of the gas thrust phase was unable to carry the fountaining mass more than a few hundred meters above the vent. Instead of initially discrete currents collapsing from column margins at heights of a few kilometers, there may have been a broad low fountain flooding Novarupta basin and disgorging radially from it. Rapid excavation of a flaring vent funnel 1–2 km wide, further confined between Broken Mountain and Trident, may have so limited radial air entrainment relative to mass discharge that fountain collapse was atypically low. There is no evidence for asymmetrical collapse during Episode I, as all nine ignimbrite packages are distributed radially into the Knife Creek and Lethe arms (as well as the narrow valley separating Baked and Broken Mountains, fig. 26), including even Package 9, which was laid down by short-lived lobate currents of low mass flux.

The north and south sides of Novarupta basin are hemmed in by highlands (fig. 70), but the slopes west and east of the vent were gently sloping valleys broadened by Pleistocene glaciation. The western outlet now slopes only ~1.7° W toward the Lethe arm (fig. 18) and is unlikely to have been much steeper before the eruption. The eastern outlet now slopes 5°–6° NE toward the Knife Creek arm, but its (probably

Figure 70. Aerial view southeastward over Novarupta toward Trident and Mount Katmai. In foreground, vent depression (commonly called Novarupta basin) extends 2.5 km from pumice-covered fissured scarp of Broken Mountain at left to 400-m-high bedrock scarp of Falling Mountain dacite dome at right. Vent funnel was backfilled by ignimbrite and fallback ejecta, deformed by compaction and welding, and plugged by 380-m-wide Novarupta rhyolite dome, which is surrounded by an asymmetrical ejecta ring that consists mostly of fallout from Episodes II and III. At extreme upper left, Katmai caldera is centered 10 km east of Novarupta. Tephra-covered Knife Creek Glaciers (1 to 3) descend at left from four ice-sculpted peaks of Trident group. Partly hidden at right is black cone of Southwest Trident (1953–74) and its lava-flow apron in Katmai Pass and Mageik Creek. In far right distance is Katmai River floodplain and Katmai Bay. On remote horizon lies Kodiak Island, beyond Shelikof Strait.

gentler) preeruptive gradient is hard to estimate because of the thick Episode II and III deposits in that sector. Densifying underflows encountered no barriers emerging east or west from fountains collapsing in Novarupta basin, and the gentle slopes away from the vent would have reduced their initial momentum. Coarse lithic lag breccias, representing vent-funnel excavation and collapse of the Falling Mountain scarp (fig. 71), are presumed to have been dumped in Novarupta basin during Episode I, but none are exposed owing to the thick proximal deposits of Episodes II and III. Dense clasts lifted by the gas thrust but unsupported by the fountain would have been dropped promptly by initially dilute currents emerging on such gentle slopes. Clasts in traction-stratified HEPI (and sparse ballistics) on circum-vent ridgecrests include dense lithics as big as 25–35 cm and abundant pumice blocks as big as 40–60 cm.

During most of Phase B, a pulsating fountain filled Novarupta basin and discharged concentrated pyroclastic underflows into both arms of the VTTS. The laterally expanding fountain may have been little thicker than the highest HEPI deposits, a few hundred meters above Novarupta. Concurrently, a plinian convective column partially survived and deposited equivalently zoned Layer B fallout downwind as far as Kodiak Island. More speculatively, the column might have been augmented by thermal plumes rising from the periphery of the fountain, their resurgent buoyancy owing to peripheral ingestion of air during collapse. Another possibility is alternation of collapsing fountains and convecting plumes, perhaps punctuating the ignimbrite packages, owing to variations in discharge rate, in gas content released from changing proportions of the different magmas, or in fluctuating exit pressures caused by vent-wall instabilities. In any event, persistently low fountaining during Phase B is consistent with the low momentum of VFI packages, which failed to ride far up sidewalls in the upper VTTS or to overrun the low valley-blocking moraine at Three Forks (fig. 64). Progressive excavation of a

Figure 71. View southwestward across Novarupta rhyolite dome from crest of Turtle to truncated face of Falling Mountain dacite dome, a distance of ~2.2 km. Novarupta is 65 m high and Falling Mountain ~400 m. Ignimbrite as thick as 13–15 m is plastered on east (left) shoulder and part of steep face of Falling Mountain, indicating syneruptive loss of vent-facing side of dome early in 1912 eruptive sequence. Dacite of Falling Mountain makes up a modest fraction of lithic fragments among Episode I ejecta, as does andesite of Trident volcano, lava flows of which were truncated by the vent-margin scarp to left of Falling Mountain. Ice-capped Mount Mageik rises 1,000 m higher and 8 km beyond Falling Mountain. Mount Cerberus, a dacite dome much like Falling Mountain, stands 1 km to its west (right). Both are of late Pleistocene age, and both are lapped by 1912 valley-filling ignimbrite in upper Lethe arm at right.

wider vent and increasing proportions of dacite and andesite, at the expense of gas-richer rhyolite, would favor extensive fountain collapse during Phase B, when ignimbrite greatly exceeded fallout, following rhyolitic Phase A, when only a modest fraction of the ejecta contributed to lateral currents.

Particle Concentration Gradient

Proximity of (1) thinly stratified bedsets of variably sorted HEPI draping ridgecrests and (2) massive poorly sorted VFI on valley floors a few hundred meters downslope provides good evidence for vertical gradients in the concentration of solid particles in the currents that discharged laterally from the collapsing fountain(s) in Novarupta basin (fig. 43). General equivalence of the succession of compositional proportions making up the sequence of HEPI and VFI packages (and intercalated fall layers that locally avoided ingestion and scour) attest to contemporaneity of deposition. Graded pyroclastic density currents have been called "stratified" by some, following Valentine (1987).

In a turbulent suspension current, a concentration gradient inevitably develops, as differential settling of particles experiencing a wide range of turbulent support should rapidly produce vertical gradients in the masses transitioning from fountain collapse to outflow currents (Sparks and others, 1978; Druitt, 1998; Choux and others, 2004). Not clear, however, is whether stratification is in part intrinsic to the collapsing fountain nor how rapidly the emerging currents develop an interface between a dilute turbulent cloud and a denser underflow. Nonetheless, even at high radial velocity around Novarupta, sedimentation from an expanded turbulent current should soon have produced a poorly sorted dense underflow, which could be diverted or channelled by topography, while the radially expanding dilute cloud concurrently laid down a thin-bedded set of variably sorted, lenticular, stratified deposits upon ridges only a few hundred meters higher (fig. 43). Certainly, such a dichotomy had developed here within a few kilometers of outflow, as massive VFI deposits bank around the bases of Mount Cerberus (fig. 72) and Whiskey Ridge (fig. 73), which are capped by stratified HEPI that lie 250 m and 400 m higher, respectively, than VFI on the adjacent valley floors (but only 250 m higher than Novarupta). After proximal HEPI deposition, turbulent ash clouds may have become so dilute that much of their remaining fine ash lofted and dispersed or was convectively drawn back inward and upward to join what survived of the buoyant plinian column.

Currents that deposited the most voluminous of the nine ignimbrite packages may have been sustained for an hour or more. However, viewed as a whole, the Phase B sequence was episodic rather than continuous, as suggested by (1) scour and truncation of successive HEPI bedsets, (2) varied range and distribution of successive VFI packages, (3) intercalation of fall layers between some VFI packages, and (4) compaction swales that had time to develop between successive packages. Such episodicity is likely to have reflected spasmodic fountain collapse and vent-wall spallation that caused repeated discharge fluctuations, including radial blasts during deposition of Packages 1 and 7 (Fierstein and Wilson, 2005).

Alternating HEPI beds range from fines-rich and poorly sorted (much like thicker massive VFI) to fines-poor and better sorted layers rich in lithics and coarse pumice (figs. 22, 41, 42). They reflect fluctuations in velocity, capacity, competence, shear rate, tractional effects on bedload, and degree of turbulence in the pulsating currents. Sedimentation was from turbulent suspension currents with or without traction at the depositional surface. Some matrix-richer layers may have been virtually plastered by current pulses breaking like waves over the ridgecrests, the roughness of which influenced fluctuating degrees of local turbulence. Alternation of massive beds and lensoid traction-dominated layers on a decimeter scale suggests abrupt rapid fluctuations in solids concentration—vertically, laterally, and temporally—in the turbulent currents passing.

Densification of Flow Units

The dense granular underflows (with solids concentrations probably up to several tens of volume percent), retained most of the particulate mass, flowed gravitationally into the different arms of the valley, were channelled and diverted by surrounding terrain, filled and masked the prior topography, and continued much farther than the dilute lofting clouds from which each had separated. The latter deposited thin HEPI veneers only out to ~6 km from vent, whereas some of the VFI-depositing currents extended at least 23 km. Turbulent support was soon replaced by particle interactions and partial fluidization in each underflow. Currents were not "ponded" but, instead, were topographically guided down gently sloping, open-ended valleys with steep sidewalls.

Although gradually deflating and losing momentum along valley floors, each current retained modest mobility owing to reduction of interparticle friction by partial fluidization promoted by excess pore pressure and hindered settling associated with upward-escaping interstitial gases (Fenner, 1923, p. 63; Sparks, 1976; Wilson, 1980, 1984; Druitt and others, 2004, 2007; Girolami and others, 2008). High temperature and internal shear during transport would have helped maintain expansion of the gas-rich particulate currents. In addition to ingestion of air by the collapsing fountain and continuing diffusive exsolution of magmatic gas from glassy particles, gas content of the currents would be partially replenished by pumice attrition, burning vegetation, boiling of overrun surface waters, and perhaps by additional frontal entrainment of air where currents encountered rough terrain, all helping to sustain mobility.

The high fines contents of the VFI-depositing currents would have enhanced partial fluidization and hindered settling (self-fluidization not requiring gas replenishment) by reducing permeability, thus extending gas-retention times. Predominance in the ignimbrite of rhyolite pumice and ash, which had high magmatic gas content and thin fragile bubble walls, no

Figure 72. Mount Cerberus, late Pleistocene dacite dome, 1,500 m across and 360 m high, at foot of 2,165-m Mount Mageik compound dacite stratovolcano. High-energy proximal ignimbrite (HEPI) deposits cap both ridges of Mount Cerberus as well as south ridge of Baked Mountain in foreground. Where valley-filling ignimbrite laps against base of Cerberus, its tan (oxidized) margin is sharply defined against pale-gray pumice-fall scree. Some ignimbrite packages flowed around both sides of Mount Cerberus en route to Katmai Pass, which lies 2 km to left of image and is drained by prominent arroyo at left. East Mageik Lake, barely visible at right center, was occupied in 1912 by the glacier now hanging ~1 km above it.

Figure 73. Scarp of Whiskey Ridge, 400 m high, and panorama along Knife Creek arm of VTTS toward ignimbrite terminus 23 km away. Atop great wall of Jurassic siltstone at upper right is 5-m-thick intact section of HEPI deposits capped by Layer C (see figs. 23, 42).

doubt helped produce a large initial fraction of fines. Fines replenishment by attrition of fragile pumice during outflow would have compensated in varying degree for the elutriation of fines by escaping interstitial gas.

Currents in the VTTS propagated 9–23 km downvalley along low gradients, infilling the prior topography to produce generally planar sheets that slope only 0.9° in the Lethe arm and 1.2° in the Knife Creek arm. Beyond the constriction where the VTTS is narrowest, ~10 km from Novarupta (fig. 74), the ignimbrite surface steepens to ~1.5° as far as the Three Forks moraine, probably because some of the later packages terminate near the head of that narrow reach (Fierstein and Wilson, 2005). Such modest slopes are gentler than normal angles of repose for granular materials, underlining that excess mobility was imparted by retention of interstitial pore pressure, despite the low momentum indicated by failure of most packages to run up sidewalls (fig. 75) or to surmount the cross-valley moraine (fig. 76).

Deposition of the valley-filling VTTS ignimbrite was an incremental process that lasted 14–16 hours (figs. 2, 34, 60; Hildreth, 1983; Hildreth and Fierstein, 2000; Fierstein and Wilson, 2005). Each of the nine compositionally distinguishable packages consists of the deposits of several discrete currents (flow units), and each or most of the currents may have deposited material incrementally at its base by progressive aggradation (Fisher, 1966; Branney and Kokelaar, 2002), at least in the upper VTTS. As stressed by Wilson and Walker (1982) for pyroclastic currents in a partially fluidized state, the thickness of a deposit bears no simple relation to the thickness of the current that passed, and the processes of sedimentation and particle segregation that take place during outflow cause the grain size and componentry of a deposit at any point to bear no simple relation to either those of the original eruptive mixture or to those of the current that passed.

Progressive densification of currents downvalley, however, increased intergranular friction, leading to unusually

Figure 74. Aerial view of VTTS downvalley from above its narrowest reach (3 km wide), extending ~11 km to Ukak lobe terminus at foot of Mount Katolinat. At left, ignimbrite wraps nose of Buttress Range into Windy Creek lobe. In center distance, narrow Ukak River corridor connects Ukak lobe with 5-km-wide ignimbrite sheet that was elsewhere blocked by moraine belt crossing lower valley. Four stream gorges are prominent: River Lethe at left margin, mid-valley fork of Knife Creek, main fork of Knife Creek (with gray overbank alluvium), and Juhle fork at right. Chains of pinkish ignimbrite hills along Lethe and Juhle fork margins contrast with elsewhere planar surface of sheet; hilly marginal strips have 10–15 m of primary local relief that formed during Episode I ignimbrite emplacement (see figure 81) before being draped by Episode II fallout.

sluggish medial-to-distal advance. Densification and early depletion of interstitial gas may likewise account for failure of the depositing currents to loft buoyantly at their runout extremities or to produce a volumetrically important cloud of coignimbrite ash. As each current flowed out into the valley, its expanded condition would have gradually deflated, and basal sedimentation would have progressively reduced its moving mass. After outflow of several kilometers, each densifying ash-flow current would first have developed a sharp density discontinuity and eventually have left behind most of its turbulently dilute upper clouds. Development of free surfaces on some currents is implicit in the intercalation of pumice falls (but no fine ashfall layers) between ignimbrite packages at medial runout distances in upper Knife Creek and the Katmai Pass/Mageik Creek area. Such intercalations are absent more proximally, where the turbulent currents would have incorporated contemporaneous fallout.

In its progress from fountain collapse to the lower VTTS, each current would thus have transitioned from turbulent suspension to fluidized laminar flow to shearing flow with partial gas support by hindered settling to shearing granular mass flow with increasing matrix strength and density (Druitt, 1998; Freundt and Bursik, 1998; Iverson and Vallance, 2001). On reaching the central VTTS, most currents had evidently lost enough of their interstitial fluid pressure that intergranular friction greatly reduced velocities. Downvalley, some discrete currents took on the characteristics of model flow units (Sparks and others, 1973) with fine-grained and/or lithic-rich layers at the base, buoyant pumice concentrated at or near the top, and free surfaces attended only by billowing ash clouds produced by ongoing elutriation. Reduction of internal agitation as fluid escape lessened would have increased particle collisions (leading ultimately toward interlocking), increasing rigidity toward plug flow

Figure 75. Ignimbrite margin at southwest corner of VTTS, where tuff banks against Buttress Range adjacent to West Mageik Lake. Dark ignimbrite ledge near rim is illustrated in figure 38, where it overlies thick rhyolite-rich ignimbrite that extends beneath lake level. Light-colored deposits above dark ledge include dacite fallout but are mostly pumice-rich scree sloughed from steep bedrock slope above. Gray material veneering cream-colored tuff above lakeshore is till left by glacier that occupied lake basin in 1912. In distance, ignimbrite bench stands as high as 60 m above valley floor axis at right, owing to differential compaction-welding of thick valley-filling tuff.

and promoting alternation of stalling and surging of the flow front, perhaps stimulated by the ongoing seismicity and/or slight resurgence of pore pressure by gas flux from wet substrate. Such densifying flow units, though perhaps still aggrading basally, might ultimately glide to a halt, not quite en masse but by internal laminar shear. Even at the extremity of outflow, such flow units probably would not undergo plug flow but move instead by laminar flow with organized shear, perhaps pulsatorily. Each distal current may have ground to a halt from the base up, shear propagating upward as the granular mass stalled and its upper levels moved onward in unsteady pulses (fig. 77). At the ultimate Ukak terminus, the VFI feathers out in tapering wedges less than a meter thick (fig. 78).

Rhyolite-rich distal ignimbrite in the lower VTTS is not strongly crystal enriched relative to phenocryst-poor (0.5–3 wt percent) rhyolite pumice, indicating that loss of fine vitric ash was limited. Small fractions of phenocryst-rich (25–42 wt percent) andesite and dacite pumice render calculation of free-crystal enrichment uncertain distally and nearly impossible upvalley, where proportions of the three pumice types are large and varied. Nonetheless, componentry yields only 2–4 wt percent free crystals in rhyolite-rich nonwelded Packages 3 and 4 in the lower VTTS, 14–20 km from vent (Fierstein and Hildreth, 1992). Both rhyolite-rich and compositionally more heterogeneous ignimbrite packages in the VTTS contain 30–50 wt percent ash finer than 63 µm, supporting the inference that elutriative loss of fines was not large during downvalley flow. Moreover, the all-rhyolite plinian pumice-fall deposit, Layer A, likewise contains 2–4 wt percent free crystals at medial downwind distances of 7–13 km. (Crystal content of Layer A is greater than this on Kodiak Island, 170 km downwind, by which distance the deposit is entirely ash-grade and aeolian fractionation had kept a larger fraction

Figure 76. Top of main VTTS ignimbrite (Package 4) wedging out against cross-valley moraine east of Three Forks (see figure 64). Within 50 m upvalley (to left of image), ignimbrite is 10–20 m thick. Ruler, 16 cm long, rests against brown soil containing charred willow roots. Oxidized ignimbrite here is overlain by thin Layers C–D, fine-grained pale-gray Layer E, and thicker Layers F–G at top. Ignimbrite lacked sufficient momentum, 16 km from Novarupta, to run up modest slope of soil-covered moraine.

Figure 77. Thinly laminated ignimbrite on gorge wall of "Moraine Creek" just upstream of cross-valley moraine in northeastern part of VTTS (Hildreth and Fierstein, 2003). Exposure here is 7–8 m high, but gorge is locally as deep as 20 m, cut entirely in rhyolite-dominant packages of Episode I ignimbrite. Note that laminae, mostly 3–10 cm thick, continue to top of deposit. Distal deposit here, 15–16 km from Novarupta, represents entire transporting current, no higher part of which existed to pass over moraine barrier. It is thus inferred that lamination developed within dense sluggish current itself, not by incremental deposition at base of a current that continued farther downvalley.

Figure 78. Ukak lobe terminus of ignimbrite at foot of Mount Katolinat, ~20 km from Novarupta. Gray-brown flats at center are posteruptive deposits of ashy mudflows. White deposits are primary rhyolite-dominant nonwelded ignimbrite that feathered out into brush with little or no run-up.

of vitric fines aloft). The similarly modest losses of fine ash in plinian fallout and distal ignimbrite inferred from equivalently small degrees of crystal enrichment suggests that most of the fines were lost by violently agitated fluidization of the dilute mass around the vent (Wilson, 1980) as it transitioned from collapsing fountain into radial currents. If so, and if subsequent elutriation of fines from downvalley currents was unimportant, this further implies only weak fluidization, in accord with the inference of low velocity and low momentum of nearly all VTTS ash flows.

Further support for very slow distal advance comes from the scarcity of ground layers enriched in crystal and lithic components. Widely interpreted as deposited from flow fronts diluted by turbulent entrainment of air (Walker and others, 1981), basal ground layers (2–4 cm thick) have been observed just beyond Katmai Pass and along the Ukak River corridor, both rough, steepening pitches where turbulent entrainment might provide exceptions to otherwise sluggish advance of the distal currents. The only other ground layers observed are in thin all-rhyolite flow units that banked into the cul-de-sac at the snout of Glacier 5. Consistent too with slow and halting progress of flows near the downvalley terminus is absence in their well-exposed basal deposits of any bones of large mammals, several species of which would be expected here in June. They got away.

Episodes II and III: Small Density Currents from Column Margins

The several pyroclastic density currents (PDCs) produced during Episodes II and III (fig. 47) were very different from the sustained ignimbrite-depositing currents of Episode I. Each current was only minutes in duration, trivial in volume (≤0.01 km^3), and generally sector-confined. The deposits crop out in five packages, ~0.1 m to ~3.5 m in aggregate thickness, that each consist of several thin layers or flow units; none extends more than 3 km from vent, though a few pass laterally into ashfall layers a few centimeters thick. The laterally emplaced PDC deposits are intercalated with the proximal parts of regional plinian Layers C through G and with a few locally dispersed sector-confined wedges of andesite-rich proximal fallout (Fierstein and others, 1997; Houghton and others, 2004). Like the ultraproximal fall deposits, the PDC deposits are enriched (relative to the coeval dacite-dominant plinian falls) in andesitic and banded pumice and in vent-wall lithic fragments. As their total aggregate thickness averages ≤3 m, total volume of the five packages is probably <0.05 km^3, even accepting that a modest fraction of each current may have been lost as dilute turbulent ash clouds overwhelmed by synchronous plinian fallout. Altogether, then, the PDC deposits make up less than 1 percent of the Episode II–III ejecta, which include ~8.2 km^3 of regional plinian fallout (Fierstein and Hildreth, 1992).

During Phase B of Episode I, the main VTTS-filling multipackage ignimbrite accounted for ~80 percent of the material being erupted and the subordinate convective plume that produced Layer B only ~20 percent. No large-scale collapse into a radially spreading fountain took place, however, in either Episode II or III, during each of which the high convective column was sustained. Rather than from a fountain, PDCs were initiated by spasmodic sedimentation from external margins of the emerging jet and/or lowest margins of the convective plume, each in a limited sector and from probably less than 1 km above the vent. Collapsing spasmodically and highly asymmetrically, the PDCs are inferred to have resulted from column-margin instabilities caused by annular overloads rich in andesitic and lithic ejecta that emerged in irregular short-lived pulses. Such instabilities took place on a time scale of minutes, such that many of the multipulse currents were separated by short intervals of fallout, producing deposits of falls and flows that alternate on a thickness scale of decimeters to meters.

The PDC deposits range widely in thickness, grain size, sorting, fraction of fines retained, and in abundances of andesite scoria and vent-wall lithics, showing such contrasts even within packages emplaced simultaneously in different sectors. As documented in detail, illustrated, and interpreted by Houghton and others (2004), the Episode II and III PDC deposits include at least four facies. (1) Channel-confined matrix-rich, multipulse ignimbrites with sharp lateral and terminal boundaries and containing abundant pumice blocks. Largely confined to the western outlet of Novarupta basin, these probably originated as laterally ejected slugs or small-scale local collapses of the lower column margin. (2) Thinly bedded ash-rich deposits with pinch-and-swell lamination and a wide range of sorting, interbedded with thin pumice-fall layers that share their pumice-composition proportions. These are distributed widely and probably represent dilute suspension currents that lofted distally and graded out into thin ashfall deposits. (3) Channel-confined matrix-poor lenses of pumice lapilli and blocks largely confined to slopes of the Turtle, interpreted as local grain-flow avalanches of pumice-fall material remobilized by seismicity and/or compaction-induced sagging and faulting of the vent-fill.

Finally, (4) a radial blast-like deposit (PDC 4 of Houghton and others, 2004) that extends symmetrically as far as 2.5 km issued from the same vent but during a temporary shutdown of the high plinian plume. Proximally, it is temporally equivalent to regional Layer E, deposited during the interval between Episodes II and III. The deposit is a 1–3.5-m-thick, matrix-supported, massive, landscape mantling, poorly sorted ignimbrite that has as many as five flow units with pumice-concentation zones and elutriation pipes. Although more even in thickness and distributed more radially than the other PDC deposits, it is nonetheless sectorially varied in grain size, fines retention, lithic fraction, and degree of welding. Its abundant welded vitrophyre lithics (including some ballistic blocks) indicate major vent erosion, and its 350-m runup height on the slopes of Trident suggests energetic lateral transport. Its radial outburst was probably the result of sudden clearance of the vent blockage that had shut down plinian Episode II.

Physical Features of VTTS Ignimbrite Sheet

Aspect Ratio

Walker and others (1980) proposed that aspect ratio is a measure of "violence"—lateral dispersal by momentum—and defined it as the ratio of average thickness to the diameter of a circle with the area of the deposit. The VTTS sheet (fig. 74) has commonly been cited as a prime example of a high-aspect-ratio ignimbrite. The fact, however, that the 1912 sheet is a composite product of nine packages (and certainly dozens of flow units) that merged into the main valley from its three arms means obviously that such citation addresses the whole deposit and does not concern the thickness, velocity, or runout of its many component currents. Masking of topography by early packages and deposition of smooth degassing surfaces would clearly have enhanced the mobility and runout of later currents. Duration of the multipackage eruption also influenced aspect ratio, because maximum runout area was achieved by Package 3, and Packages 4–9 only served to thicken the whole, thereby progressively increasing the aspect ratio. Moreover, even neglecting HEPI veneers and VFI near valley margins, the thickness of axial VFI ranges at least 5-fold and possibly 20-fold, rendering average total thickness rather difficult to estimate. As an exercise, nonetheless, for an area of 120 km^2 for the main VFI, a (very uncertain) average total thickness of ~50 m gives an aspect ratio of 4×10^{-3}. By comparison, the (likewise composite) HEPI veneers, which drape rather than fill topography, have an average thickness of ~5 m over ~50 km^2 (or ~2 m over ~100 km^2), giving arguable aspect ratios of $2–6 \times 10^{-4}$—an order of magnitude thinner, for facies of the same eruption.

Flow Units and Progressive Aggradation

Numerous flow units can be identified within the VTTS ignimbrite sheet, some of which are thin with well-defined contacts and some of which are thick and structureless and may have been deposited continuously or incrementally at the base of passing currents. Conceptually, we consider a flow unit to be the deposit of a single vent-derived current, initiated by a discrete eruption, collapse pulse, or major fluctuation in discharge during a sustained eruptive interval, while further recognizing that some currents can subdivide laterally or vertically during outflow. In practice, we follow Wilson and Hildreth (1997) in mapping "packages" of ignimbrite with similar components, compositional proportions, and lithologic characteristics (exclusive of welding style) that were emplaced as one or many pulses or flow units. This usage skirts the uncertainty in interpreting whether such common features as discontinuous partings or stringers of pumice or lithics represent unsteady deposition from a continuous current or flow units separated by time breaks between discrete episodic currents. Thick sections of massive ignimbrite, either structureless or with a few such discontinuities, have been variously interpreted in terms of progressive aggradation or as a succession of flow units at intervals so short that batches amalgamated by shearing along their contacts before each lower batch had time to degas and consolidate sufficiently to provide a firm substrate for the next. These depositional modes may themselves essentially merge if discrete pulses follow one another rapidly enough. The better demarcation of flow-unit contacts distally, as commonly observed, may reflect longer setup times between batches and/or more advanced defluidization. Mixing and merging at boundaries between successive fluidized pulses may so obscure the contact that a seamless join is either indiscernible or marked only by subtle textural changes and particle trains hard to distinguish from products of unsteady deposition from a continuous current.

Flow-unit boundaries (Sparks, 1976; Wilson, 1980; Druitt, 1998) are fairly unequivocal where marked by (1) layers of intercalated fallout, (2) truncation of degassing pipes, (3) inversely graded fine-grained basal layers, or (4) ground layers strongly enriched in dense components—crystals and lithics. In addition, (5) major pumice swarms and (6) normally graded lithic-rich zones are usually clear indicators of upper and lower parts, respectively, of flow units. Small clusters or trains of pumice or lithics, discontinuous on a meter scale, and vague impersistent stratification, however, are more ambiguous. Not all such discontinuities need reflect discrete pulses with time breaks, as opposed to short-lived irregularities in deposition (or nondeposition) at the base of a continuing current (Branney and Kokelaar, 2002).

The nine successive packages of Episode I ignimbrite are the cumulative product of 14–16 hours of eruptive activity (Hildreth and Fierstein, 2000; Fierstein and Wilson, 2005). Flow-unit boundaries between and within packages are well marked in the upper VTTS and Mageik Creek by pumice-fall intercalations (fig. 37) or stratified intervals between massive units of contrasting composition and induration (fig. 38). The lack of coignimbrite fine-ash layers at flow-unit contacts, however, means either that they were scoured, that intervals between currents were short, or that the fallout of plinian pumice was uninterrupted. As illustrated by Fierstein and Wilson (2005; their fig. 6), sharp contacts between flow units are variously marked by compactional swales filled by later units or erosional channels cut by the units that fill them (figs. 31, 67), and, in the central and lower VTTS, by fine-grained basal layers (fig. 68) that occur commonly (but not exclusively) between nonsuccessive packages. Andesite-rich Package 8 may have been partially welded before the thin block-rich flow units of postplinian Package 9 overran it (fig. 35).

Striking compositional changes (fig. 79) that define packages usually also coincide with flow-unit boundaries, but exposures in the central and lower VTTS are common in which a compositional change, either sharp or gradational within a meter or two, is accompanied by no visible discontinuity or grain-size change (fig. 39) to indicate a depositional hiatus between currents. In particular, transitions between voluminous rhyolite-rich Packages 2, 3, and 4 in the lower VTTS are commonly featureless except for abrupt upward increases in proportions of andesite pumice. Some of the discontinuous stratification and clast trains may well reflect incremental aggradation within thick flow units that by mid-valley had become dense grain flows that were extensively defluidized, advancing in pulses, and halting layer-by-layer. Rather than stopping en masse as plug flows ~10 m thick, distal flow units apparently sheared out laminarly, depositing progressively or spasmodically and feathering out thinly at their termini—in the Windy Creek embayment, beyond saddles in the Three Forks moraine, and at the base of Mount Katolinat. Engulfment of charred but upright trees in distal ignimbrite several meters thick suggests creeping velocity and near-exhaustion of momentum.

Remobilized Flow Units

Stringers of pumice or lithic fragments and subtly discontinuous layering that extends only a few meters, not uncommon features on gorge walls in the central and lower VTTS, may represent traces of amalgamated contacts between closely successive flow units or, alternatively, products of unsteady sedimentation at the base of a unitary current depositing incrementally. Distally, however, where Packages 3 and 4 bypassed

Figure 79. Distal ignimbrite exposure, ~7 m high, on wall of lower "Moraine Creek" gorge (Hildreth and Fierstein, 2003), ~18 km from Novarupta and 2 km east of Three Forks, showing contact of rhyolite-rich Package 4 overlain by compositionally heterogeneous Package 6.

the Three Forks moraine and flowed down the narrow Ukak River corridor (fig. 30) to the terminal lobe, numerous meter-scale flow units are well developed (fig. 80). Called "segregation units" by Hildreth (1983) and "pseudo-flow-units" by Fierstein and Hildreth (1992; their fig. 10a), these have no visible equivalents upstream and appear to have originated distally by budding or surging of massive unstratified ignimbrite that was by then halting or advancing very slowly. Many of these strata are inversely graded, with fine-grained basal layers and coarse pumice preferentially buoyed toward the top. Downstream from the Three Forks moraine, the slope of the ignimbrite surface (where uneroded) increases from ~1.5° to ~2.1° for 3.5 km and to as steep as 3.2° for one 500-m-long reach (figs. 29, 30).

After 16 km runout from vent, the internal fluid pressure may have been so depleted that intergranular friction increased matrix strength sufficiently to resist further advance. After burying a stream valley and beginning to stall on a slope of a few degrees, however, upper parts of the deposit may have been intermittently remobilized gravitationally by slight renewal of pore gas pressure or perhaps by the strong seismicity continuing during caldera collapse. Although developed within the same ignimbrite package and similarly associated with slowing or stoppage of a granular mass flow, these meter-scale remobilized flow units (fig. 80) appear to reflect a process rather different from the systematic internal shear thought to have produced the centimeter-scale lamination (figs. 65, 66, 77) where the flow was stopped and diverted by the Three Forks moraine (fig. 64).

There are at least three places in the VTTS where the upper few meters of ignimbrite were remobilized after the sheet had clearly come to rest, leaving concave scarps and drainaway channels in nonwelded tuff that had developed sufficient rigidity to preserve them. Each location is adjacent to the margin of the main VFI: (1) along the southeast margin of the Ukak terminal lobe; (2) the northeast margin of the central VTTS along the Juhle fork; and (3) the southwest margin of the central VTTS where the River Lethe hugs the bend in the Buttress Range (fig. 74). The last involves a kilometer-long area, the Lethe Hills (fig. 5), where the ignimbrite piled up in a cluster of mounds and ridges as much as 15 m higher than the elsewhere smoothly planar surface of the sheet (fig. 81). It seems likely that the hills were deposited where lobes of Packages 3 and 4 that flowed out of the Knife Creek arm collided with the Buttress Range. Pronounced narrowing of the VTTS at this location may also have played a role (Hildreth, 1983), as the lobes coming down the Lethe arm converged with those from Knife Creek. Directly across the valley constriction, where the Juhle fork hugs the northeast wall of the VTTS, the Lethe Hills have a counterpart in a cluster of ridges and hummocks showing 7–10 m of primary local relief on the ignimbrite surface. Both hilly areas were affected by local drainaway, the upper few meters of ignimbrite having been remobilized within hours of emplacement, before the fall of dacite Layers C–H, which drape them.

Whatever the origin of the hills, the internal friction and rigidity suggested by such primary relief indicates that some of the granular currents were poorly fluidized by the time they reached the central VTTS. At several places in the Lethe Hills, scalloped embayments provide evidence for drainaway of ignimbrite that remobilized obliquely downstream toward the lateral margin of the VFI. Both the hills and the drainaway scarps (which added only slightly to the hilly relief) are draped by the dacite fall deposits of Episodes II and III, showing that hills are primary constructional features and that remobilization took place within a day after Episode I ignimbrite deposition.

Diffusely Plane Laminated Ignimbrite

In the lower VTTS, dense flow units of at least two Packages (3 and 4) ran straight into the Three Forks moraine, banking against it (fig. 76), diverting laterally into low gaps through it (figs. 29, 30) but failing to override the barrier that lay only a few meters higher than the surface of each current (fig. 64). The deposits, there pervasively marked by diffuse parallel stratification (figs. 65, 66, 77), had previously condensed into a dense sluggish current with a free surface. Condensation had concluded earlier and, though the mass was still moving slowly, no separate transporting current was left to continue across the moraine. The evenly spaced, well-organized, centimeter-scale stratification (fig. 65) is exposed from the uneroded top surface to the bottom of gorges 10–15 m deep. Occurring within packages of elsewhere-massive ignimbrite, the thin layering is best developed within 300 m of the moraine barrier, but subtler manifestations are also exposed as far as 1 km upstream of the moraine and down on the terminal Ukak lobe. Any incremental aggradation had been complete farther upvalley, so the systematic lamination of these densely granular avalanche-like flow units was apparently accomplished by internal shear as they banked into the barrier and ground fitfully to a halt nearby (Hildreth, 1983).

"High Sand Mark"

The elevated bench around the upper VTTS where the ignimbrite sheet is thickest was called the "high sand mark" by Griggs (1922) and Fenner (1923, 1925), who thought it was either a veneer or a valley-margin runup. However, it was deposited neither by passage of an inflated current nor a high-velocity current that drained away downvalley. Instead, it represents a marginal bench (figs. 75, 82) around most of the Lethe arm that formed in thick ignimbrite by differential welding-compaction (Fenner, 1923; Hildreth, 1983). The benches change gradually into side slopes in the central VTTS and then fade out completely as the sheet thins downvalley and becomes nonwelded. Around the upper Lethe basin, however, an elevation difference of 30–60 m between the valley floor and the fissured hingeline of the bench suggests an axial thickness as great as 150–200 m, in agreement with Curtis' (1968)

Figure 80. Meter-scale distal flow units that developed near Ukak River terminus from what had been thick massive ignimbrite just upvalley. Hammer is 30 cm long. Budding of such distal secondary flow units is interpreted to have taken place within densified, extensively defluidized currents that were discontinuously stalling and pulsing, preferentially near their tops.

Figure 81. Lethe Hills: primary relief on Episode I ignimbrite surface draped by pale-gray stratified dacite fallout of Episodes II and III. Hills have as much as 15 m relief above planar surface of sheet. Such hills developed only along margins of narrowest reach of VTTS, here only 3 km wide (see figure 74). View is northeastward across gorge of River Lethe at base of Buttress Range, 11 km northwest of Novarupta.

estimate based on comparative stream profiles. Although less well developed, comparable benches and inward slopes around the upper Knife Creek arm (fig. 26) suggest 100–150 m of variably welded ignimbrite there as well. Some have speculated that the benches might conceal lateral moraines, but Neoglacial moraines in the VTTS lie 15 km downvalley near Three Forks, and glacier-fed valleys that parallel the VTTS nearby (Ikagluik Creek, Rainbow River) lack such proximal lateral moraines (Hildreth and Fierstein, 2003).

Compaction faults with decimeters to meters of displacement are common on the benches in the upper VTTS, especially along hingelines, where indurated ignimbrite adjacent to antithetic faults is commonly tilted 15º–30º toward the valley axis (fig. 83; Fenner, 1923, p. 40). Such faults and fissures intercepted surface precipitation, runoff, and groundwater, leading to vigorous fumarolic discharge and alteration along the benches (as observed when active in 1916–1923; Griggs, 1922, p. 216, 225, 232) and to explosive phreatic excavation of craters and fissures (fig. 84) concentrated along the hingelines (probably all before 1916).

Because many of the faults displace the dacite fall deposits as well as the ash-flow tuff, compaction deformation of the ignimbrite sheet certainly persisted for at least 2 days after Episode I deposition. From another perspective, however, because some of the later ignimbrite packages were channelled along broad axial swales atop the thick packages emplaced only hours earlier, axial compaction was clearly underway almost immediately after emplacement.

Valley Margin Runups

Not to be confused with the marginal benches that marked the deposit surface in the upper VTTS before compaction and welding, there are also a few locations where momentum-driven runup deposited thin wedges of nonwelded ignimbrite still higher than that surface. Apart from the initial blast-like radial emplacement of the all-rhyolite Package 1, however, the subsequent packages underwent significant runup in only a few topographically special situations. The sparsity of evidence for runup is consistent with the sluggishly halting advance inferred for the ash flows that reached the lower VTTS. The most voluminous packages, 3 and 4, which alone flowed all the way to the Ukak terminus, exhibit marginal runups of less than 10 m where exposed in the upper VTTS (fig. 75).

Figure 82. "High sand mark"—ignimbrite bench along Buttress Range margin north of West Mageik Lake. Fissured hingeline along rim of bench reflects bending of indurated ignimbrite sheet owing to differential compaction and welding. Elevation difference between bench and valley-floor axis is variously 30–60 m. Valley-filling ignimbrite along valley axis is more densely welded and thicker than along its margins. Lake was occupied in 1912 by snout of glacier from Mount Mageik. Recession of ice front by ~1 km has exposed 3.1-Ma rhyolite sill complex (white ledge at lower left) cutting stratified Jurassic siltstone. Lake outlet is origin of River Lethe.

Package 1 was uniquely mobile in veneering the vent-facing slopes of Trident to elevations >700 m higher than Novarupta basin and in sweeping across the slopes of Mount Griggs as high as 450 m above the margins of subsequent VFI packages that filled the Knife Creek arm. A massive to vaguely stratified tongue of pink nonwelded rhyolite tuff (fig. 25) that extends ~5 km up the narrow Griggs fork canyon (to ~500 m higher than the valley floor at Knife Creek) was deposited by unobstructed outflow directly from Novarupta, which lies 6 km from the canyon mouth. Along the Griggs fork, Package 1 ignimbrite is as thick as 13 m, but it feathers out upslope on Mount Griggs and thins northwestward across slope to <1 m at the Juhle fork, 9 km from vent, where it then vanishes. A similar veneer of pink all-rhyolite ignimbrite, vaguely stratified and typically thinner than 20 cm, mantles the lava-flow benches of Mount Mageik, west of Katmai Pass and upper Mageik Creek (figs. 26, 28); the veneer extends as much as 250 m higher than the tongue of Package 1 ignimbrite (4–6 m thick) that fills the axis of Mageik Creek. These Package 1 veneers on Trident, Griggs, and Mageik have a runup quality to them, but they were all probably deposited from turbulent overcurrents rather than the dense undercurrents that deposited ignimbrite along the canyon floors.

Package 2, predominantly rhyolitic, also ran ~3 km (and 300 m in elevation) directly up the Griggs fork but did not veneer the adjacent slopes of Mount Griggs as Package 1 had. None of the subsequent packages left evidence for significant runups in the upper VTTS. However, the southwest margin of the (composite multipackage) ignimbrite sheet that fills the upper Knife Creek arm extends smoothly along the Broken Mountain slope, paralleling the northwesterly flow direction 30–50 m higher than the valley floor (fig. 26). This feature, only 2 km downhill from the vent, is certainly not a result of runup but is, instead, a combined product of deposition from proximally expanded currents that passed and subsequent differential welding, as in the Lethe arm. Similarly, for Package 7 to be limited to the upper VTTS (fig. 27) but to have poured through Katmai Pass more voluminously than any other package requires not runup around a left-turn corner but some process otherwise unique in the 1912 eruptive sequence. Much of its fountain collapse is likely to have enveloped Falling Mountain, allowing Package 7 currents to go both south and north from that obstacle (figs. 70, 71).

Where the Lethe and Knife Creek arms merge in the central VTTS and the valley floor narrows to only 3 km, there are runups of Packages 3 and 4 on both walls of the valley (fig. 74). At this constriction, adjacent to the Lethe Hills and just downstream from fadeaway of the marginal bench, a wedge of pink to terra cotta nonwelded ignimbrite (fig. 85) extends up the Buttress Range sidewall to elevations 30–60 m higher

Figure 83. Fissured and faulted hingeline of the bench ~1 km north of West Mageik Lake seen in figure 82. Bending in response to differential compaction produced dips as great as 25° toward valley axis. Fissures focussed intense fumarolic discharge and alteration for a decade after eruption. Visible relief 1–3 m along trough, now partly filled with reworked pumice and ash.

A

B

Figure 84. Overview of Fissure Lake. Views westward along trough occupied by Fissure Lake (Griggs, 1922, pp. 212, 244) for the first decade after eruption: *A*, 1917 and *B*, 1982. Trough was excavated by phreatic explosions along hingeline of compactional bench at southwest edge of VTTS. Fumarolically altered ignimbrite dips ~20° N at right but is almost horizontal at left and in distance where banked against Buttress Range. Scarp at right is ~20 m high; note two men on rim in 1917 image. Lake silted in by 1923. Terminus of glacier, ~1 km from camera, occupied present-day site of West Mageik Lake. Since 1951, glacier has retreated ~1 km upslope to left, exposing rhyolite sill complex that intrudes Jurassic siltstones. White sill at upper left in 1982 image is 10 m thick, dips 15° NNE along with enclosing Jurassic strata, has ~77 percent SiO_2, and yielded a K-Ar age of 3.08±0.07 Ma (Hildreth and Fierstein, 2003; Lowenstern and Mahood, 1991; Lowenstern and others, 1991). 1917 photo courtesy of National Geographic Society; 1982 photo by authors.

than the hills and the planar surface of equivalent units on the valley floor. The massive unstratified deposit is ash-rich and rhyolite-dominant and contains abundant pumice lapilli where it is still a few meters thick, but it becomes lapilli-poor where it feathers out upslope to <20 cm thick (fig. 86). This elevated fringe extends along the Buttress Range for ~1.5 km but is absent along the same slope immediately upstream and downstream. Only 4 km farther downvalley, the ignimbrite filled the 1912 channel of Windy Creek (fig. 87), but it was moving so sluggishly that the sidewall and overbank turf, though charred, was not scoured away.

A similar but less well preserved runup fringe of pink ignimbrite is present directly across the narrows of the VTTS above the Juhle fork (fig. 74), where the main VFI also exhibits a cluster of ridges and hummocks mirroring those of the Lethe Hills. The runups apparently represent a local phenomenon related to narrowing of the VTTS.

Because Package 3 and 4 currents failed to surmount the low moraine only ~5 km farther downvalley, their evidently

Figure 85. Thin runup fringe of pink rhyolite-dominant ignimbrite along Buttress Range margin at narrowest part of VTTS, 11–12 km from Novarupta. Brush was killed up to 60 m above planar surface of valley-floor ignimbrite. Deposit overlies dark brown organic-rich soil. Shovel is 58 cm long.

poor mobility makes it unlikely that, when they passed the narrows, they could have been inflated to twice the thickness of their deposits in this part of the valley. From the general lack of significant runups elsewhere in the VTTS, it can be inferred that valley-confined currents here were seldom much thicker than their nonwelded deposits. More likely, currents from the two converging arms of the VTTS either crossed paths and ran directly up the opposing walls or promoted a temporary elevation of the confluent flow surface, limited to the narrow VTTS constriction (Hildreth, 1983; Fierstein and Hildreth, 1992; Fierstein and Wilson, 2005).

Induration and Welding

In the lower VTTS, the rhyolite-dominant ignimbrite is wholly nonwelded where exposed top to bottom in gorges 10–20 m deep (figs. 31, 68, 79, 87). Mechanical compaction nonetheless promoted induration by matrix packing (with or without traces of fumarolic or groundwater precipitates) sufficiently strong to permit the massive nonwelded tuff (bulk density 1,100–1,300 kg/m^3) to stand in vertical cliffs for decades. Within 3 km upvalley from Three Forks, lower walls of 30-m-deep gorges begin to exhibit weak sintering (figs. 32, 88), and in the central and upper VTTS most surface exposures (figs. 35, 36, 69, 89) become moderately sintered to partially welded (1,400–1,700 kg/m^3). Because sintering represents fusion of point contacts among vitric shards without macroscopic deformation, sintered ignimbrite remains permeable and subject to vapor percolation. As a result, there is widespread overlap between near-surface zones of vapor-phase crystallization and sintering (or weak welding) in the VTTS. Along the mid-valley forks of Knife Creek, such zones developed slender cooling-joint columns that extend down the gorge walls to depths of 5–20 m, where they grade down into massive sintered or nonwelded tuff. Columnar jointed vapor-phase-indurated tuff is also widely exposed on the benches fringing the upper Lethe arm.

Partial welding (1.7–2.0 g/cm^3), marked by incipient to moderate flattening of pumice, is exposed in relatively few areas of the upper VTTS, where stream cuts into the ignimbrite are only 3–7 m deep (figs. 35, 89). Most of the dark-gray surface exposures there, largely of rhyolite-poor Packages 8 and 9, are strongly sintered or incipiently welded (fig. 35), but Package 8 displays flattened pumice near West Mageik Lake and has a strong eutaxitic (planar foliation) texture near Knife Creek Glacier 1 (fig. 33), ~2 km east of Novarupta basin. Blocks of welded tuff ejected from the many phreatic craters in the upper Knife Creek arm (Hildreth and Fierstein, 2003) range from tack-welded to eutaxitic, some having measured densities of 1.8–1.9 g/cm^3 (Curtis, 1968; Hildreth, 1983). Most such blocks are of rhyolite-poor ignimbrite, indicating shallow sources, but some are rhyolite-rich, evidently blasted out from earlier packages concealed deep in the deposit. Similarly, the vitrophyric blocks of welded vent-fill recycled as the principal lithic component of Episodes II and III range

Figure 86. Close-up view of 15-cm-thick pink ignimbrite runup along Buttress Range margin, as viewed in figure 85. White base chilled against substrate of dark brown soil, avoiding oxidation. Note that ignimbrite coarsens upward. Pink rhyolite-dominant ignimbrite is overlain by white dacite fallout of Episode II. Knife is 8.5 cm long.

Figure 87. Nonwelded ignimbrite that filled pre-1912 course of Windy Creek, forcing stream to incise new channel (toward left) through hyperconcentrated sandflow deposit of latest Pleistocene Lethe assemblage (Hildreth and Fierstein, 2003). Streambank here is 8–10 m high. Note that sluggishly emplaced ignimbrite did not scour mat of charred vegetation from upland surface it overran (at left) and only partially scoured it from sidewall of channel it filled. Mount Katolinat on skyline.

from weakly to densely welded (figs. 62, 63, 90), presumably reflecting a range of excavation depths.

Densely welded ignimbrite (2.1–2.4 g/cm^3) is nowhere exposed in the VTTS, but it may underlie axial parts of the upper Lethe and Knife Creek arms, as suggested by 30–60 m of differential welding compaction. It is inferred that rhyolite-rich Packages 2, 3, and 4, which extend farthest downvalley, make up the bulk of the unexposed welded tuff beneath the upper and central VTTS. The partially welded or strongly sintered material capping much of the upper VTTS, however, belongs to later rhyolite-poor ignimbrite packages dominated by dacite and andesite (Fierstein and Wilson, 2005). Because magma temperature of the andesite was at least 100°C higher than that of the rhyolite (Hildreth, 1983) and because the gas-poorer andesitic ejecta collapsed from generally lower heights (with less air entrainment) than rhyolite and dacite ejecta (Houghton and others, 2004), the andesite-richer flow units were intrinsically more subject to welding. During the break between Episodes II and III, a pair of andesite-rich flow units ~2 km east of Novarupta welded moderately (PDC 4 of Houghton and others, 2004), even though each is only ~1 m thick and overlain only by ~8 m of pumice-fall deposits.

Welding of ignimbrite entails the viscous deformation of vitric fragments under load. Glass has its viscosity reduced (and is thus made more susceptible to welding) by higher emplacement temperature, greater alkalinity, and more retained (or resorbed) water and halogens. It has been shown experimentally that (1) high-silica-rhyolite ignimbrite (comparable to that of the 1912 eruption) requires temperatures of 550°–600°C to permit welding and that (2) under load pressure of 5 MPa (50 bars) with 0.4 weight percent water, such material can achieve partial welding with fiamme in 2-day

Figure 88. Gorge along mid-valley fork of Knife Creek (see fig. 74) , ~10 km from Novarupta. Upper ~5 m is nonwelded, weakly sorted ignimbrite with vague meter-scale layering. Lower 10–12 m is sintered and jointed, grading to partial welding at depth. Stream originates in central VTTS and is fed in part by mid-valley thermal springs of Keith and others (1992).

Figure 89. Former natural bridge in partially welded tuff along main fork of Knife Creek, 7 km north of Novarupta. Later ignimbrite packages (5 through 9), rich in andesite and dacite pumice, are typically sintered or welded in upper and central VTTS and seldom incised deeper than 8 m. Bridge collapsed at unknown time in interval 1986–1989.

runs (Grunder and others, 2005, and references therein). These conditions were probably met or exceeded by those inferred for the thick rhyolite-dominated ignimbrite buried in the upper VTTS. Rhyolite magma temperature was 800°–850°C, and fumarolic gases were still as hot as 645°C in 1919. H_2O retained by dense dome rhyolite at Novarupta is ~0.25 weight percent and was certainly greater for rhyolite pumice within the ash flows. And 100–200 m of overlying compacting tuff would have applied loads similar to that of the experiments cited.

Rapid onset of welding in thick upper-valley ignimbrite is inferred from the compaction swales that must have formed within an hour or two after emplacement of Package 4 on the first day of the eruption in order to have partially channeled subsequent packages. Hingeline faults along the compactional benches, however, are observed to offset Package 9 as well as some pre-Episode II reworked deposits and the dacite fall Layers C-H, demonstrating that welding compaction also continued beyond the third day of the eruptive sequence (Hildreth, 1983; Fierstein and Wilson, 2005). Many of the long fissures that controlled fumaroles in the upper and central VTTS are transverse to the valley axis, suggesting that viscous downvalley creep at depth influenced the brittle fracture in sintered tuff at the surface. For voluminous Packages 2 through 4, the fines-rich crystal-poor rhyolite-dominant ash flows that issued as dense currents from a low fountain (and overran streams, ponds, marshes, and snowbanks in the upper VTTS) probably accumulated, compacted, and lost permeability rapidly enough to retain heat and pore fluids (Sparks and others, 1999) adequate to promote dense welding at depth on a timescale of only hours to days.

Degassing Pipes

Gas escape pipes, variously known as elutriation pipes, segregation pipes, or degassing pipes, are remarkably uncommon in the 1912 ignimbrite. Where small clusters of them are exposed near Mageik and Windy Creeks and along the Griggs fork, they are typical of pipes widely reported in massive nonwelded ignimbrite—nearly vertical, 5–15 cm in diameter, and 0.5–3 m high. Rarely, some are inclined, branching, or curtain-like in three dimensions. Gas escape has flushed them of fine ash, leaving openwork concentrations of crystals, lithics, and pumice lapilli, much better sorted than their host tuff. Because gas can be gathered internally from the host, no external fluid source (wet substrate or burning vegetation) need be

Figure 90. Block of vitrophyric welded tuff, 56 cm long, atop north ridge of Mount Cerberus, 3 km southwest of Novarupta and ~250 m above valley floor. Fiamme are all rhyolite, and abundant small red lithics are oxidized Jurassic siltstone. Material is welded vent fill equivalent to ignimbrite Package 1. Block was excavated and probably transported in blast-like current that punctuated emplacement of ignimbrite Package 7 (Fierstein and Wilson, 2005). Knife is 8.5 cm long.

required; the few pipes observed are rooted either within the ignimbrite or in permeable fall material at its base. Nonetheless, all of those found in the 1912 sheet are in nonwelded tuff less than 10 m thick that had clearly buried streams and willow thickets. None of the pipes observed had leached, oxidized, or otherwise visibly altered their walls; none had precipitated incrustations at their upper ends; and none have been found in sintered or welded tuff. In these respects, they contrast strikingly with the "ten thousand" fumaroles discussed in chapter 9.

How can scarcity of elutriation pipes be reconciled with the abundant fossil fumaroles so ubiquitous in the 1912 ignimbrite? Although less abundant and shorter-lived than in the thick ignimbrite of the upper and central VTTS, fumaroles were indeed present in the nonwelded tuff of the lower VTTS, where some persisted for a decade. If the granular currents had lost most of their interstitial gas before distal deposition, thus rarely producing elutriation pipes, were distal fumarolic gases almost entirely meteoric water that became acid by leaching the hot tuff? Distal fossil fumaroles appear to be more common over buried stream courses, but this is likewise true for the sparse elutriation pipes.

Emplacement Over and Against Ice

The five Knife Creek Glaciers lie within 1.5 to 6 km east of the Episode I vent rim, their termini adjoin the valley floor, and their lower reaches are at elevations lower than Novarupta. The snouts of none have receded significantly since 1912. Two glaciers from the north side of Mount Mageik also extended in 1912 to the valley floor, where their snouts were 4 km and 6 km southwest of Novarupta and 100 m lower in elevation. Both have receded more than 1 km since 1912, leaving their former terminal reaches occupied by East and West Mageik Lakes, each bordered by bluffs of ignimbrite that had banked against the ice.

Around East Mageik Lake, the sheet of partially welded ignimbrite has slumped back toward the lake, opening a set of arcuate fissures (fig. 91), presumably in response to withdrawal of the ice front against which it had been deposited. West Mageik Lake is enclosed on three sides by ignimbrite that is widely altered fumarolically and pitted with phreatic craters, has collapsed into aprons of ignimbrite talus, and is locally lapped by lake sediments (fig. 92). The lake periphery is littered with brightly oxidized blocks of ignimbrite, some

Figure 91. East Mageik Lake and fissured welded ignimbrite of Package 8, which banked against glacier that occupied lake basin in 1912. Snout of glacier from Mount Mageik has now receded ~1 km (see figure 72), withdrawing support for tuff margin, which has slumped back toward lake, opening complex fissure system. View is northwest across Lethe arm of VTTS to Buttress Range margin, where "high sand mark" bench stands as high as 60 m above valley axis. Thin black veneer of alluvium in middle of Lethe arm overlies ignimbrite and is largely derived from andesite-dacite lavas of Mount Mageik.

prismatically jointed, that were expelled explosively (fig. 93). Many such blocks are associated with the fissured hingeline of the compactional bench nearby (figs. 84, 92), but others may have been ejected from the ice-contact zone at the glacier front itself. Lakeward tilting, fissuring, and slumping of the ignimbrite bluffs may also reflect melting away of buried snowbanks and ice-cored moraine.

When first visited in 1917, a narrow lake extended from the glacier front for 600 m northeast of the present lake along a defile excavated phreatically along the fissured hingeline. Called "Fissure Lake" by Griggs (1922), this extension had silted up by 1923, and today its 10-m-thick deposit of ashy lacustrine and alluvial fill has been incised by a stream flowing back southwest into the modern lake (fig. 94), which was not fully formed by glacial recession until the 1950s. The lower reaches of the two Mageik glaciers have now disappeared, but photographs from 1917 show that the lower kilometer of the western one was still mantled by a few meters of eruption products, thick enough to require an ignimbrite veneer as well as fall Layers C–H. The lower kilometer or two of the eastern glacier was surely veneered with ignimbrite, as shown by the thin runups (largely of Package 1) still present on Mageik lava benches adjacent to the deglaciated trough.

All five Knife Creek Glaciers remain heavily mantled with 1912 fallout (fig. 95), which is as thick as 10 m atop their lower lobes and still 6.5 m at the Katmai caldera rim. Much of it has been disrupted by ice movement and wasting, but sections complete from Layer A to Layer H are locally intact. Intercalated within some supraglacial fall sections are (1) layers and pods of pyrite-bearing hydrothermal mud ejected from Mount Katmai during caldera collapse (Hildreth, 1991) and (2) meter-thick deposits of pumiceous debris flows that were remobilized down snowclad slopes during the eruptive sequence.

At the western edge of Glacier 1, an ice-marginal stream has cut a narrow slot ~20 m deep, along which at least 12 m of stratified fines-poor ignimbrite rich in andesite is exposed beneath 10 m of dacite fall deposits, both of which had banked against the ice and are still in local contact with it. Across the chasm to the east, 3–5 m of ignimbrite are locally visible extending across the glacier under a hummocky mantle of disrupted fallout. In a few places atop the steep flow front of Glacier 1, supraglacial ignimbrite is seen to consist of several meters of rhyolite-dominant material overlain by a few meters of compositionally heterogeneous darker material, capped in turn by the stratified dacite fall deposits. The supraglacial

Figure 92. Southwest margin of VTTS viewed eastward from crest of Buttress Range. Outlet of West Mageik Lake at lower left is origin of River Lethe. Trough across center of image was excavated by chain of phreatic explosions along hingeline of compaction-related bench similar to that seen in figure 82. Wide silt-floored central segment of trough was site of Fissure Lake (figs. 84, 94), which ultimately drained toward modern lake at right. Three discrete phreatic craters mark foreground peninsula. Blue East Mageik Lake is visible at ignimbrite margin in left distance, 2 km from seasonally muddy West Mageik Lake in foreground. Ignimbrite packages banked against dark slopes of glacially scoured lavas of Mount Mageik with hardly any run-up. In left background, black cone and lava flows that erupted at Southwest Trident in 1953–1974 rise above Katmai Pass, 10 km from camera.

pumice-fall deposits, though now widely chaotic, are little altered; most are gray or tan, but they locally exhibit red or yellow staining caused by diffuse steaming, which was variously promoted by the heat of the pumice deposits themselves or that of the Katmai mud layers or the thin ignimbrite veneers intercalated near the base of the fall section.

A few kilometers farther east, on lower reaches of Glacier 3, thin remnants of disrupted ignimbrite are exposed within thick fall material chaotically disarrayed by severe wasting of the glacier, which had been beheaded by caldera collapse. On the cleaver separating Glaciers 3 and 4, ~3 km east and 670 m higher than the snout of Glacier 3, a complete fall section has three all-rhyolite ignimbrite flow units of Package 1 intercalated within Layer A. Separated by fall intervals, the three are successively 30, 60, and 60 cm thick, and all are matrix-supported but fines-poor, perhaps owing to excess gas flux during transport over snow and ice. The site, 6 km east and 500 m higher than the vent rim at Novarupta, was not reached by any of the voluminous (but less mobile) ignimbrite currents of Phase B—Packages 2 to 9.

Knife Creek Glacier 4 is laterally confined by a pair of steep cleavers (fig. 23), the tops of which are, respectively, 5 km and 7 km from Novarupta. Atop the first (Whiskey Ridge) were deposited 4 m of all-rhyolite HEPI and ~1 m of mixed HEPI; atop the second cleaver these packages had thinned to 104 cm and 9 cm, respectively. Both ridgetop deposits are ~400 m above the VTTS floor and ~200 m above the sloping surface of Glacier 4. Upper levels of blast-like pyroclastic currents passed over both steep-sided cleavers (fig. 73) and the trough-filling glacier between them. On both ridgecrests the PDC deposits are overlain directly by Layer C (fig. 42).

Fed from cirques high on Trident, Knife Creek Glaciers 1 and 2 advanced 250–300 m between 1951 and 1987 (as shown by comparing sets of aerial photographs). Fed from Mount Katmai, Glacier 4 advanced ~150 m and Glacier 3 (despite beheading by caldera collapse) ~225 m between 1951 and 1987. Moreover, examination of a 1919 photograph shows that by 1951 Glacier 4 had already advanced 500 m and Glacier 5 ~1,300 m (see section below titled "Behavior of Ash-covered Glaciers" in chapter 11). Since 1951, however, the distal 700 m of Glacier 5 has thinned and stagnated but not receded noticeably; any former ice-ignimbrite contact there was, in any case, overrun by glacial advance before 1951 or masked by alluvium.

Figure 93. Breadcrusted block of sintered, oxidized, vapor-phase indurated bulk ignimbrite ejected by phreatic explosion along trough of former Fissure Lake (figs. 92, 94). Shovel is 58 cm long.

A

B

Figure 94. Floor of Fissure Lake. Views westward along 100-m-wide trough of Fissure Lake: *A*, 1917 and *B*, 1982. Admiring its "clear green water," Griggs (1922, p. 244) called it "one of the most picturesque spots in the whole Valley." Ignimbrite scarp at right is ~20 m high and was locally still steaming in 1917; scarp at left is more than 30 m high. Sediment from the glacier (center distance, ~1 km from the camera) was already by 1919 filling the lake, which was largely silted in by 1923. Later withdrawal of glacier terminus created present-day West Mageik Lake (not in view), which now drains north (right) to the River Lethe. Fissure Lake trough then suffered drainage reversal (toward West Mageik Lake), permitting incision of a westerly outlet stream to expose ~10 m of ashy lake sediment. 1917 photo courtesy of National Geographic Society; 1982 photo by authors.

The termini of Glaciers 1, 2, 3, and 4 have thus all overrun margins of the 1912 ignimbrite sheet. No further changes in terminal positions have been noticed since 1987 (Hildreth and others, 2003a), but small ones could easily have escaped detection. It can be inferred that most of the ignimbrite packages had banked against the first four glaciers and that some currents flowed a short distance up their surfaces. Only all-rhyolite Package 1, however, was energetic enough to climb the glaciers and reach the Trident saddles and the middle slopes of Mount Katmai.

Advance of the glaciers and incision of streams along their fronts have removed most evidence for direct ignimbrite-ice interaction. At one exposure across the stream from the snout of Glacier 3, ~4 m of cross-bedded but poorly sorted tuff rich in andesite and dacite pumice rests directly upon sintered ignimbrite of Package 8 and probably represents pyroclastic-flow material thrown back from the ice front by steam blasts. Many phreatic craters were blasted through the ignimbrite sheet in the upper Knife Creek arm, a few of them close to the glacier termini (Hildreth and Fierstein, 2003); because their surge-bedded deposits (discussed in chapter 10) overlie the dacite pumice falls, however, these explosive interactions between meltwater and hot tuff took place at least a few days after ignimbrite emplacement.

Figure 95. View eastward across Knife Creek Glaciers 1 (right), 2 (center), and 3 (left), all heavily ash-covered except in upland cirques where ice has cleared in a century of accumulation and flow. Glaciers 1 and 2 issue from opposite flanks of East Trident volcano (peak 6010). Glacier 3 descends trough from skyline saddle, which is caldera rim of Mount Katmai. Twin peaks on center skyline are Peaks 6128 and 6200 on west rim of caldera. In foreground, undisturbed complete 7-m-thick section of dacite fallout of Episodes II and III overlies ~8 m of Episode I ignimbrite; same deposits have been chaotically disarranged on glacier surface. Fine-grained ribs of Layers E and H are conspicuous in intact fallout remnant. In July 1923, Yori and Fenner contoured across these glaciers at about this elevation to reach Glacier 3 and Katmai caldera rim, achieving first ascent from VTTS side of Mount Katmai.

Katmai Project campsite on south bank terrace of Griggs Fork of Knife Creek in June 1953. View northward toward Knife Peak (now Mount Griggs). Geological parties were based here in 1953 and 1954, supplied by fixed-wing aircraft. Garniss Curtis bends over the camp stove. Others standing, from left to right, are John B. Lucke (University of Connecticut), bush pilot Schmiel, Howel Williams, and Ronald Kistler (Johns Hopkins University). Curtis and Williams were faculty members at the University of California at Berkeley, and Kistler soon after became a graduate student there. Photo courtesy of Ron Kistler, U.S. Geological Survey.

Chapter 7

Vent Structure at Novarupta

The failure of the first generation of investigators to define clearly the source vent for the 1912 pyroclastic deposits is attributable to late-stage backfilling of nested vent craters by their own ejecta and to poorly developed notions of magma storage and pyroclastic eruption. Griggs and Fenner assumed incorrectly that most of the fallout came from the summit crater of beheaded Mount Katmai, but they did observe that the ignimbrite had not descended from that edifice and inferred correctly that it originated in the VTTS. Both were impressed by the valley-filling fumarole field, leading Griggs to postulate a batholith beneath the VTTS and Fenner to imagine a rhyolite sill extending under much of the valley. Although both acknowledged Novarupta to be one of the chief sites of emission, they thought that the swarms of fumarolic fissures throughout the VTTS also ascended directly from a subjacent magma body. Even as late as 1953, a new generation of observers attributed the ignimbrite to fissure swarms at the head of the VTTS, thought to extend from the base of Mount Mageik as far as Falling Mountain (Williams and others, 1953). An extended field season in 1954 allowed Curtis (1968) to appreciate the structural significance of the fault scarps that surround Novarupta basin and to propose that the postulated fissure vents were restricted to that 2-km-wide hollow (fig. 70).

Rather than fissure vents, we envisage an open flaring funnel-shaped vent for the great Episode I outpouring, which released plinian Layers A and B, all of the main VTTS ignimbrite, and ~70 percent of the magma erupted in 1912. A cylindrical vent created by piston-style subsidence is very unlikely here because of the small diameter, silo-like geometry required, and evidence against a shallow, directly subjacent, magma chamber. The presumed vent funnel was flared explosively, was backfilled by Episode I ejecta toward the close of that episode, was partly reamed out a day later to form the smaller nested vent for Episodes II and III, and was then concealed deeply by the ejecta of those episodes—Layers C through H and the ring of complex proximal deposits that encircles the still-later lava dome (fig. 96).

The rim and steeply inward sloping upper walls of the Episode I vent funnel are defined by truncated scarps of Falling Mountain, distal Trident lava flows, and southwestern Broken

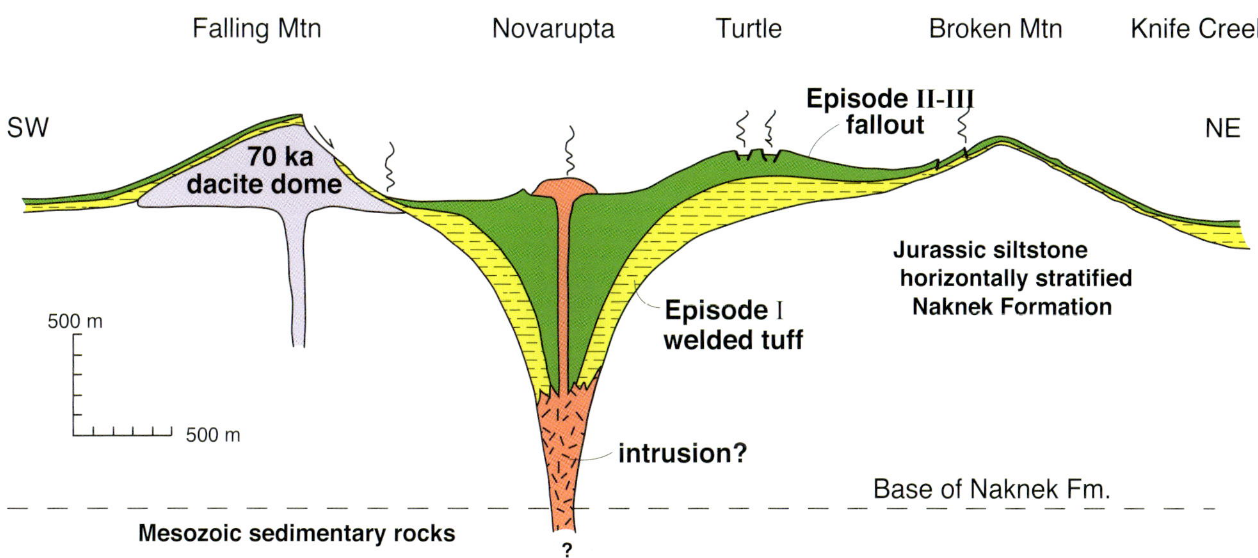

Figure 96. Conceptual cross section, to scale, of Novarupta vent area. Representation is based on surface exposures, geophysical interpretations, regional Mesozoic stratigraphy, and estimates of volumes of basement lithic ejecta and recycled welded ejecta. No vertical exaggeration.

Mountain (figs. 70, 71). The northeast rim is defined by a set of arcuate fissures and a shallow graben that probably reflect compaction and welding of the thick multiepisode ejecta blanket where it drapes the buried rim (fig. 97). The western rim of the funnel is more subtly defined by a gentle monocline that dips ~5° inward toward Novarupta (fig. 18), probably reflecting differential compaction of the thick vent fill. Just north of Falling Mountain, the monocline occupies the topographically lowest side of the vent complex, a pre-1912 valley excavated by Pleistocene ice that truncated the southern spurs of Baked Mountain and flowed into what is now the Lethe arm of the VTTS. A 1917 photograph (p. 194 of Griggs, 1922) illustrates abrupt lessening of fumarolic discharge outside the buried west rim, just north of the toe of Falling Mountain. The Episode I vent funnel is ~2.5 km wide at its ejecta-mantled rim and 1.7-2.0 km wide at its present-day floor (fig. 14 of Hildreth, 1983; Wallmann and others, 1990).

The truncated wall is best exposed as the shattered 425-m-high north face of Falling Mountain, where, as recognized by Griggs and Fenner, a large slice of the dacite dome had spalled and disappeared into the ignimbrite early in Episode I (fig. 71). Frequent rockfalls persisted from the face at least as late as 1923 (Fenner, 1938a), inspiring the name of the dome (Griggs, 1922). Ample post-1912 talus that accumulated below the face includes great blocks that disintegrated into hummocks as tall as 12 m, which have repeatedly been mistaken by casual observers for independent mini-domes (Hubbard, 1932, p. 90); the largest (fig. 55) was already there in 1917. Fumaroles high on the face of Falling Mountain persisted until the 1980s, but most issued from talus or from the thick sheet of ignimbrite draping its east end (fig. 71; also see photo on p. 136 of Griggs, 1918f) and only a few came from cracks in the dacite lava itself. A row of craters and troughs, with 5–8 m of local relief and rimmed by crossbedded surge deposits, was excavated by phreatic explosions through talus and tuff along the toe of the scarp, marking the buried structural margin of the vent funnel (see photo-fig. 8 of Fenner, 1925).

The depth of the backfilled Episode I funnel is poorly known but can be roughly limited by the volume and types of lithic fragments ejected. Episode I deposits are unusually rich

Figure 97. Aerial view east-northeastward over Novarupta dome and its ejecta ring, which rises to form the Turtle on far side. In distance, floor of upper Knife Creek valley lies 4 km from dome. Mutually perpendicular sets of graben cut Episode II fall deposits across summit of Turtle, resulting from viscous spreading of thick welded-tuff pile. Beyond Turtle, ~1 km from dome, elongate arcuate fissures and graben define outer margin of backfilled vent structure, which is inferred to be funnel-shaped.

in lithics, dominantly fragments of the sedimentary Naknek Formation, with lesser fractions of Trident andesites and Falling Mountain dacites, the three bedrock units truncated by Novarupta basin. Layer A contains 9–24 weight percent lithics on the mainland, rhyolite HEPI 9–14 weight percent, and all-rhyolite VFI 5–15 weight percent. Layer B has 6–17 weight percent lithics on the mainland, heterogeneous HEPI 1–16 weight percent, and the (main) heterogeneous VFI 1–6 weight percent (Fierstein and Hildreth, 1992). On Kodiak Island, however, Episode I fallout has only ~1 weight percent lithics, and the fraction no doubt continues to diminish still farther downwind (out to sea). Because the lithics in all these deposits are predominantly Naknek Formation lithologies, derived from a well-studied regional formation predicted to extend to a depth of ~1,500 m here (Detterman and others, 1996), and because Mesozoic formations that regionally underlie the Naknek are absent in the 1912 lithic suite, the reamed-out vent funnel likely does not extend deeper than 1.5 km (and could be much shallower). Sparse granitoid cobbles among the 1912 ejecta are from conglomerate members of the Naknek Formation, not from any subjacent pluton.

If the volume fraction of lithics were a third of its weight fraction in the ejecta, then lithics would make up an average of ~1 volume percent of all Episode I fallout (8.8 km^3) and heterogeneous VFI (~8 km^3) and ~3 volume percent of all HEPI, rhyolite VFI, and concealed proximal VFI (~3 km^3). Such approximations, based on volume estimates and measured lithic fractions (Hildreth, 1987; Fierstein and Hildreth, 1992), yield an Episode I lithic volume of only ~0.26 km^3, the volume of an inverted cone only 350 m deep, if 1.7 km in diameter at the surface (the floor of Novarupta basin). If, instead, the volume fraction of lithics were half its weight fraction in the pyroclastic deposits, similar calculations yield a lithic volume of ~0.39 km^3 and a cone depth of ~500 m. Calculation of such modest volumes suggests either strong near-surface flaring of a shallow vent funnel above a narrow conduit or a much larger proximal fraction of (concealed) lithics. The latter is not observed recycled into the ejecta ring (fig. 98) close to the dome nor in the recycled welded-tuff material (figs. 62, 63, 90) reamed from the Episode I vent fill that dominates the lithic fraction in Episode II and III ejecta.

Rim benches north and south of Novarupta appear to be pre-1912 surfaces, ejecta-mantled but little deformed, that were sharply truncated by the Episode I eruption. The benches are now ~200 m higher than the floor of Novarupta basin west of the dome. This 200 m of visible relief could account for half of the lithic volume calculated, rendering the apparent depth of the backfilled Episode I crater shallower still.

Gravity and magnetic surveys of Novarupta basin (Goodliffe and others, 1991), tied to a geodetic grid with a 200-m spacing, support a strongly flaring, rather shallow backfilled crater. The crater is not defined at all by the gravity contours, owing in part to the small density contrast between Naknek Formation basement and welded vent fill. A subtle gravity low (–1 mgal) over the Turtle is likely to reflect the great tephra thickness of that asymmetrical pile of ejecta. The buried crater is well defined by the magnetic contours, owing to the enormous susceptibility contrast between Naknek Formation sandstones and welded andesite-dacite ejecta. The magnetic contours define a high that dips radially outward on all sides of Novarupta basin, reflecting the shallow radially inward dip of the base of the high-susceptibility vent-filling ejecta. Lack of a magnetic low over the Turtle disproves the hypothesis that it is underlain by a siltstone slide mass from adjacent Broken Mountain. Three modest magnetic highs (+200 nT) in the vent area, notably one located 500 m northeast of the dome and 500 m south of the Turtle top, are likely to reflect, not cryptic intrusions, but sector-confined emplacement of andesite-rich fallout and PDC lobes as well as local variations in degree of welding (Fierstein and others, 1997; Houghton and others, 2004).

Compaction and welding of vent-filling ejecta hundreds of meters thick provide the best explanation for the fault and fissure patterns atop the Turtle as well as for the arcuate graben and fissures in the thick Episode II and III deposits rimming Novarupta basin (fig. 97). The Turtle is an integral part of Novarupta's ejecta ring (figs. 55, 97), stratigraphically continuous with it but 200 m thicker owing to strong asymmetry in accumulation of proximal Episode II and III fallout (fig. 96). The Turtle is clearly not a structural uplift. The mutually perpendicular sets of extensional faults crossing its summit probably reflect shallow spreading of the steep-sided 1-km-wide mound of ejecta during viscous deformation accompanying the welding of its interior (Hildreth, 1983). Although wispy fumaroles and warm ground are still present along some of the Turtle faults, the Turtle was never a major focus of fumarolic discharge. Photographs taken in 1917–1919 (Griggs, 1922) show only modest emissions on the Turtle (fig. 55*A*), compared to vigorous plumes then issuing nearby, variously from Novarupta, its lower ejecta ring, the Falling Mountain scarp, and many of the arcuate fractures marginal to the wider depression. Relatively weak fumarolic activity only 5 years after the eruption is inconsistent with existence of a cryptic intrusion beneath the Turtle (Hildreth, 1983; Goodliffe and others, 1991).

Within Novarupta basin, extreme variations in conductive and convective heat flow were measured over horizontal distances of only a few meters (Ballard and others, 1991), and the several surviving warm areas (25°–90°C) are largely confined to faults and fractures, as are the areas of former acid alteration (Keith, 1991b). Such warm areas dominate heat flow in Novarupta basin, and a few still emit near-neutral wet steam, weakly from tiny orifices. It is inferred that the fractures provide high-permeability pathways that guide ascent of water vapor (previously also acid gases), which in turn dominates heat transport and arises from depths where infiltrating groundwater is heated to boiling by proximity to the still-cooling magmatic conduit.

The lithic fractions of Episodes II and III consist predominantly of welded ejecta that had backfilled the wider vent during Episode I (Hildreth, 1987). Fragments range in size from centimeters to several meters, in composition from rhyolite to andesite, and in texture from dense eutaxite to tack-welded pumice agglutinate (figs. 62, 63, 90). Most of the fragments are poorly sorted tuff that consists of a variety of juvenile (and lesser lithic) components; some are breadcrusted or have prismatic joints, and others have a hackly fracture pattern characteristic of densely welded vitrophyres. Virtually all are glassy, and most contain angular to rounded fragments of Episode I lithics—andesite, dacite, and Naknek Formation lithologies, the last commonly oxidized to brick-red. Although these three lithic types are also sparsely present as discrete clasts, the rarity of free basement lithics among the Episode II-III ejecta shows that these episodes contributed little to further excavation of the bedrock envelope. Their conduits were nested well inside the broader backfilled vent for Episode I. Low H_2O contents of matrix glass in the vitrophyre ejecta (0.2±0.1 weight percent; Westrich and others, 1991) confirms that they consist of material that decompressed to one atmosphere before falling back, welding, and being re-ejected.

Residence time of the recycled vent fill was evidently long enough for it to weld but too short to devitrify before being blasted out during excavation of the smaller nested vent for plinian Episode II. It is not known whether the Episode II vent was thickly backfilled before plinian Episode III, but similar radial distribution of their products does suggest that their vent sites were the same. Episode III strata dip radially inward and enclose a low-relief moat ~450 m in diameter that surrounds the lava dome. The volume of welded recycled material for Episodes II and III combined is estimated to be ~0.05 km^3 (Fierstein and Hildreth, 1992). Such a volume could represent a 450-m-wide cylinder about 300 m deep, or an inverted cone three times as deep, or a geometrically less regular structure of intermediate depth.

Asymmetry of the Proximal Ejecta Ring: The Turtle

In the northeast sector of Novarupta basin, the ejecta ring surrounding the inner vent plugged by the Novarupta dome grades into a northeast-elongate hill of coarse Episode III ejecta, 225 m high and 1 km long, first called "the Turtle" by Howel Williams in 1953. The hill is cut by sets of mutually perpendicular graben (and other faults) that cross its crest (figs. 70, 97). The faults (fig. 99) were fully developed when the Turtle was first visited in 1917, and they still influence the distribution of wispy fumaroles (as warm as 90ºC) today (Keith, 1991a, b). Photographs taken in 1917–1919 (Griggs, 1922) show, however, that the Turtle was not a major focus of fumarolic discharge (fig. 55*A*). Compared to the vigorous

Figure 98. Stratified fall deposits that form inner south wall of ejecta ring only 25 m from margin of Novarupta dome. Section exposed is ~10 m thick, consisting mostly of coarse fall deposits and a few thin fines-rich pyroclastic density current (PDC) deposits; juvenile material in all layers is rhyolite-poor and rich in dacite, andesite, or both. Lithics are predominantly recycled welded-tuff vitrophyre along with sparse basement siltstone and rare fragments of Trident lavas. Section was deposited entirely during Episode III. A few large dacite blocks scattered on top surface are part of Episode IV block bed.

plumes then emitted from other nearby sites, emissions atop the Turtle were unimportant.

The Turtle is not a structural uplift but, instead, a stratigraphically continuous sector of an extraordinarily asymmetrical ring of ultraproximal ejecta (figs. 55, 97, 98). The ejecta ring ranges in exposed height from ~7 m in the southwest sector to 225 m in the northeast. Inside the ring, the narrow moat surrounding the dome is inferred to represent the vent active during Episodes II and III. On the Turtle itself, erosive incision is minimal and only the last few proximal fall layers of Episode III are exposed, including fine-grained Layer H, which caps the fallout sequence. However, outer fringes of the ejecta ring are seen to rest on a thin andesite-bearing ignimbrite equivalent to the ashy parting between plinian Layers C and D (Fierstein and others, 1997; Houghton and others, 2004). Most or all of the ring deposits therefore accumulated during the last full day of the explosive sequence (the latter half of Episode II and all of Episode III), and they consist of the proximal facies of regional Layers D, F, and G, along with ultraproximal intercalations of several sector-confined fall beds and pyroclastic density current deposits (summarized on p. 104-105 of Houghton and others, 2004).

Why the overthickening should here be so asymmetrical and why such asymmetrical deposition should have been sustained toward the northeast sector for ~24 hours remains uncertain. Overthickening of ultraproximal (but nonballistic) pumice deposits may generally be attributable to (1) vortices in the emerging jet; (2) inefficient reentrainment of ejecta released from greater heights and swept inward by surrounding air drawn toward the buoyant convective plume; (3) roughness of conduit walls, decelerating margins of the emerging jet; and (4) excess andesite and lithics partitioned into a boundary layer around the core of the jet. The last two could create an unstable collar of low-fountaining ejecta marginal to the higher-velocity core feeding the plinian column (fig. 53). Such peripheral fountaining evidently produced the several sector-confined, locally dispersed fallout lobes and small pyroclastic density current deposits (mostly andesite-enriched) close to Novarupta (Fierstein and others, 1997; Houghton and others, 2004).

There are at least two components to the northeastward overthickening of the ring around Novarupta: (1) In contrast to southeastward dispersal of all the earlier regional fall units, isopachs for Layers F and G exhibit a lobate bulge to the northeast (Fierstein and Hildreth, 1992), and (2) intercalated, locally dispersed, fall and flow units are more numerous to the northeast and tend to be thicker in that direction (Houghton and others, 2004). Factors that might have contributed to asymmetry of the ring deposits include: (1) Low-altitude winds drawn into the base of the convecting column preferentially from the Lethe arm and Katmai Pass on the wide-open southwestern side of Novarupta basin (fig. 18), potentially resulting in (2) slight bending of the plume and preferential sedimentation from the lower northeast part of the column; (3) asymmetry of the flaring vent funnel, distorting axisymmetrical emergence of the jet; and (4) asymmetrical semiannular distribution of andesite magma in the conduit during flow that was otherwise dominated by dacite magma during Episodes II and III. A strongly asymmetrical ultraproximal deposit was also produced by the plinian rhyodacite eruption of Quizapu in 1932 (Hildreth and Drake, 1992), but there a compositionally contrasting component like the andesite at Novarupta is not conspicuous in the main-stage crater-rim deposit.

It has occasionally been suggested that the Turtle might be a displaced slump block of Broken Mountain basement draped by 1912 ejecta. This idea may well have inspired the name Broken Mountain, and it is clear that Griggs (1922, p. 244) intended the name to include both hills. However, there is no amphitheater-like scarp on Broken Mountain that could be the source of a 1-km-long block. Although blanketed by ejecta, the south face of Broken Mountain, which rises more than 600 m above the saddle separating it from the Turtle, appears to have been sculpted by glacial erosion in the same manner as adjacent Baked Mountain and the other sides of Broken Mountain. Moreover, a Bouguer anomaly map (Goodliffe and others, 1991) shows a weak gravity low (–1 mgal) over the Turtle, more likely to reflect the thick pile of low-density tephra (1,000–2,000 kg/m^3) near the surface there than a slide block of basement siltstone (2,500–2,600 kg/m^3). Absence of a magnetic low over the Turtle (Goodliffe and others, 1991) casts severe doubt on the hypothesis that it is underlain by a slide mass of Naknek Formation siltstone (of low magnetic susceptibility: ~1.1x10^{-3} SI units) surrounded by andesite-rich 1912 ejecta (14-30x10^{-3} SI units).

Another idea advanced by Goodliffe and others (1991) was that a shallow cryptodome was emplaced beneath the Turtle after the ejecta had accumulated. Their speculation was based on magnetic data, but in fact their magnetic contours run parallel and undeflected across most of the Turtle. They identified three modest magnetic highs (+200 nT) within Novarupta basin, the closest of which is centered 125 m lower and 500 m south of the top of the Turtle. Rather than intrusions, the highs are likely to reflect the concentration of proximal fall and flow lobes in the northeast sector that are rich in andesite (30x10^{-3} SI units) as well as local variations in degree of welding (Fierstein and others, 1997; Houghton and others, 2004). The sets of extensional faults crossing the summit of the Turtle (fig. 97) probably reflect shallow spreading of the steep-sided mound of ejecta during viscous deformation accompanying the welding of its interior (Hildreth, 1983). Some of the faults continue well beyond the Turtle to the east margin of the vent funnel, where they likewise represent extensional spreading in vent fill that was compacting and welding where it drapes the rim. The fault displacements atop the Turtle (fig. 99) range from <1 m to ~20 m, fairly modest had a cryptodome caused the extension. Finally, the rather subdued fumarolic activity on the Turtle only 5 years after the eruption, compared to the abundant high-temperatures plumes all around it, is more consistent with a thick impermeable lens of welded tuff than with a cryptodome intruding a shallowly flaring vent funnel.

Figure 99. Extensional faults atop the Turtle. Views toward northwest along southern set of northwest-striking faults near crest of the Turtle, ~1 km northeast of Novarupta dome: *A*, 1919 and *B*, 1982. Taken from near intersection with northeast-striking fault set (see fig. 97). Lower scarp represents ~3 m displacement, upper scarp ~20 m. Dacite-dominant pumice fall exposed is proximal deposit of Episode III, locally capped by a thin remnant of fine-grained Layer H. Degradation of scarps by 1982 reflects 63 years of sloughing induced by wind and snowmelt. For a profile perpendicular to these scarps, see figure 6 of Fierstein and others (1997). 1919 photo courtesy of National Geographic Society; 1982 photo by authors.

Chapter 8

Caldera Collapse and Seismicity

Collapse of the summit of Mount Katmai during the eruption at Novarupta was one of the few caldera-forming events of the scientific age. Owing, however, to syneruptive and summer-long ash clouds, lack of nearby observers, and the obstructed perspectives of remote observers, no one witnessed the collapse. During the summer investigations of 1912 and 1913, Mount Katmai was obscured in cloud, so the first record of its beheaded summit is Griggs' photograph of July 1915 (fig. 100*A*). Nonetheless, study of circum-caldera deposits intercalated with Novarupta pumice-fall layers (Hildreth, 1991) and analysis of the seismic record for June 1912 (Abe, 1992) together permit a fairly good reconstruction of the fitful collapse sequence (Hildreth and Fierstein, 2000).

Before 1912, the Katmai edifice was a compound stratovolcano that consisted of two large contiguous cones (fig. 5), both beheaded by the caldera collapse (fig. 101). On the caldera walls are exposed separate foci of hydrothermal alteration and abutting stacks of lavas and ejecta that dip toward each other from discrete northeast and southwest centers. A coastal survey in 1908 (Griggs, 1922, p. 270–275) sketched four summits (fig. 102), three of which were lost in 1912. The westernmost of the four survived the collapse and forms present-day twin peaks 6128 and 6200. These rise above the western edge of a marginally preserved remnant of a pre-1912 filled crater of the southwest cone; the eastern margin of that crater apparently included the main summit, former peak 7500. The two other pre-1912 summits (peaks 7360 and 7260; fig. 102) belonged to the northeast cone and were probably high points on the rim of its summit crater, now wholly destroyed.

Katmai caldera, described by Griggs (1922), Fenner (1930), Hildreth (1983, 1991), and Hildreth and Fierstein (2000, 2003) has an area of 4 km^2 at the floor and 8.8 km^2 at the rim. Its floor lay at an elevation ~990 m in 1916, and its jagged rim has a highest elevation of 2,047 m and a lowest of about 1,488 m. Caldera walls exposed above lake level generally have average inward slopes of 52°–62°, interrupted by slump-block benches in two sectors and by a 35° chute in a western scallop, which alone provides (hazardous) foot access to the lakeshore (fig. 103). Internal volume of the caldera is today about 4 km^3, and the foundered superstructure amounted to an additional 1.1 to 1.5 km^3, giving a total collapse volume of ~5 to 5.5 km^3. The Horseshoe Island dome and numerous fumaroles on the caldera floor (Fenner, 1930) were by 1929 covered by rising lakewater, which is now deeper than 200 m and more than 1 km^3 in volume.

Seismicity Accompanying the 1912 Eruption

There were no seismic instruments within 1,000 km of Katmai in 1912. Abe (1992), however, studied the records of about 40 seismological stations worldwide and was able to assign magnitudes to 50 earthquakes associated with the eruption. Of these 50 shocks, 33 had surface-wave magnitudes (M_S) greater than 5.5, and 14 were in the range M_S 6.0 to 7.0. A plot of M_S vs cumulative frequency for the sequence (fig. 8 of Abe, 1992) shows a deficiency of recorded events smaller than M_S 5.5 and suggests that there could actually have been as many as 100 Katmai earthquakes with $M_S \geq 5.0$. If so, about half either failed to register on 1912 instruments or did not meet Abe's selection criteria.

Many precursory earthquakes were said to have been felt locally for at least 5 days before the eruption (Martin, 1913; Fenner, 1925), but none of them were recorded instrumentally. Similarly, following the eruption, although shocks were felt (at a village 60 km southwest of the caldera) on 50 of the next 70 days (Martin, 1913, p. 158), Abe's search detected only two Katmai earthquakes later than 10 June and none after 17 June. Since the detection threshold of the station at Victoria, B.C., was about M_S 5.0 (Abe, 1992), the absence of an instrumental record of the precursory and posteruptive felt seismicity suggests that all such earthquakes were of lesser magnitude. (Because large events so dominate the total energy released, an additional 50 earthquakes of M_S~5.0 would add only 0.22 percent to the cumulative energy of the 1912 seismic sequence.)

Figures 60 and 104 show the sequence of earthquakes evaluated by Abe in relation to our analysis of the eruptive sequence. The first of the 50 instrumentally recorded earthquakes (M_S 5.4) took place at 0508 (local time) on 6 June, about 8 hours before the plinian eruption began. Seven additional shocks (M_S 4.8 to 5.5) were recorded during the next 7.5 hours, before the first sighting of the plinian column at about 1300. During the opening 10 hours of the eruption, eight more shocks (M_S 4.8 to 5.6) were recorded, culminating in the 11th hour with the first major seismic event (M_S 6.5) at 2356 on 6 June. This event may have signalled the first strong pulse of caldera collapse, which was probably associated with ejection

A

B

Figure 100. Mount Katmai beheaded. Views northward from alluvial floodplain 11 km south (1915) and 13 km south-southwest (1989) of caldera rim. Rhyodacite crag in south rim notch at center; Peak 6200 on west rim at left. *A,* 1915 photo was taken from Katmai River floodplain, ~2 km east-northeast of *B,* 1989 photo site on Mageik Creek alluvial fan. Ghost forests have today largely been destroyed by aggrading gravels of both streams. 1915 photo courtesy of National Geographic Society; 1989 photo by authors.

of the first phreatic mud layer from Mount Katmai (intercalated in the Novarupta fallout at the B_2/B_3 boundary; fig. 60). Moreover, the debris avalanches described by Griggs (1922) at Katmai Canyon, Noisy Mountain, and Martin Creek ("Mageik Landslide") all broke loose about this hour, as each deposit is draped directly by all or part of Layer B (see section below titled "Syneruptive Landslides" in chapter 10. About 8.5 km^3 of magma had by then been released at Novarupta (Fierstein and Hildreth, 1992), so, if our correlation is correct, it indicates an average magma discharge rate of ~5 x 10^8 kg/s (8.5 km^3 x 2,400 kg/m^3 x 11 h^{-1}) for the initial 11 hours leading up to the onset of major collapse-related seismicity.

Soon after the M_S 6.5 event at 2356, three more major shocks took place early on 7 June (M_S 6.3 at 0035, M_S 6.1 at 0223, and M_S 6.5 at 0825). The seismic energy released during this 8.5-hour interval amounted to 10.2 percent of the total for the 50-event 11-day series (fig. 104). After 0825 on 7 June there was a 6.5-hour interval marked by four earthquakes of $M_S \geq 5.1$ but none greater than M_S 5.6. This interval of lesser seismic-energy release ended with a M_S 5.8 event at 1457, opening a remarkable 6-hour series of earthquakes of magnitudes 5.8, 5.7, 5.8, 5.9, 5.4, 5.8, 5.7, 6.0, 5.8, and 6.3 (at 2049), culminating with a 7.0 at 2136. This powerful event was succeeded an hour later by a magnitude 6.8 earthquake (at 2248 on 7 June), followed by events of M_S 6.2 at 0029 and 6.6 at 0300 (8 June). The five big earthquakes between 2049 and 0300 on the night of 7–8 June account for 61.3 percent of the cumulative seismic energy released by the 11-day series (fig. 104), the M_S 7.0 shock alone representing 37.4 percent of the total seismic energy. Much of the structural collapse of Mount Katmai probably took place during these hours.

Surprisingly quiet by contrast, a 28-hour seismic respite (interrupted by modest shocks of M_S 5.4 at 2057, 5.1 at 2045, and 5.6 at 2241) ensued from 0300 on 8 June until 0714 on 9 June, when the "lull" was broken by another M_S 6.0 event. This was followed by another extended interval of relatively low seismic-energy release, 23 hours long and marked by only four recorded earthquakes (M_S 4.8 to 5.7), that ended with a magnitude 6.9 shock at 0606 on 10 June.

This earthquake, second largest of the entire series, took place 24-30 hours after the main eruptive activity ceased. Relocated by Abe (1992; p. 177) about 12 km west of Mount Katmai (apparently near Novarupta if located accurately), this major shock could conceivably have reflected posteruptive failure of a distal part of the plumbing system or (less likely) fault dislocation of vent fill around Novarupta, instead of additional caldera subsidence at Mount Katmai. The faults atop the Turtle 0.7–1 km northeast of Novarupta have vertical displacements as great as 20 m (fig. 99; also Griggs, 1922, p. 244; Fenner, 1925, p. 203). The seismic moment (M_O) implied by a M_S 6.9 earthquake demands, however, that the rupture surface area or the vertical displacement have been at least an order of magnitude

Figure 101. Southwestward aerial view of Mount Katmai caldera with ignimbrite-filled upper VTTS at right. Snow patch on wall across lake is where Yori and Fenner descended to caldera floor from west notch in 1923. Lake level is ~250 m above 1912 caldera floor, but sedimentation may have reduced present water depth to ~200 m. West wall now rises as little as 275 m above lake, whereas northeast wall (hidden by summit ridge at right) is ~800 m high. Several small glaciers still drape outer slopes despite loss of summit in 1912. Ice shelf on southwest rim predates caldera, but it took decades for its narrow ice appendage to extend down west wall. Beyond caldera, two tongues of Pacific sea fog are about to envelop Trident group, the more distant tongue invading Katmai Pass at toe of iceclad Mount Mageik.

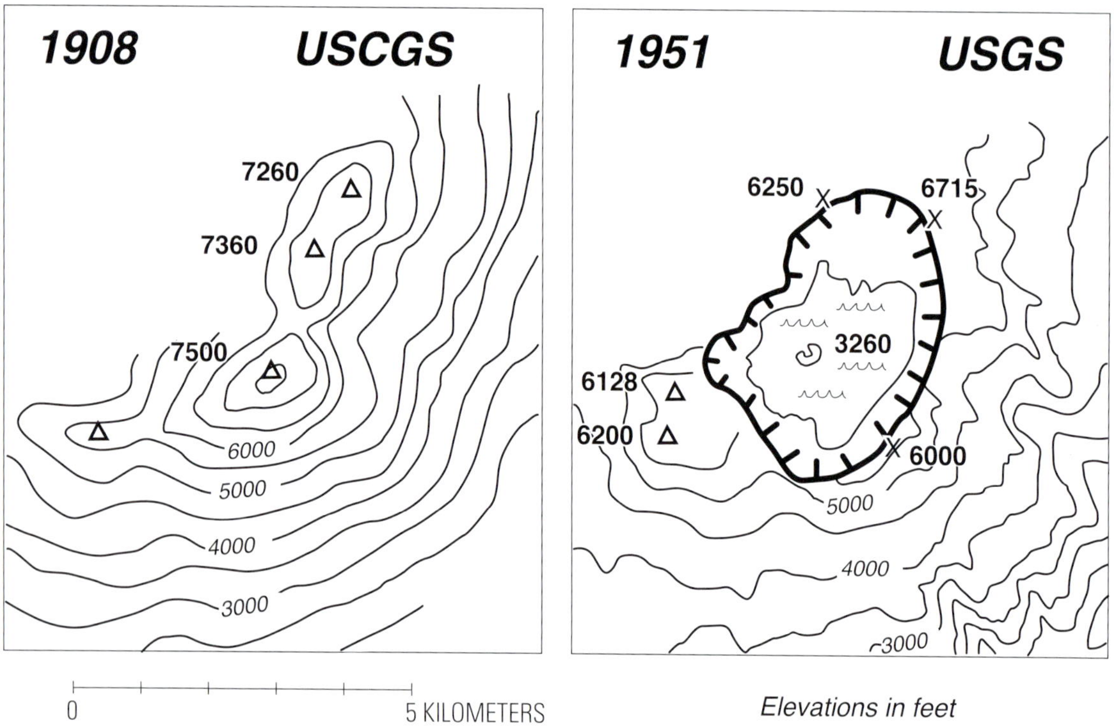

Figure 102. Mount Katmai before and after the caldera collapse of 1912. Left panel is adapted from Coast Survey Chart 8555 (U.S. Coast and Geodetic Survey, 1908; also reproduced by Griggs, 1922, and Fenner, 1923); summit elevations were triangulated from the coast 25–42 km south of Mount Katmai, and contours were apparently sketched visually with little additional control. Right panel is adapted (for the same area at the same scale) from U.S. Geological Survey quadrangles Mt. Katmai A–3, A–4, B–3, and B–4 (1:63,360 series, 1951), which were prepared by photogrammetric methods from aerial photographs taken in 1951. Elevations in feet. Horseshoe Island dacite dome, on sublacustrine caldera floor at an elevation of about 3,260 feet, is now covered by as much as 820 feet (250 m) of sediment and still-deepening lakewater.

Figure 103. Aerial view northwestward over Katmai caldera and down ignimbrite-filled VTTS, terminus of which is 29 km from center of caldera. Muddy deposit mantling shelf on northwest rim consists of caldera-collapse phreatic ejecta and ablation diamict left by beheaded glacier that has receded >500 m from rim since 1912. Narrow intracaldera glacier draping west wall (at left) and glacier that mantles slump-block bench at base of north wall (at right) both originated in 1920s. Isolated snow shelf just right of narrow glacial tongue marks descent route from west rim notch taken by Yori, Fenner, and Motyka.

greater than discernible for any posteruptive faults within or surrounding the Novarupta depression (fig. 70). The event is likely to have taken place at Mount Katmai or elsewhere in the plumbing system.

Unlocated adjustments in the vent, caldera, caldera-fill, or magma-storage and plumbing systems continued for more than two months, as recorded by a magnitude 6.1 earthquake at 2104 on 11 June, a magnitude 6.2 earthquake at 0128 on 17 June, and an abundance of felt seismicity (not recorded instrumentally) that persisted at least through mid-August (Martin, 1913).

Comparison with Other Seismically Recorded Caldera Collapses

Figure 104 illustrates the earthquake sequences for the three caldera-forming eruptions for which adequate seismic records exist. Table 2 of Hildreth and Fierstein (2000) summarizes comparative volcanological and seismological data for the three eruptions, collapses, and their host edifices. Although the volume collapsed and the volume of magma erupted at Katmai were each only about twice as large as at Mount Pinatubo (Philippines) in 1991, the cumulative seismic energy released at Katmai was 250 times greater. No fewer than 24 Katmai earthquakes were as large or larger than the greatest shock at Pinatubo (magnitude 5.7). Likewise, during the 1968 caldera subsidence at Isla Fernandina (Galápagos Islands), which was similar to Pinatubo in volume collapsed, more than 600 earthquakes were assigned magnitudes (Filson and others, 1973), but none was larger than M_S 5.2, and the total seismic energy released was 800 times smaller than at Katmai. What factors can explain such discrepancies?

In attempting to account for the far greater seismic energy released during the collapse at Mount Katmai, factors to consider logically include the structure and strength of the precaldera volcanic edifice, the location, depth, and configuration of the magma reservoir, the conduit geometry and flux of magma removal, and perhaps the viscosity and volatile content of the

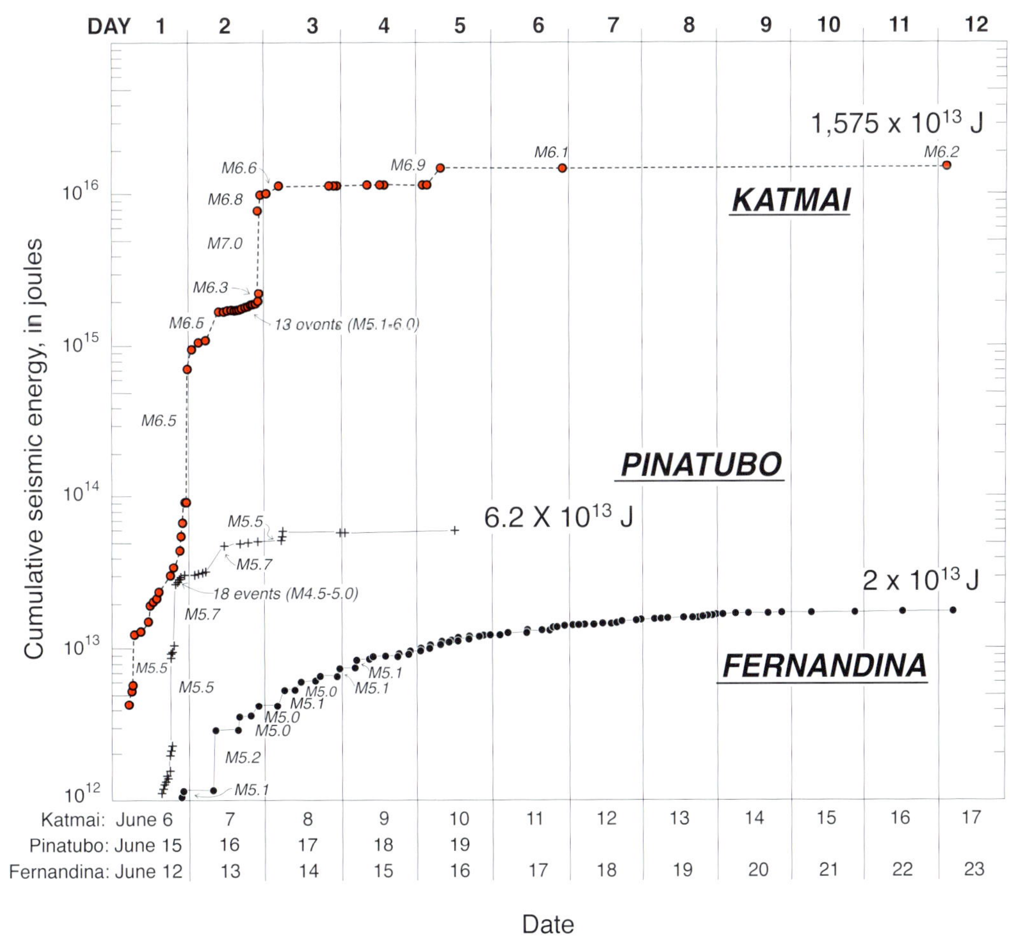

Figure 104. Cumulative seismic energy release (in Joules) during June earthquake sequences accompanying caldera collapse at Katmai in 1912, Pinatubo in 1991, and Fernandina in 1968. Note log scale. Energy for each event recalculated by revised energy-magnitude relation of Kanamori and others (1993) as employed by Mori and others (1996). Data from Abe (1992), Scott and others (1996), and Filson and others (1973). Sequence of Katmai earthquakes (with magnitudes shown here) is plotted relative to eruptive sequence in figure 60.

magmas. At Fernandina, as at Mount Katmai, magma was withdrawn from beneath the collapsing edifice rather than erupting through it, but the Fernandina magma was basaltic and the edifice a basaltic shield (Simkin and Howard, 1970). At Pinatubo, however, the climactic eruption issued from vents at the site of caldera collapse, expelled predominantly dacitic magma, and took place at a long-lived andesite-dacite stratovolcano (and dome complex) somewhat larger than Mount Katmai and marked by numerous ignimbrite-forming eruptions in the past ~35 kyr (Newhall and others, 1996).

Because the volumes of collapse differ by at most a factor of three and the bulk densities of collapsed material are unlikely to differ substantially, we speculate that the following factors were implicated in producing the seismic-energy discrepancy. (1) In contrast to Pinatubo and Fernandina, both of which had older calderas, Mount Katmai lacked preexisting caldera faults and shear zones, requiring that all breakage be new. Many earthquakes may have taken place on older faults at Pinatubo and Fernandina. (2) Subsidence may have been buffered at Pinatubo by buoyant viscous magma vesiculating and rushing upward to escape at fissure vents surrounding or within the foundering block. To the extent that the cauldron block was broken up and the subsidence piecemeal, such cushioning might have been still more effective in limiting the seismic energy generated. At Katmai (and Fernandina), because the magma was withdrawn from below, no such subsidence-opposing magma flow took place. At Katmai there were no magmatic ring dikes serving to lessen friction around the subsiding cauldron block. (3) At Katmai, about 8.5 km^3 of magma had erupted before the first cluster of large earthquakes ($M_S > 6$) 11–13 hours into the eruption, and nearly 12 km^3 had been released by the time of the second major cluster 32–38 hours after the eruption started (fig. 104). The strength of the previously unfaulted Katmai edifice may have permitted greater stress accumulation during withdrawal of magmatic support, resulting in individually greater stepwise vertical dislocations of the ultimately collapsing block(s) and, accordingly, fewer but larger earthquakes than at the already faulted Pinatubo and Fernandina edifices. Moreover, the horizontally layered Mesozoic siltstones and sandstones, at least 5 km thick beneath Mount Katmai (Detterman and others, 1996), could have contributed to greater average stress drops for each event by initially bending (with interbed slip, phonebook-like) before finally breaking. (4) The high-silica rhyolite magma that dominated the 1912 eruption, probably at least as viscous as crystal-rich Pinatubo dacite, certainly more so than Fernandina basalt, and proportionately more retentive of the excess volume contributed by vesiculation, may have played a role in slowing magma transfer within the plumbing system and slowing withdrawal of support, thereby contributing to the stress accumulation and overstepping that resulted in larger individual dislocations.

At Katmai in 1912, seismic-energy release was concentrated into a few short intervals (fig. 104), reflecting principally the distribution of the few M_S 6.5–7.0 earthquakes. The three largest earthquakes alone accounted for 76.5 percent of the total seismic energy released: The M_S 7.0 shock in the 33rd hour of the eruption contributed 37.4 percent of the cumulative seismic energy, the M_S 6.8 event 72 minutes later about 15.2 percent, and the M_S 6.9 event on 10 June about 23.8 percent. By comparison, the largest earthquake at Fernandina (M_S 5.2) contributed only 8.9 percent of the total seismic energy for the swarm, and the largest at Pinatubo (M_S 5.7) about 27.1 percent.

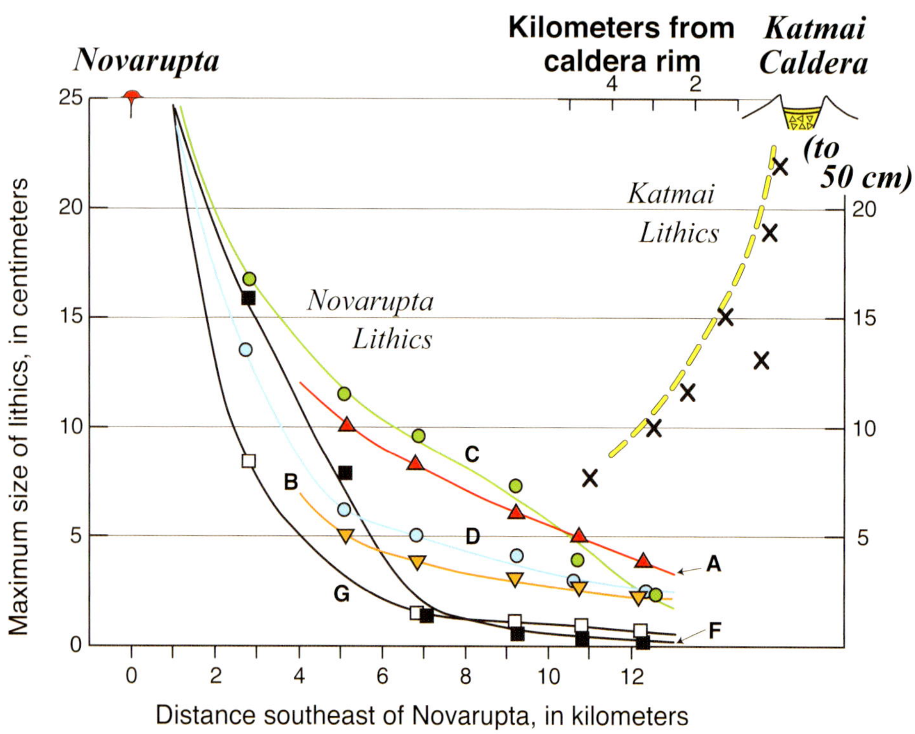

Figure 105. Variation with distance from source of average maximum sizes of lithic fragments expelled concurrently from vents at Novarupta and Mount Katmai. The hydrothermal/phreatic mud layers that contain Katmai-derived lithics were expelled during caldera collapse and intercalated within sequence of Novarupta-derived pumice-fall layers (A–G). "Maximum size of lithics" is average of three axes of five largest lithic clasts at each station. Katmai data are largely from south and southwest slopes of Mount Katmai edifice.

Three main intervals of seismic activity at Katmai produced 95.3 percent of the total seismic energy. (1) Between 2356 and 0825 on the first night (11 to 19.5 hours into the eruption), 10.2 percent was released. (2) Between 2049 and 0300 on the second night (32 to 38 hours into the eruption), 61.3 percent was released. And (3) the lone shock (M_S 6.9) at 0606 on 10 June (89 hours after the eruption began and 24–30 hours after it ended) produced 23.8 percent. Hildreth and Fierstein (2000) further elaborated the timing of seismic-energy release as spread over the multiday earthquake sequences, and their figure 7 shows the percentages of total seismic energy released for successive 4-hour intervals during the three caldera subsidences.

As discussed by Abe (1992), the total moment estimated from the collapse geometry of Katmai caldera can be approximated as $M_O=\mu\Delta V$. Taking ΔV=5.0–5.5 km^3 as the caldera volume displaced and taking μ, the shear modulus for shallow events, to be 3×10^{10} N/m^2, then ΣM_O= 1.5–1.65x10^{20} Nm, a satisfactory approximation to the cumulative seismic moment, 1.4x10^{20} Nm, estimated from the magnitudes of the 50 Katmai earthquakes interpreted. There is thus no need to invoke major additional release of seismic energy or large displacements elsewhere in the plumbing system or at Novarupta. The cumulative seismic moment for the Fernandina collapse was roughly equivalent to that of a single earthquake of *M* 5.9, the Pinatubo seismic sequence to one of *M* 6.2, and the 1912 sequence at Katmai to one of *M* 7.4.

Hydrothermal Explosion Breccia and Phreatically Ejected Mud Layers

On and near the caldera rim at Mount Katmai, several layers of massive, poorly sorted, unconsolidated "mud," studded with Katmai-derived lithic fragments, were ejected by phreatic-hydrothermal explosions during caldera collapse and intercalated within the sequence of widespread plinian fallout layers from Novarupta (Hildreth, 1991). No juvenile material has been recognized in the Katmai-derived mud layers, which thin from 10–200 cm near the rim to ≤1 cm at radial distances of 5–8 km from the rim. In contrast, the pumiceous plinian layers (A through G) thicken and coarsen toward Novarupta, away from Mount Katmai (fig. 105). We call them "mud layers" because under conditions prevailing at Mount Katmai these fines-rich deposits are always wet (or frozen), even gooey, and contain a hydrothermal clay component that imparts a faintly sulfurous odor. They are best exposed on the southern slopes of Trident and Mount Katmai, but a few intact remnants have been identified on and adjacent to the glaciers west of the caldera. Abundant crystals and centimeter-scale lithics occur within the dominant muddy matrix, and sparser large lithics (as big as 17 cm, as far as 5.8 km from the rim) were probably emplaced ballistically (Hildreth, 1991). The several mud layers identified are interpreted as a series of explosive ejection events released by unloading the hydrothermal system inside the cone during stepwise caldera collapse.

The first (and most voluminous) of the Katmai-derived mud layers lies between Novarupta plinian layers B_2 and B_3, providing the earliest evidence for onset of caldera collapse. The chronology of fallout reported at Kodiak (Martin, 1913) and of seismicity reevaluated by Abe (1991) supports correlation of this stratigraphic horizon with the first major syneruptive earthquake (M_S 6.5), at 2356 h on 6 June, 11 hours into the eruption (fig. 60). By that time, ~8.5 km^3 of magma had vented at Novarupta, suggesting an average mass eruption rate of ~5x10^8 kg/s for the first 11 hours. Some or all of this had been withdrawn from beneath the Katmai edifice, which then began to collapse.

Two mud layers near the B_2–B_3 contact are each a few centimeters thick on lower southern slopes (where the upper one

Figure 106. Hydrothermal explosion breccia as thick as 12 m on northwest rim of Katmai caldera. Ice axe (lower right) is 81 cm long. Derived from altered zones inside Mount Katmai edifice, material was ejected at onset of caldera collapse, thought to coincide with first major seismic event (M 6.5) in 11th hour of eruption. Deposit rests on 1.5 m of Novarupta plinian pumice fall (Layers A through B_2), which in turn rested on glacial ice, which has since melted (fig. 107).

contains accretionary lapilli), but they thicken and merge into a common layer 0.5–2 m thick where it overlies a breccia sheet (described next) on the west rim. This mud layer is 65–85 cm thick on a 4,600-ft lava-flow shelf ~2.5 km west of the rim. A thin mud layer is locally present at the B_3–C contact, and as many as eight mud layers (each 3–30 cm thick) are intercalated within the dacite pumice-fall sequence near the caldera rim. The thickest of these is within Layer F, where its thickness is 30–80 cm on the south rim, 14–45 cm near the west rim, and 6 cm on the 4,600-ft western shelf. At least three more mud layers, rich in sand and granules, are intercalated around the rim near the top of Layer G and within what appears to be an extraordinarily thick (~3 m) accumulation of Layer H. Capping the 1912 section, this near-rim fines-rich accumulation is apparently composite, consisting in part of Novarupta-derived vitric ash (normal Layer H), in part of contributions from continuing phreatic outbursts, and in part of deposits from dustclouds attending caldera collapse and continuing slumps and rockfalls. No ashfall layer has been identified that might correlate with eruption of the Horseshoe Island dacite dome on the caldera floor.

A wedge of coarse yellow-brown breccia as thick as 12 m is widespread on the caldera's northwest rim (fig. 106). It directly underlies the B_2/B_3 mud layer, which is here 0.5–2 m thick, though only 22 cm thick on the south rim. Confined to a 20-degree sector, the breccia was deposited atop ~1.5 m of plinian fallout (Layers A through B_2), which in turn rests on glacial ice that has since wasted away owing to destruction of its accumulation area by caldera collapse (fig. 107). Intact fallout sections have nonetheless survived disruption atop a few unmelted seracs. The breccia sheet thins westward across hummocky ablating ice and disappears about 800 m from the rim, where the glacial ice is still mobile.

Fragments in the breccia are angular to subrounded, as large as ~1 m, and include ~10 percent fresh (basaltic, andesitic, and dacitic) Katmai lavas. Most clasts, however, are hydrothermally altered lavas and tuffs, some of which contain disseminated sulfides. Identified by x-ray diffraction in the few clasts sampled were smectite, gypsum, albite, pyrite, hematite, magnetite, cristobalite, and chalcedony. Oxidized fracture films and rinds on both (internally) fresh and grey-to-black sulfide-bearing clasts attest to hot emplacement and subsequent steaming. A few hydrothermally altered blocks developed prismatic joints after emplacement of the breccia. Neither nonvolcanic nor juvenile clasts have been observed in it. The breccia deposit is unconsolidated, poorly sorted, and unstratified,; where thickest, it fines upward from >10 m of coarse clast-supported breccia (with <10 percent sandy-muddy matrix) through ~0.5 m of matrix-supported breccia into 0.5–2 m of blue-grey lithic-poor mud, which is overlain by Novarupta Layers C–H. The coarse zone thins to <1 m within 250 m northwest of the rim, but the associated mud layer extends radially as far as 8 km. Several kilometer-scale zones of hydrothermal alteration on the caldera walls extend beneath present lake level, had precollapse vertical dimensions greater than 500 m, and are an obvious source for the hydrothermal explosion breccia. Depressurization of an active hydrothermal system within the Katmai edifice where it was displaced abruptly along caldera faults caused explosive ejection of a relatively small fraction of the collapsing mass.

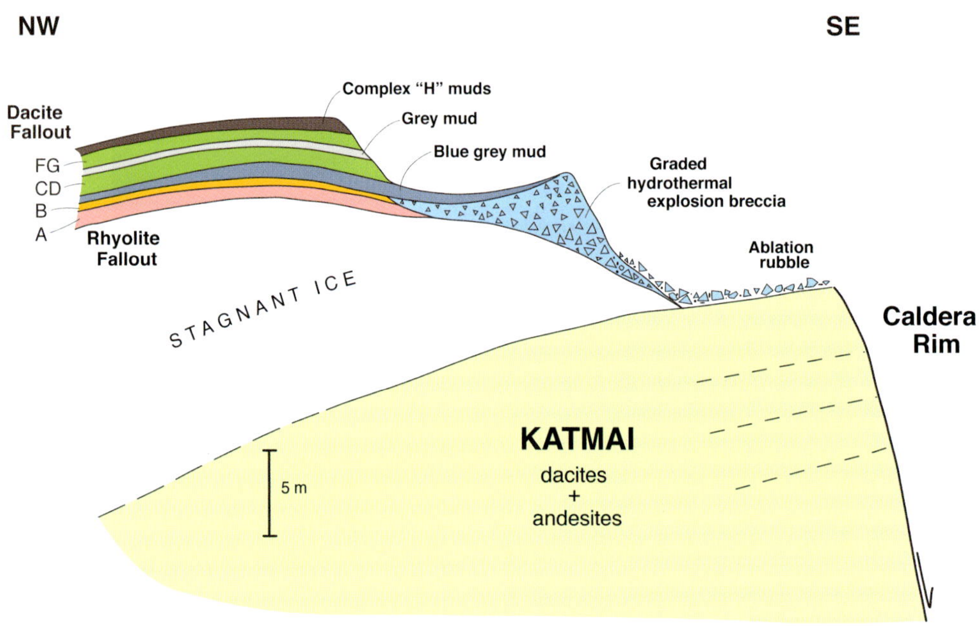

Figure 107. Sketch of stratigraphic relations on northwest rim of Katmai caldera, showing breccia (fig. 106) and mud layers expelled during caldera collapse and intercalated within the regional pumice-fall sequence from Novarupta. Deposited upon a glacier that was beheaded by collapse, section has been widely disrupted by a century of ablation of subjacent ice, margin of which has locally receded as much as 500 m from rim.

Chapter 9

The Ten Thousand Smokes

The VTTS was discovered by Griggs in 1916, who then named it the "Valley of Ten Thousand Smokes" for its myriad fumarolic plumes (fig. 108), among which he estimated 1,000 to be more than 500 feet high. He also observed fumaroles in Katmai Pass but reported none still active in the thin ignimbrite tongue farther south along Mageik Creek. During the 1917 expedition, the first to actually study the VTTS, Shipley (1920) noted fissural and point-focused orifices over the entire ignimbrite, but he also stressed ubiquitous diffuse steam emanations from surfaces with no visible orifices. Pits dug into cold surfaces everywhere reached boiling-point temperature (~97°C) within 1 m. Shipley observed that some fumaroles were largely water vapor, but many others consisted of acid gases rich in Cl, F, and S, and he was first to study the mineralogy and composition of the fumarolic incrustations. During the 1919 expedition, Allen and Zies made extensive collections of the acid gases and the incrustations (Allen and Zies, 1923; Zies, 1924a, b; 1929).

In 1918, temperatures greater than 190°C were measured for 86 vents (Sayre and Hagelbarger, 1919), the hottest found being 432°C in the central VTTS. In 1919, 54 vents were found hotter than 200°C, of which 11 were in the range 400-645°C (Allen and Zies, 1923). In the lower VTTS near Three Forks, the hottest was 274°C, while in the upper VTTS fumarole temperatures were as high as 397°C in the Lethe arm and 438°C in the Knife Creek arm. Some were as hot as 403°C near Novarupta but no hotter than 260°C on the faulted top of the Turtle. None were measured on the forbiddingly steaming Novarupta lava dome itself. Highest temperatures measured were again in the central VTTS, 1–3 km north of the noses of Baked and Broken Mountains, including all values in excess of 500°C (Allen and Zies, 1923). There was a wide range in volume and velocity of fumarolic plumes, from wispy to brisk to roaring jets and from pinhole fumes to columns 10–20 m wide that ascended >100 m and rarely >500 m. Also noteworthy in 1919, in the near-vent area, were a few feeble fumaroles that issued from bedrock shale on the south and west faces of Baked Mountain (diagram on p. 263 of Griggs, 1922) and from the face of dacitic Falling Mountain (photo on p. 202 of Griggs, 1922); the latter site continued to steam weakly into the 1980s.

Already in 1919, however, Griggs noted slackening, cooling, or disappearance of many 1917 fumaroles and conversion of many former steam vents in the lower VTTS to boiling springs. The great majority of fumaroles measured in 1918 and reoccupied in 1919 had declined in temperature (Allen and Zies, 1923). Fumarole 1 on the distal Ukak lobe of the ignimbrite cooled from >357°C in 1917 to 220°C in 1918 to 95°C in 1919. Fenner (1950) found that by 1923 hundreds of former fumaroles were either dead or had cooled to near the boiling point; hardly any survived along the Knife Creek arm, and the few superheated fumaroles remaining were either in the upper Lethe arm or around Novarupta, where some could still melt zinc (>419°C). Fenner noted that weak fumaroles were still active in 1923 high on Broken Mountain and in Katmai Pass south of Mount Cerberus, indicating that surprisingly thick ignimbrite is concealed there beneath the blanket of dacite fallout.

By 1929, only a few hundred fumaroles survived (Hubbard, 1932), nearly all near or below the boiling point. In 1930, Griggs (1933) could locate none above the boiling point, but he noted a cluster still active on the face of Falling Mountain (see 1918 photograph there on p. 280 of Griggs, 1922). When Hubbard returned in 1934, he reported that only "a few feeble" fumaroles were left, mostly in the upper Lethe arm and around Novarupta, compared to the hundreds he had seen in 1929; most conspicuous was a pair of big steamers on the western bench of Baked Mountain (Hubbard, 1935). In 1940, an aerial reconnaissance by the National Park Service reported only about 10 fumaroles still active in the VTTS. During geophysical investigations coordinated by the University of Alaska in the mid-1960s, weak fumaroles (near or below boiling) remained only on the Novarupta dome, on the west rim of its ejecta ring, along faults on and near the Turtle, and on the western bench of Baked Mountain. All but the last of these areas were still active in the 1990s, though most continued to weaken and cool; a few wispy ones remained as hot as 80°C–90°C (Keith, 1991b). The Baked Mountain bench fumarole, the lone survivor outside the 1912 vent area, kept steaming passively from a deep open fissure during the 1980s but died out in the 1990s; we obtained a temperature of 49°C in 1976 by dangling a maximum thermometer into a narrow linear orifice (fig. 109) about 6 m below its rim, and Keith (1991a) measured 32°C there in 1989. In its prime in 1917, it was photographed with a mighty plume (p. 128 of Griggs, 1918f).

Infiltration of ample rainfall and snowmelt directly into the ignimbrite, contributions of several modest side

A

B

Figure 108. Lethe arm of VTTS. Views northwest to north across Lethe arm of upper VTTS to Buttress Range (at left): *A*, 1917 image, where distance from camera on bench at northwest toe of Mount Cerberus to steaming bench across valley at left is 4 km; fissure control of many (not all) fumarolic emissions in upper valley is conspicuous. *B*, 1982 image from lower slope of Mount Mageik (~1 km west of Mount Cerberus) overlaps 1917 image and shows axial River Lethe, fissured hingeline of marginal benches (left and foreground), and post-1912 dark alluvial sheet (at right) from Mount Mageik. Distance down VTTS, entirely fumarole-free in 1982, is ~14 km to narrows where valley bends left. 1917 photo courtesy of National Geographic Society; 1982 photo by authors.

streams that disappear into it, and large contributions from the glacier-fed forks of Knife Creek and the River Lethe all helped sustain the fumaroles and greatly advanced cooling of the ignimbrite sheet. In 1919, Allen and Zies (1923) observed that total stream outflow from the sheet at Three Forks, though far greater than in 1917, was less than a tenth of the water they estimated to be entering the sheet from the various sources. Just above Three Forks, the River Lethe gave temperatures of 80ºC–85ºC in summer 1919. Allen and Zies further noted that fumaroles along the valley margins were generally of lower temperature than those near the valley axis, probably owing to direct marginal runoff from the valley walls, as well as to lateral thinning of the ignimbrite sheet. Their observation that the ratio of acid gases to steam declined downvalley can likewise be attributed to accumulative infiltration and to thinner tuff.

Gases sampled from many fumaroles in 1917 and 1919 (Shipley, 1920; Allen and Zies, 1923) were all >99.5 volume percent water vapor, but they also contained significant amounts of HCl, CO_2, H_2S, and HF (generally in that order of abundance), as well as atmospheric N_2 and Ar. Shipley mentioned SO_2, but it was not detected by Allen and Zies, which

Figure 109. Fumarolic fissure in welded tuff on southwest bench of Baked Mountain. Ignimbrite is overlain by dacite pumice-fall deposits (at surface), aeolian redistribution of which has filled or hidden many such fissures, creating an enduring hazard. This was site of longest-lived fumarolic emissions outside of Novarupta basin. It was still producing a weak vapor plume at 49ºC in 1976 and at 32ºC in 1989.

(for higher temperature gases) may reflect their analytical procedures. Sheppard and others (1992) critically reviewed the early methods of Allen and Zies and noted that the modest CO_2 contents relative to the halogens reflected the sources of the nonaqueous gases within the tuff itself. Found principally in Novarupta basin were CH_4 and NH_3, the latter precipitating white crusts of ammonium chloride within fumarolic orifices. Like Shipley before them, Allen and Zies observed a tarry black precipitate of aromatic hydrocarbon close to Novarupta dome. The presence of HF in concentrations comparable to those of H_2S and HCl was surprising, probably because gases associated with high-silica rhyolite had not been sampled previously. Fresh rhyolite pumice from Layer A contains ~500 ppm F and ~1,500 ppm Cl. Along with the eruptions of Tambora in 1815 and Hekla in 1947, the 1912 eruption was one of the few historical events to produce a fluorine spike in ice cores on the Greenland icecap (Lyons and others, 1990).

Fossil fumarole deposits and structures, though degraded, are still ubiquitously exposed on the floor of the VTTS and in Novarupta basin (Keith, 1991a, b). Several types of fumarole morphology were described by Sayre and Hagelbarger (1919), Allen and Zies (1923), and Keith (1991a). In the upper and middle VTTS, where the ignimbrite is well indurated and fractured, fissure fumaroles tens to hundreds of meters long predominated (figs. 110, 111). Many fumarolic fissures are roughly perpendicular to the valley axis, others are parallel to the valley margins (notably along the hingeline of the marginal benches produced by differential compaction of the VFI), and still other fumaroles took advantage of columnar joints and networks of variably oriented fractures in devitrified vapor-phase tuff. In the lower VTTS and Mageik Creek, where the ignimbrite is poorly indurated, slightly permeable, and thinner than ~25 m, fissure control is less common, and most fossil fumaroles are funnel shaped (figs. 112, 113), 1–5 m across at the surface, and commonly more concentrated above buried pre-1912 stream courses. Discrete altered zones—fissural, funnel-like, or irregular—are exposed on gorge walls to depths of 1–20 m (fig. 114), showing that focused gas ascent was gathered from within the tuff, where some fraction of the volatiles had been trapped interstitially during emplacement and more were liberated by secular devitrification. Many discrete pathways have light-colored leached rinds at depth in the ignimbrite, but many flare shallowly into varicolored zones of mineral precipitation near the surface of the ignimbrite and especially within the overlying stratified pumice falls (figs. 115, 116). Where well indurated by incrustations, many fumarolic orifices are still marked by mounds and

Figure 110. Fractures in sintered ignimbrite that concentrated fumarolic emissions for about a decade after the eruption, leaching tuff of mobile constituents. Lethe arm of VTTS near course of River Lethe. Backpacks for scale.

ridges resistant to erosion (Keith, 1991a). Precipitates at some orifices included mounds of amorphous silica "several feet high" (Shipley, 1920; Allen and Zies, 1923), a few of which have survived a century of erosion.

Fumarolic Deposits

Most fumarolic deposits are complexly zoned in mineralogy, color, and degree of alteration of the host ignimbrite or fall deposit (Keith, 1991a). An axial throat of high-temperature acid-gas release was surrounded on a meter scale by successive zones of boiling and condensing steam that permeated the tuff. As each fumarole cooled, such zones collapsed inward, converting the orifice to a hot spring or mudpot within a few years and replacing in full or in part the original fumarolic minerals with lower-temperature assemblages. Most of the fluorides, chlorides, and sulfates, originally ubiquitous, were dissolved and removed by aqueous condensate and by secular weathering in a cold wet climate.

Figure 111. Close-up view of fumarolically altered crack seen in figure 110. In well-indurated (sintered) ignimbrite, strong leaching was commonly limited to narrow zones only 5–10 cm wide; but note wider grayish zones of lesser alteration outboard of cream and pink strip. Notorious Katmai winds have widely scoured indurated ignimbrite surface, especially along Lethe arm of VTTS, stripping away formerly overlying fall deposits. Knife is 8.5 cm long.

Higher temperature vents produced anhydrite, corundum (Shipley, 1920), ammonium chloride, ammonium silicofluoride, molybdates (Zies, 1929, p. 28), and abundant vapor-phase magnetite (much of which was gradually replaced by hematite). Ramdohr (1962) later identified bornite, chalcopyrite, zincite, molybdenite, pyrrhotite, and pyrite as inclusions in magnetite taken by Zies (1924a) from mid-VTTS Fumarole 148 (239°C in 1919). In 1923, Fenner found that Fumarole 148 had cooled to 97°C and that the abundant clusters of magnetite crystals had been replaced by hematite.

Every incrustation studied contained abundant F, analyses yielding 0.1–15 weight percent F. Hard opaline mounds at Fumaroles 108 (238°C) and 134 (298°C) were impregnated by fluorides (as well as chlorides and sulfates, including barite), and the latter site (in the middle VTTS) had 12 weight percent F (Zies, 1929). Extensive bluish deposits along a high-temperature fissure just north of Baked Mountain (Fumarole 127; 264°C) contained 700–3,500 ppm Mo (Zies, 1929), a remarkable degree of concentration from the 3 ppm Mo in 1912 rhyolite pumice (Hildreth, 1987). Borate minerals were not identified, and B contents of gases were not measured. But incrustations at 22 fumaroles (97°C to 414°C) contained 0.01–0.21 weight percent B_2O_3, and those at three additional high-temperature fumaroles had 0.72–2.17 weight percent B_2O_3 (table 9 of Zies, 1929); these values can be compared with 31–58 ppm B in the rhyolite pumice, 28–41 ppm in dacite

Figure 112. Funnel fumarole in nonwelded ignimbrite east of Three Forks in lower VTTS. Orifice flares to ~2 m wide at surface. Note concentrations of lapilli along irregular cracks, from which fine ash was flushed by gas flux.

Figure 113. Orifice of funnel fumarole in nonwelded ignimbrite just upvalley from cross-valley moraine in lower VTTS. Note contrast between pervasive mild oxidation of tuff and restriction of white and bright red alteration to margins of fossil fumarole.

Figure 114. Large focus of fumarolic alteration, 5–7 m wide, exposed on wall of River Lethe gorge, 12–15 m deep here, in lower part of central VTTS. Rhyolite-dominant ignimbrite grades down from nonwelded to weakly sintered.

Figure 115. Natural cross section through hemicylindrical zoned fumarolic incrustations developed in dacite fall deposits, above a long-lived fumarolic fissure in underlying ignimbrite (exposed in foreground). View westward toward Buttress Range from draw that drains northward from Katmai Pass between Mount Cerberus and Falling Mountain (see fig. 72). Fumarolic structure is ~20 m across. Backpack and shovel at left center for scale. See Keith (1991a) for more on this site.

Figure 116. Hemispherically zoned fumarolic incrustations formed in permeable dacite fall deposits directly atop sintered to partially welded ignimbrite. Fissure in brown indurated tuff that fed sustained gas flux is exposed at base of structure. Pale-gray fall deposit is ~3 m thick here in upper Lethe arm of VTTS.

pumice, and <4 ppm in andesite pumice (Hildreth, 1987). Shipley, Allen, and Zies were well aware that some fraction of most metals they detected in the fumarolic deposits (B, Cu, Zn, Pb, Ti, Mn, Fe, Co, Ni, Mo, Ag, Cd, Sn, Sb, Te, Ga, As, Se, Bi) had been extracted from the pyroclastic deposits by the acid gases, but the great enrichments of such elements as Mo, Tl, and Bi persuaded them to accept the view of Griggs and Fenner that the gases and much of their metallic cargo issued directly from a magmatic intrusion beneath the valley.

Lower temperature fumarolic vents had deposits of elemental sulfur, realgar, orpiment, gypsum, halite, alunite, tarry aromatic hydrocarbons, and multicolored oxides, chlorides, and hydroxides of iron (Shipley, 1920; Allen and Zies, 1923). Sulfur was not present in vents hotter than 97°C but was conspicuous along lower temperature fissures and ubiquitous in fringing zones within ~1 m of hotter orifices. Pyrite was widespread in mudpots, and vivianite (hydrous iron phosphate) was found in a few (Shipley, 1920). Methane, ammonia, and tarry hydrocarbons were detected mostly within Novarupta basin, site of the deeply reamed vent structure and not a site where exceptional stands of pre-1912 vegetation might be expected to have grown or to have been buried. Perhaps these compounds were distilled out of the fossiliferous Mesozoic siltstone that surrounds the vent and spalled into it, a possibility raised by Zies (1929).

Keith (1991a) investigated in detail the mineralogy of the deposits that remained seven decades after extinction of most of the fumaroles (figs. 117, 118). Black zones typically reflect surviving magnetite or hematite that replaced magnetite. White zones may be enriched in silica, alunite, fluorite, barite, clay minerals, and hydrated aluminum hydroxyfluoride (AHF), but some are simply the leached ashy protolith with little added. Dark red and purple zones are usually enriched in hematite and goethite, and pink zones contain a little of these. Orange to yellow zones typically reflect abundant limonite, goethite, or sulfur, in addition to varied amounts of the white-zone minerals just cited. Within the coarse pumice-fall deposits, fine-grained Layers E and H served as low-permeability barriers, causing rising vapors to spread out laterally, creating diffuse zones of red, orange, and yellow precipitates in the pumice and thin multicolored crusts of

Figure 117. Severe fumarolic alteration in sintered ignimbrite and overlying stratified dacite fall deposits exposed along River Lethe in upper Lethe arm of VTTS. Black and red lineation on stream-scoured surface of ignimbrite reflects closely spaced cracks that governed gas emission. Blackening of originally near-white dacite fallout reflects high-temperature fumarolic deposition of magnetite (now widely retrograded) where gases emerged from fissures in ignimbrite into cold permeable fall deposits of pumice and crystal-rich ash. Black fall section is ~2 m thick here.

silica, limonite, and amorphous Fe-hydroxide directly beneath the barrier layers (Keith, 1991a).

Chemical studies of fossil fumaroles have shown intense leaching of most elements adjacent to fissures in indurated ignimbrite, resulting in residual enrichment of Si and Ti. In contrast, the far more permeable stratified pumice falls that overlie the ignimbrite provided favored sites for precipitation of incrustations and enrichment of most elements transported by the fluids (Papike and others, 1991a; Keith, 1991a). Lovering (1957) studied a sample transect collected by Ray Wilcox in 1952 from roughly concentric zones around Fumarole 1 in nonwelded rhyolite-rich ignimbrite on the Ukak lobe of the distal VTTS; the fumarole had been hotter than 357°C in 1917, 220°C in 1918, 95°C in 1919 (Allen and Zies, 1923), and was cold and dead in 1952. Because the ignimbrite protolith was rhyolite-dominant, gas expelled had been halogen-rich and sulfur-poor, in contrast to gases from andesite-richer tuff in the upper VTTS, where S may locally have exceeded F + Cl. Lovering documented complex multistage remobilization of many elements and concluded that, in general, halogen-acid alteration promotes greater loss of silica relative to alumina than does sulfuric-acid alteration.

It is hard, however, to extract systematic generalities about chemical redistribution in the several fumarolic deposits investigated (Papike and others, 1991a,b; Papike, 1992; Keith, 1991a; Lovering, 1957) because of (1) the temporal succession through decadal cooling of the acid-gas phase to aqueous boiling and eventually liquid-water alteration; (2) instantaneous lateral and vertical gradients in the thermal zonation implicit in those stages; (3) the multipackage (fig. 27) compositional (andesite-dacite-rhyolite) complexity of the VTTS ignimbrite, much of which exhibits great heterogeneity on a 10-m scale; and (4) downvalley (upwind) fining of the supra-ignimbrite dacite fallout (Layers C–H) from coarse pumice to crystal-rich medium ash. Slight enrichment in alumina and loss of alkalies were fairly general, and most metallic elements were transported from ignimbrite to surficial incrustations—many to be

Figure 118. Orifice of fumarolic fissure vent near East Mageik Lake. Ignimbrite is severely altered, but overlying white dacite pumice fall deposit only weakly so. Hammer is 28 cm long.

dissolved and lost subsequently to lower temperature alteration and secular weathering.

Fumaroles in Novarupta Basin

Within the vent area of Novarupta basin, fumarolic activity was longer lasting but no hotter than in thick parts of the VTTS ignimbrite sheet. In 1918, Sayre and Hagelbarger (1919) measured temperatures of 166ºC and 264ºC atop the Turtle and 294ºC west of the dome; in 1919, Allen and Zies (1923) measured 238ºC to 414ºC at nine fumaroles in Novarupta basin, where exposures are predominantly of the thick coarse fallout of Episodes II and III with thin ignimbrites intercalated. Keith (1991a, b) studied the fumarolic deposits and alteration around Novarupta in detail, describing the overprinting of high-temperature vapor-phase suites (magnetite-hematite-silica), first by lower temperature (<400ºC) vapor-phase minerals (fluorides-chlorides-sulfates-goethite-orpiment-NH_4Cl-silica) and then by argillic suites (<200ºC; alunite, kaolinite, pyrite, smectite, goethite, gypsum, and silica). Argillically altered areas are mostly aligned along faults and fissures where surface waters were acidifed by fumarolic gases, attacking the ash and pumice from the surface downward and creating impermeable clay-rich caps mostly <50 cm thick (fig. 1 of Keith, 1991b). The clayey areas are mostly white, yellow, pink, or red (fig. 119), but the few that are dark gray typically have visible pyrite. They overlie brightly colored fall deposits, coarse and permeable, that had been altered only modestly during diffuse high-temperature gas flux.

Residual warm areas today (25ºC–90ºC; fig. 10 of Keith, 1991a) lie along the fissures that governed major fumaroles in 1917–1919 and subsequently became sites of argillization. From the warm areas, numerous wispy fumaroles still emit wet steam, which at pH 5.7–7.0 precipitates a little silica from its condensate but accomplishes little further alteration. Although no cinnabar or other Hg-bearing sulfide has been recognized, elemental mercury was measured (~0.1–7 ppm) in Novarupta basin by Kodosky (1989), concentrated principally in the warm wet areas mapped by Keith, where it is probably adsorbed on clay minerals. On the Novarupta lava dome itself, thin films of vapor-phase hematite and silica remain along cracks and faces of glassy blocks, but subsequent alteration is not discernible, and the wisps of vapor still sometimes seen atop the dome (on foggy mornings) are water vapor and warm air—pH 6–7 in 1978 and 49ºC in 1986 (Keith, 1991a, b).

Thermal springs (21ºC–30ºC) in the middle VTTS discharge on gorge walls deep within the ignimbrite. As far as 11 km from Novarupta and certainly diluted by cold infiltration, their century-long persistence suggests either underflow from the vent region or residual heat in the thick welded tuff of the upper VTTS (Keith and others, 1992).

Figure 119. Advanced clay alteration along arcuate fissure system northeast of Novarupta dome (see fig. 97), where weak vapor emissions and near-boiling temperatures persist a century after the eruption. Broken Mountain at upper left, mantled by dacite fall deposits; Mount Griggs stratovolcano on distant center skyline.

"Glacier Priest" Bernard Hubbard and his malamute "Katmai" at a fumarolic pit just west of Novarupta in September 1934, the third summer he camped in the Valley of Ten thousand Smokes with dogs and football players. With experience accrued, he came well equipped this time, even though he knew the fumarolic gases had by then diminished greatly. Photo courtesy of University of Santa Clara archives.

Chapter 10

Secondary Deposits

Emplacement of ignimbrite and fall deposits upon rugged terrain—barren or brushy, much of it snowclad or ice-covered, and drained by braided glacier-fed streams in broad valleys—inevitably produced rapid and sustained alluvial, phreatic, and sedimentary responses. Here, we describe secondary effects that took place during or soon after the eruption, and in chapter 11 we summarize longer lasting secular responses.

Syneruptive Runoff and Remobilization

In early summer, the upper VTTS usually retains extensive snowbanks, and the north-facing slopes of Trident, Mageik, and Katmai are almost always snow covered. Deposition of warm pumice falls and hot ignimbrite on surfaces of snow and ice thus inevitably led, on 6–8 June 1912, to widespread remobilization of eruption products by meltwater, principally as ashy mudflows and pumice-rich slurries. Little evidence for this survives at the north base of Mount Mageik, where fallout was thin, later phreatic cratering was extensive, and the two large glaciers have receded ~1 km since 1912, with attendant fluvial erosion.

However, at the foot of the Knife Creek Glaciers, which were blanketed with many meters of ejecta, the evidence for remobilization is extensively preserved. In a deep narrow gorge incised along the lower west wall of Glacier 1, ~12 m of stratified pumice-rich alluvium is sandwiched between the ignimbrite and a complete section of Layers C–H. In a coarse matrix of crystal sand, the pumice blocks and lapilli are mostly dacitic and andesitic, derived from final ignimbrite Packages 8 and 9. At the snout of Glacier 3, remnants of a small proglacial pond include dropstones and accretionary mudballs (2–25 cm in diameter) in a very poorly sorted massive deposit rich in vitric silt, crystal sand, pumice lapilli and blocks, and angular lithics as big as 20 cm. Just downstream, massive exposures of hyperconcentrated crystal sand are as thick as 5 m. And for 2 km downstream from Glacier 3 to its confluence with the Griggs fork, the proglacial stream provides extensive exposures of plane- and cross-stratified high-energy fluvial beds of pumice and crystal sand (figs. 35, 44), in layers 1–20 cm thick and altogether 1-8 m thick. Like the similar reworked section adjacent to Glacier 1, rhyolite pumice is sparse and small, dacitic pumice dominates over andesitic, and successive sets of steep cross beds truncate earlier ones. Because the fluvial section here overlies the sintered ignimbrite (Packages 8 and 9) and underlies (typically complete) sections of Layers C–H, it was clearly deposited during the hours-long pause between eruptive Episodes I and II. Where the main proglacial stream reemerges from a 300-m-long tunnel under the edge of Glacier 4, the fluvial beds also overlie a 50-cm-thick remnant of Layer B_3 pumice fall, confirming that meltwater runoff from the ejecta-mantled glaciers commenced only after the end of eruptive Episode I.

Below the snout of Glacier 5, secular stream erosion has removed any deposits reworked syneruptively from the glacier, but slopes on the north side of its outlet stream provide a record of several decimeter-scale debris-flow deposits, all dominated by rhyolitic pumice and ash, that ran back downslope to intercalate with Layer B and lowermost Layer C. There is also evidence here that fluvial erosion had cut a sidewall into an A–B fall-and-flow section ~7 m thick before the unconformity was draped by Layer C.

Along the lower Griggs fork, similarly, a few pink mudflow deposits, 0.5–2 m thick, are intercalated within the A–B section of fallout and valley-fringing ignimbrite. These were remobilized by snowmelt on the slopes of Mount Griggs and, largely confined by swales, flowed down to interfinger with the primary pyroclastic deposits still accumulating on the valley floor. Juvenile components in the earliest such mudflows were almost exclusively rhyolitic pumice and ash, but later ones intercalated with the dacite fallout also contain a little andesitic and dacitic pumice and lenses of crystal sand. Small sparse lithics are mostly Jurassic shale originally emplaced with ignimbrite Packages 1 and 2, from which the mudflows were remobilized.

Along the Noisy Mountain Fork of Katmai River, on the southeast side of Mount Katmai, ~15 km from Novarupta, remnants of several matrix-dominated pumice-rich debris flows crop out downstream from the glacier front. The earliest is pale gray, rhyolite-dominant, and 4–6 m thick; it truncates Layer A and is capped by Layer B_3. Another, 1–5 m thick, contains both rhyolite and dacite pumice and is overlain directly by the upper half of Layer C. A third, 5–6 m thick, is terra cotta, contains all three pumice types, and overlies Layer G. The timing of these ashy debris flows is roughly consistent with that inferred from seismicity and fall stratigraphy for major pulses of caldera collapse (see chapter 8, above). The Noisy Mountain Glacier just upstream, though actively crevassed, is still heavily pumice covered, right to its snout. The pumiceous debris flows were remobilized, not from the glacier, but from the steep snowclad southeast slope of the Katmai edifice.

The Katmai River Debris-Flow Deposit.—This deposit at the south toe of Mount Katmai was recognized by Griggs (1918c, p. 124, 130) as a secondary pumiceous diamict emplaced at low temperature atop the stratified dacite fallout of Episode II and III, but several subsequent observers interpreted it as a hot pyroclastic-flow deposit, which it resembles (fig. 7 of Hildreth, 1983). The deposit is an unstratified sheet of poorly sorted 1912 ash and pumice lapilli. As thick as 18 m, the sheet slopes ~3°S and today covers an area of ~0.75 km^2, but before erosion it was at least twice as extensive (fig. 120). It debouched from a single steep gorge fed from Metrokin Glacier, which mantles the south and southwest slopes of the Katmai edifice all the way to the caldera rim (Hildreth and Fierstein, 2003). The deposit overlies nearly complete fallout sections of Layers A through G, although G is partly scoured in places. Layer H is widely preserved atop the primary uneroded surface of the debris-flow deposit, indicating that the flow was emplaced right at the end of eruptive Episode III. Field pumice counts yield a 7:3 ratio of dacite to rhyolite lapilli, similar to that of the bulk fallout mantling Mount Katmai and unlike that of any 1912 ignimbrite package (Hildreth, 1983; Fierstein and Wilson, 2005). Andesitic and banded pumice are rare, consistent with their extreme scarcity in fallout farther than ~4 km from Novarupta. In several exposures the deposit contains uncharred branches and twigs, as well as rooted bushes that are bent over downhill and retain intact bark (fig. 121). The source area was a snow-mantled glacier ~4 km^2 in area that sloped 11°–14° southward and was blanketed by at least 5 m of pumiceous fallout. Such supraglacial fall deposits, representing at least 0.02 km^3 of material, would be nearly double the volume originally constituting the debris-flow deposit (Hildreth, 1983). Beneath the main dacite-rhyolite debris flow, an all-rhyolite

Figure 120. Katmai River pumiceous debris flow (lower right center) emerged from canyon fed by Metrokin Glacier on south slope of Mount Katmai (summit in cloud, upper left). Pink deposit remnant, much eroded in a century, is ~2.5 km long and as thick as 18 m; dark stream alluvium (lower center) has replaced about half of the original deposit. Repeatedly mistaken for a pyroclastic density current deposit, which it resembles, it contains uncharred wood and consists of Novarupta fall deposits that were remobilized from snowclad slope of Mount Katmai and emplaced as a cold wet debris flow sandwiched between fall Layers G and H. For a close-up view, see figure 7 of Hildreth (1983). At left and center, thick Katmai andesitic lavas; Katmai Canyon in right distance. Tephra-mantled knolls at lower right are Tertiary granitoid pluton.

pumiceous debris flow left a deposit 0.5–1 m thick (Hildreth, 1991) that is intercalated within the pumice-fall section at the level of Layer B_2 (fig. 122).

Syneruptive Landslides

Seismicity attending caldera subsidence triggered collapse of at least three substantial extracaldera landslide masses (fig. 5), in addition to many minor slumps and rockfalls. Numerous pumice-rich debris flows (described in the previous section) during or soon after the eruptive sequence may likewise have been shaken loose from fallout deposited on steep snowy slopes by the severe earthquakes that continued at least as long as 10 June.

The Mageik Landslide.—This feature, described at length by Griggs (1920, 1922), broke loose from a glacially eroded stack of dacite lava flows on the south side of Mount Mageik, leaving behind a 120-m-high scarp with a rim at the 3,000-ft level. The rugged deposit extends ~6 km down the valley of Martin Creek to an elevation of 650 ft and consists predominantly of angular blocks of fresh dacite, along with sparse chunks of basement sandstone. It has millions of dacite blocks larger than 1 m, slabs as long as 20 m, and hummocks as high as 20 m. It locally left 30-m-high superelevated trimlines and impounded drainages along its steep margins. About 800 m wide proximally, the deposit spread out to ~1.5 km medially and covers an area of ~6 km^2. Griggs (1920) estimated an area of 10 km^2, but this included parts of pre-1912 avalanche deposits mapped separately by Hildreth and Fierstein (2003). Although the thickness (5–30 m) is hard to average, the volume is probably in the range 0.05 to 0.1 km^3. The avalanche took place during eruptive Episode I, on the night of 6/7 June, because the deposit is overlain by only the upper part of Layer B and later fall units (Hildreth and others, 2000).

The Katmai Canyon Landslide.—This deposit likewise broke loose on the night of 6/7 June, after deposition of Layers A and part of Layer B. It is overlain directly by Layers B_3 through H. Near the south toe of Mount Katmai, the headwall scarp extends ~500 m above the canyon floor, exposing a stack of andesite-dacite lavas with thin scoriaceous interlayers. The chaotic 0.08-km^3 slide mass covered ~1.2 km^2, was as thick as 125 m, and consists entirely of shattered debris from the Katmai edifice. The deposit blocked the narrow gorge of Katmai Canyon (fig. 123), impounding a 5-km-long lake, which breached by 1915 (Griggs, 1922); see section below titled "Lakes Impounded."

The Noisy Mountain Landslide.—This deposit in upper Katmai Canyon is avalanche debris consisting of Tertiary(?)

Figure 121. Katmai River pumiceous debris-flow deposit at its distal margin, where it overran willows and alders. Branches retain uncharred bark. Shovel (center) is 50 cm long.

andesite from Noisy Mountain, on the south slope of which the headwall scarp extends as high as 900 m above the valley floor. The 6.5-km^2 deposit is studded with numerous barren hummocks 10–20 m high (Griggs, 1922, p. 128) and a few shattered megablocks as wide as 30–80 m. Like the other landslides, large parts of this avalanche broke loose during eruptive Episode I, because the lowest unit of the pumice-fall section overlying it is the upper part of Layer B_2, even though Layer A is 30 cm thick directly adjacent to the landslide deposit. Griggs (1922) noted that, when his party camped there in 1917, rockfalls large and small continued on the scarp every few minutes, inspiring the name Noisy Mountain. It was quiet when we investigated the area in the 1990s. The deposit impounded Upper Katmai Lake, which was ~2 km across in 1917 (Griggs, 1922) but only a shallow pond ~600 m long in 1997 (fig. 124) and still shrinking as a pumice fan gradually encroaches on its eastern end (see next section, titled "Lakes Impounded").

Lakes Impounded

Windy Creek was blocked by the ignimbrite at the north end of the Buttress Range. An unpublished National Geographic Society photograph taken from the top of the Buttress Range in 1917 shows a shallow lake extending from the blockage for ~1.4 km upstream, where it was fed by a broad braided stream much like today's. It was called "Windy Lake" in the 1918 diary of Jasper Sayre and mentioned by Griggs (1918c, p. 128). The pre-1912 channel of lowermost Windy Creek was filled by the ignimbrite (fig. 87) and can be traced as a gentle swale on the tuff surface for 1.3 km northward to its former

Figure 122. Katmai River pumiceous debris-flow deposit (KRdf) resting on nearly complete Novarupta fallout section (Layers A–G). Layer H is locally preserved atop the debris flow deposit, indicating that the flow took place within days after eruptive Episode III. At base, 1.2 m of Layer A rests on bouldery alluvium and is overlain by Layers B_1 (3 cm) and B_2 (48 cm). Brown boulder-bearing next layer is a syneruptive mudflow deposit (105 cm) in which all pumice clasts are rhyolite (rhy df). This is overlain by 2.8 m of stratified fall deposits (Layers B_3 through G), capped by ~10 m of main compositionally heterogeneous debris-flow deposit, which includes two principal flow units. Geologist at center for scale. Barrier Range on skyline to east.

A

B

Figure 123. Katmai Canyon landslide deposit. A chaotic 0.08-km^3 slide mass broke loose from south toe of Mount Katmai on night of 6/7 June, after deposition of Layer A and part of Layer B. It is overlain directly by Layers B_3 through H. Headwall scarp (outlined in white) extends ~500 m above canyon floor, exposing a stack of andesite-dacite lavas with thin scoriaceous interlayers. Covering ~1.2 km^2 and as thick as 125 m, the shattered andesitic debris blocked the narrow gorge of Katmai Canyon, impounding a 5-km-long lake, which breached by 1915 (Griggs, 1922, chapters IX, X). *A,* 1917 photo courtesy of National Geographic Society; *B,* 1998 photo by authors.

confluence with the River Lethe. The former channel was never reoccupied, as Windy Creek cut a gorge ~250 m long through the blockage and established a new course around the western margin of the ignimbrite, ~1 km west of its original one. The ignimbrite is only ~10 m thick and wholly nonwelded in this area, and it is presumed that the outlet gorge was cut to lake level and Windy Lake was drained within a few years after 1918, but the actual date is unknown.

Fissure Lake.—Named by Griggs (1922, p. 212), this feature extended ~600 m east from the site of present-day West Mageik Lake (fig. 92), which was then still occupied by the snout of West Mageik Glacier (fig. 84). The lake was as wide as 100 m, and in 1917 it was filled with "clear green water." Despite being fed by meltwater from the nearby glacier, the lake was warm in 1917–1919, owing to its steep walls of steaming ignimbrite. The broad "fissure" had been excavated as a chain of coalescent phreatic explosion craters along the hingeline (fig. 92) of the compactional bench marginal to the Lethe basin (see "High sand mark" in chapter 6). Accordingly, its south rim is at least 10 m higher than its north rim, and both sides are strewn with multicolored tuff blocks that had been sintered or welded and fumarolically altered before phreatic ejection. June snowbanks and the glacial outlet stream buried by the ignimbrite may have promoted exceptionally vigorous fumarolic and explosive phreatic activity here. It was noted by Griggs (1922, p. 244) that a delta then building out into Fissure Lake from the ice front would lead to its early extinction. Indeed, the lake silted up by 1923, and today its 10 m of ash-rich lacustrine and alluvial fill (fig. 94) have been incised by a stream that flows back into the modern lake (fig. 92), which was not fully formed by glacial recession until the 1950s.

East Mageik Lake and West Mageik Lake.—These lakes filled the bowls left by recession of the termini of glaciers at the north toe of Mount Mageik (see "Emplacement over and against ice" in chapter 6). When the earliest set of aerial photographs was taken in 1951, the eastern lake had fully formed but the western one was still a pair of ponds not yet coalesced into the present-day lake. The lakes are now, respectively, 400 m and 700 m across, and their color varies from blue-green to mud-brown, depending on the influx of silt-laden seasonal runoff from the nearby glaciers (figs. 91, 92).

Griggs Fork Lake.—This feature may have been the largest of the new or temporary lakes in the VTTS. Its deposits are 10–15 m thick and crop out as steep bluffs along the Griggs fork for ~1.5 km, from north of the snout of Glacier 4 nearly to the confluence of the forks of Knife Creek. The lower half of the lake-sediment section (fig. 125) consists

Figure 124. Hummocky avalanche deposit from slopes of Noisy Mountain (out of view to right) that blocked upper Katmai Canyon and impounded upper Katmai Lake (foreground). View westward down canyon, 6 km to narrows where Katmai Canyon landslide deposit blocked gorge. Both these avalanches took place during eruptive Episode I; their deposits overlie Layer A and part of Layer B, and both are overlain by part of Layer B and Layers C-H. Barrier Range at left; Mount Katmai lavas at right; snowclad Mount Mageik on central skyline.

of plane beds of gray to tan massive mud, poorly sorted and 50–200 cm thick, separated by thinner layers of pumice-bearing sand (1–10 cm) or gray silt (1–3 cm). The upper half is thin bedded to laminated mud with pumice-rich lenses and layers of alluvium and mudflow deposits. We envisage shallow still-water ponds and seasonal mudflats that intermittently received ash-rich mudflows and pumice-rich slurries, some of which slightly scoured their muddy substrate. Sparse pumice blocks and coherent clusters of pumice lapilli (probably frozen when deposited) may have been emplaced as dropstones from floating ice or from mudflows or snowbank slurries. Water depth probably fluctuated seasonally from several meters to empty or frozen.

Along this reach of the Griggs fork, which is now a gravelly braided stream ~100 m wide, the stratigraphically lowest exposures are of dark gray ignimbrite, sintered to partially welded, which is widely disrupted, faulted, tilted, or shattered into a chaos of blocks. Because the disrupted welded tuff is separated from the snout of Glacier 4 by terraces of undisturbed deposits, it seems clear that the disruption was caused by phreatic explosions, not by advance of the ice front. Surge-bedded phreatic deposits are indeed locally exposed between the ignimbrite and the overlying lake sediments. We think the trough along which the sediments were deposited was a combined product of axial compaction of the welded tuff, proglacial high-energy fluvial erosion of the overlying dacite fallout section, and blockage of the downstream end by a thick pile of phreatically ejected rubble.

The lake deposits variously rest directly on the sintered ignimbrite (Package 8) or upon overlying remnants of cross-bedded stream gravels, surge-bedded phreatic deposits, or the stratified dacite pumice falls of Episodes II and III. At the upstream end of the lake-basin deposits, there are more outcrops of disrupted welded tuff, overlain by interstratified deposits of lake mud, stream alluvium, and pumiceous mudflows, into which had calved rotated blocks (several meters across) of the layered dacite fallout. The lake deposits are overlain by 5–20 m of later, post-Episode III, explosive phreatic deposits that issued from many nearby craters and by 1–5 m of wavy-bedded terrace gravels. Parts of the sedimentary pile are fumarolically altered, particularly in their more permeable pumice-bearing lenses, attesting to enduring fluid flux from the hot underlying ignimbrite. The lacustrine section, ignimbrite, and varied deposits between them (but

Figure 125. Ash-dominant lacustrine and mudflat sediments deposited during first year or two after eruption and exposed along Griggs fork of Knife Creek opposite snout of Glacier 4. Exposure is 10 m thick and also includes pumice-rich sandy alluvium and mudflow deposits. Top 1–2 m is alluvial terrace deposit. Ephemeral or seasonal lakes were impounded by phreatic explosion deposits that blocked drainage just downstream near confluence of forks. Abrupt breakout transported debris flows and trains of welded-tuff blocks to lower VTTS before incision of deep gorges into midvalley ignimbrite.

not the overlying terrace gravels and upper phreatic deposits) are cut by several small faults, probably the result of welding deformation and/or adjustment to nearby phreatic cratering.

The Griggs fork lake was the apparent source of an outburst flood that scoured the central VTTS and laid down ashy mudflow deposits all the way to the Ukak River terminus (Hildreth, 1983). Distal deposits are typically 1–8 m thick, poorly sorted, and massive; although the thickest Ukak sections do consist of a few conformable flow units, emplacement appears to have largely been a continuous event. The mudflows overlie both the ignimbrite and Episode II–III fall deposits; they poured beyond the ignimbrite terminus, engulfed upright uncharred trees, and feathered out into the brush (fig. 126). The deposits are pumice-rich diamict with pumice blocks as big as 35 cm, diorite and andesite cobbles up to 15 cm, and abundant fragments of Jurassic shale, all in a muddy matrix of crystal-vitric ash. Also conspicuous in the deposit are blocks of dark-gray sintered or welded tuff, material exposed only in the upper VTTS and widely disrupted and phreatically shattered along the Griggs fork. Energetic runoff might also at times have issued from the glacier-fed streams and the Juhle fork, but the only location satisfactory as a source of the pre-1912 clasts, the welded-tuff blocks, and the abruptly released volume of water required by high-energy transport for 20 km to the Ukak terminus is the Griggs fork lake. The River Lethe arm is not nearly as widely scoured of dacite fall deposits as parts of upper Knife Creek and the broad Knife Creek floodplain of the central VTTS, where both the Episode II-III fallout and the top meter or two of the sintered Episode I ignimbrite were stripped. The right margin of the flood path can be tracked for kilometers along the central and northeastern VTTS as a 10-m-wide mudflow veneer strewn with blocks of welded tuff (fig. 127). Gorges on several forks of Knife Creek were already more than 20 m deep in the central VTTS by 1917 and would have intercepted and confined the flood flow had it not taken place earlier. Moreover, a 1917 photograph of the Griggs fork shows it much as it is today (fig. 128). Rapid sedimentation along this short-lived proglacial trough is not surprising because the rugged landscape was everywhere mantled with unstable unconsolidated eruption products. Breakout and drainage of the lake took place in one of the first summers—1912 to 1916.

Katmai Canyon Lake.—This ephemeral feature was impounded behind the Katmai River landslide deposit, which blocked the narrow gorge at the south base of Mount Katmai (fig. 123). The 0.08-km^3 slide mass consists of shattered lavas from the Katmai edifice, covered ~1.2 km^2, and was as thick as 125 m; its headwall scarp extends ~500 m above the canyon floor. Although a complete pumice-fall section (Layers A–H) is intact alongside it, the slide mass is overlain by only part of Layer B, all of Layers C–H, and minor secondary slide debris; clearly therefore it was emplaced during the seismically

Figure 126. Upright uncharred trees buried by ash-dominated debris-flow deposits as thick as 6 m, adjacent to Ukak terminus of ignimbrite, 20 km downvalley from Griggs fork—the inferred source of the flows (see fig. 125).

vigorous eruptive Episode I. Canyon blockage led to growth of a 4-km^2 lake, ~5 km long, as wide as 800 m, and as deep as 70 m, which was never witnessed but was inferred from field evidence by Griggs (1922, chapter IX). The strandline illustrated by Griggs (1922, p. 120) stood ~64 m above a small lake remnant in 1917 (called "Lower Katmai Lake" by Griggs). The lake remnant has since vanished, the strandline has been obscured by scree, and scant trace of any lake sediment remains. When the landslide dam breached, the outburst and subsequent fluvial erosion cut a steep-walled trench through the blockage (Griggs, 1922, p. 116), removing about a third of the deposit (fig. 123). Nonetheless, north-bank terrace remnants ~1 km upstream of the blockage preserve ~15 m of plane-bedded sand-dominant strata with intercalated pumice lenses and layers of laminated mud. Foreset bedding is locally exposed, but most of the strata are horizontal (unlike alluvium of the steep present-day streams adjacent), supporting their interpretation as lakeshore deposits. Their elevation is ~60 m higher than the present-day floor of the breach, thus acceptably consistent with Griggs' estimate of lake depth.

When Griggs (1922, chapters VIII, IX) first landed at Katmai village in June 1915, he was so impressed with the alluvial devastation that he assumed the outbreak flood had just occurred. He illustrated swamped barabaras, bent-over brush, trees uprooted or snapped off, and the sandy lowland at the village site covered by a sheet of wet mud all the way to the toe of the adjacent mountain. The village chapel had floated loose from its foundation and was marked by a high-water mark 3 m above its floor (Griggs, 1922, p. 104). The river-mouth estuary, which had extended 3–4 km inland to the village dock and up which Martin (1913) had sailed in a 35-foot schooner as late as August 1912, had by 1915 completely silted up. The 8-km-wide floodplain of Katmai River was a sheet of ashy mud and pumice with wide expanses of shallow water and quicksand, rather than the braided plain of coarse gravel restored today. Griggs reported flood runups of 3–8 m along the floodplain margins, a bedrock ridge athwart the lower-canyon flood path that had been stripped of its soil and pumice mantle, and a train of huge boulders extending 3–5 km out from the canyon mouth. His evidence was convincing for recent passage of a great flood, and his inference of its occurrence in 1915 (three years after the blockage) is fairly persuasive. Of course, aggradation of the Katmai River floodplain continues today, a century after the eruption choked the whole drainage system with pumice. Comparable aggradation of tributary Soluka Creek, where there was no flood, suggests that some of the sedimentational devastation illustrated by Griggs was secular.

Upper Katmai Lake.—This lake survives today as a shallow pond ~600 m long, but Griggs (1922) mapped it as being ~2 km across in 1917. Along Katmai River canyon, between Noisy Mountain and the Barrier Range, the lake lies 6 km upstream from the slide mass that impounded the lower lake and only 2 km downstream from the snout of Princess Glacier (Hildreth and Fierstein, 2003). The lake is impounded by a 6.5-km^2 expanse of avalanche debris studded with numerous barren hummocks 10–20 m high (Griggs, 1922, p. 128) and a few shattered megablocks as high as 30–80 m. The avalanche rubble is andesite of Noisy Mountain, on the south slope of which the headwall scarp extends as high as 900 m above the valley floor. Large parts of the avalanche cut loose during eruptive Episode I, because the lowest unit of the pumice-fall

Figure 127. Alignment of meter-scale blocks of welded tuff along northeast side of VTTS, interpreted as margin of flood flow across ignimbrite surface prior to incision of gorges that would have intercepted them. Inferred source was ephemeral lake impounded along Griggs fork 12 km upvalley (see figs. 125 and 126). Gray-brown gravel veneer on surface of ignimbrite is lag deposit left by deflation of debris-flow deposits and of nonwelded top of ignimbrite by perennial Katmai winds. Backpack at lower center provides scale.

Figure 128. Griggs fork of Knife Creek. Views eastward from north-bank rim at lower end of Griggs fork up its 3-km-long braided floodplain: *A,* 1919 and *B,* 1982. At head of valley in center distance, Glacier 5 has advanced 1,300 m. Glacier 4 at right has advanced 650 m. Both glaciers remain heavily tephra-covered a century after 1912 eruption. Cutbanks consist of ashy fines-rich ponded sediments overlain by terrace gravels and phreatic surge beds. Floodplain has widened modestly at expense of south-bank terrace (right, middle distance). Foreground terrace at right served as airstrip and Katmai Project campsite in 1953–54, occupied by Curtis, Juhle, Williams, Kistler, and others. 1919 photo courtesy of National Geographic Society; 1982 photo by authors.

section overlying it is the upper part of Layer B_2, even though Layer A is 30 cm thick directly adjacent to the avalanche deposit. Griggs (1922) noted that, when his party camped there in 1917, rockfalls large and small continued on the scarp every few minutes, inspiring the name Noisy Mountain. The lake had no surface outlet when we examined it in 1997, all drainage being through the permeable subsurface of the blocky avalanche (fig. 124). Lake area was shrinking, owing to encroachment of a pumice-rich alluvial plain at its eastern end. The only strandline observed was ~1 m above the August lake level, presumably representing seasonal fluctuation.

"Horseshoe Pond."—This feature, illustrated and so named by Griggs (1922, p. 142), is a shallow lake ~500 m long that was impounded on the south side of Martin Creek by the Mageik Landslide, which (like several other mass movements) was emplaced during eruptive Episode I in 1912 (Griggs, 1922, chapter XI; Hildreth and Fierstein, 2003). The lake has changed little in size or configuration since studied by Griggs in 1917. What he assigned to the Mageik Landslide, however, includes two older avalanche deposits as well as that of 1912 (Hildreth and others, 2000; also see section above titled "Syneruptive Landslides").

Two temporary lakes also formed along the braided plain of the Ukak River (fig. 6), as documented by unpublished National Geographic Society photographs taken in 1917. One was ~5 km long, only a few hundred meters wide, and headed ~4 km downstream from the ignimbrite terminus. The other, 6–8 km farther downstream, was more equant but larger in area, flooding the brushy marshland just above the mutual delta of the Ukak and Savonoski Rivers. Summer discharge down the Ukak in 1917–19 was less than a tenth of what it is today, owing to boiling away of its principal inputs (Allen and Zies, 1923), and it was probably nearly dry in winter. Pileup of sandy pumice-rich alluvium along the braided courses of both rivers (and lower Ikagluik Creek), temporarily severely overloaded with eruption products, are presumed to have caused the transient impoundments.

Phreatic Craters and Deposits

Phreatic explosions within the VTTS ignimbrite produced numerous craters surrounded by tuff rings and aprons of poorly sorted, cross-bedded, remobilized pyroclastic debris (Hildreth, 1983). Many craters are circular, most are 20–60 m across, but some are as wide as 150 m. Crater depths were 3–20 m in 1919 (Allen and Zies, 1923, p. 87, 108), but most have since been partly filled with sediment (fig. 129). About 60 craters are identifiable along a 6-km-long swath adjacent to upper Knife Creek, and conspicuous clusters of craters are also present near the snout of Glacier 3 and along the compactional bench at the north base of Mount Mageik (Hildreth and Fierstein, 2003). A few small craters also formed between the Buttress Range and the River Lethe, opposite Baked Mountain. Most craters are in areas where some level of the ignimbrite was welded or otherwise well indurated, suggesting the importance of an impermeable seal adequate to permit buildup of steam overpressure. Although some phreatic explosions took place at glacier fronts during or immediately after ignimbrite emplacement against them, most crater formation occurred after the dacite pumice falls of Episodes II and III, indicating delays of at least two days before steam blasted through the ignimbrite. In each of the three principal areas of craters, welded tuff had buried streams and snowbanks.

Downvalley, though the tuff is nonwelded, craters formed in a few places where defeated streams poured against and over the hot tuff, inducing explosions. Resulting deposits are well preserved along the Juhle fork, along the southeast edge of the distal Ukak lobe, and where Windy Creek enters the ignimbrite. The diamictic explosion deposits at Windy Creek are 1–4 m thick, extend as high as 45 m above the ignimbrite surface on both banks, contain sparse stream cobbles, and consist predominantly of remobilized ignimbrite material.

Figure 129. One of ~60 phreatic craters rooted in ignimbrite and preserved along upper Knife Creek arm of VTTS (Hildreth and Fierstein, 2003). Craters range from 7 m to 30 m in diameter; this example is ~20 m wide, ~5 m deep, and holds ephemeral ponds that have cut encircling rings. Ejecta aprons of such craters are 1–20 m thick. View is southwestward across main fork of Knife Creek toward base of Broken Mountain, where abrupt change from brown to pale gray marks upper limit of ignimbrite, 30–50 m above valley floor; dark brown slope is equivalent to "high sand mark" around Lethe arm (see figs. 72, 82).

The principal phreatic explosion deposits in the upper VTTS are generally 2–5 m thick but 15–25 m thick near snouts of Glaciers 3 and 4 and around the confluence of the Griggs fork with Knife Creek. These deposits consist of poorly to moderately sorted thin beds and laminae that are alternately enriched in crystals or in pumice lapilli, both kinds having a crystal-vitric matrix of muddy ash. Parallel bedding is common, but dunes, antidunes, festooned cross-sets, pinch-and-swell bedding, and extensively rippled bedding are also present. Truncated bedsets provide evidence of repeated outbursts and mutal interference of deposits from adjacent craters (fig. 130).

The deposits contain abundant blocks of sintered to partially welded tuff, some of which are denser than most ignimbrite surface exposures. Such blocks are commonly 0.5 m across, and some are as large as 3 m (see photo on p. 195 of Curtis, 1968). A few show impact sags and thin zones of thermal alteration of the enclosing diamict. Most of the phreatic debris is derived from rhyolite-poor packages that were emplaced late in the ignimbrite sequence, but a few craters near Glacier 4 and along the Juhle fork expelled material that is as much as 80 percent rhyolite. Although these few suggest a somewhat deeper (or valley marginal) source, most of the outbursts ejected material from upper and intermediate levels of the ignimbrite. Nonetheless, the low-permeability welded zones that favored buildup of steam pressure (and, by ultimately failing, initiated the blasts) were generally at deeper levels than the bulk of the material ejected.

Two linear troughs were excavated from indurated ignimbrite, which was shattered and heaved by explosive formation of chains of coalescent craters, along what became Fissure Lake and along the Griggs fork of Knife Creek. These were not ice-contact features but formed, instead, hundreds of meters away from the glacier fronts, probably above outlet streams buried by the ignimbrite. Along the Griggs fork, surge-bedded deposits on both banks of the present-day braided stream are as thick as 20–25 m and dip oppositely away from the stream. At the site of Fissure Lake (now drained) and around West Mageik Lake, phreatic deposits as thick as 5 m dip away from the trough and are studded with abundant blocks of tuff, commonly as big as 1 m (fig. 93). Close to the present-day lake, however, which was occupied by the glacier front in 1912, ash-dominated proglacial pond sediments contain radially columnar blocks of indurated ignimbrite. The local evidence there suggests that blocks of hot ignimbrite calved into a meltwater-filled gap that had opened quickly between ice and ignimbrite, probably explosively.

Figure 130. Surge-bedded deposits from phreatic explosion craters near confluence of main fork and Griggs fork of Knife Creek. USGS geologist Dave Johnston stands atop primary nonwelded ignimbrite, which was source of the phreatic ejecta. Numerous craters each produced bedsets that mutually truncate and overlap each other. Johnston was swept away by the 18 May 1980 eruption of Mount St. Helens, nine months after our joint work in the VTTS.

Chapter 11

Posteruptive Evolution of the Landscape

The principal geomorphic impacts abruptly introduced by the 1912 eruption were (1) inundation of Knife Creek, the River Lethe, and upper Mageik Creek by ignimbrite; (2) mantling of the rugged landscape with fall deposits—pumice-rich tephra out to 10–15 km but crystal-vitric ash beyond; (3) caldera collapse at Mount Katmai; and (4) fluvial aggradation of the Ukak, Ikagluik, Rainbow, Savonoski, and Katmai River systems.

Only a tiny fraction of the 1912 ignimbrite transformed or was subsequently remobilized downstream into pumiceous debris flows. The Ukak River and Mageik Creek ignimbrite termini were certainly so affected, but terrace remnants of laharic deposits farther downstream are rare and small. We did find terraces with modest hyperconcentrated streamflow deposits along Rainbow River (northeast of Mount Katmai; see fig. 6), but these were derived from crystal-rich fallout, not ignimbrite. Debris flows transported from the Griggs fork and Katmai Canyon were mentioned in chapter 10, but these were abrupt events caused by lake breakouts.

The contrast with the fluvial response to the 1991 eruption of Pinatubo, which emplaced a volume of ignimbrite comparable to that of 1912, could not be more striking. Within months, the radial ignimbrite sheets at Pinatubo were intricately dissected, and within three years, a third of the ignimbrite had been swept away as lahars (Punongbayan and others, 1996; Newhall and Punongbayan, 1996). Typhoons and monsoons were important intensifiers at tropical Pinatubo, but the dearth of lahars at Katmai was a complex result of several factors: (1) Summer drizzle is frequent but downpours uncommon, and streams are frozen for half the year. (2) Despite the glacier-fed streams, more than 90 percent of streamflow over and into the ignimbrite boiled away for the first decade. (3) Upvalley the ignimbrite is welded and poorly permeable, and it was widely covered by several meters of coarse permeable pumice-fall deposits. (4) Differential compaction and welding, proportional to ignimbrite thickness, favored immediate establishment of axial trunk streams in each branch of the VTTS. These are joined intermittently by sheetwash but by only a few (mostly shallow) tributaries across the welded surface. Characteristics of the 1912 deposits and the Katmai climate thus combined to retard intricate dissection and extensive erosion of the VTTS ignimbrite. The few major drainages had been incised into its surface as narrow gorges before 1917 when the valley was first explored, and they have changed rather little ever since.

Rapid Incision of Gorges

Because three branches each of Knife Creek and the River Lethe are fed by glaciers at the head of the VTTS, streams rapidly began to establish new courses across the valley-filling ignimbrite, and the trunk streams were forced down the middle of each arm of the upper valley by the differential axial welding-compaction. When the first real exploration of the VTTS was conducted in 1917, Griggs (1918c, p. 118; 1918f, p. 132) photographed mid-valley gorges that he said were already "nearly 100 feet [30 m] deep," though those illustrated were only ~20 m deep. In 1919, the outlet channel west of Novarupta basin was at least 7 m deep northwest of Falling Mountain (Griggs, 1922, photo facing p. 240), and it is no deeper today. Also in 1919, gorges on the forks of Knife Creek in the central VTTS were said by Griggs (1922) to be "100 feet deep" in places, though the gorge he illustrated (p. 254) was again closer to 20 m deep. He noted that canyons did not cut through the tuff to the underlying substrate anywhere except in the lower valley. Fenner (1923) repeated 100-foot (30 m) estimates for some of the gorges he witnessed in 1919, but those he illustrated (p. 32, 34) were likewise no more than ~20 m deep.

The stratified fall deposits of Episodes II and III had been eroded away along the main stream channels by 1917, and incision of the underlying ignimbrite was generally only a few meters in the upper valley but widely ~20 m in the central valley. There have been no quantitative measurements of erosion rates, but there is no doubt that erosion slowed greatly after the first few years. A century after the eruption, streams in the upper River Lethe and Knife Creek arms of the valley have narrow, steep-walled gorges still only 2–8 m deep (figs. 35, 89) into sintered or welded ignimbrite. In the lower valley where the tuff is nonwelded and 10–20 m thick, the base is widely exposed (figs. 31, 68), but for only 2 km upstream from Three Forks. Today the gorges are deepest in the central VTTS, where the tuff grades from sintered to nonwelded (fig. 131). The greatest depths we measured (by rope) are 30–35 m, along the gorges of the main and Juhle forks of Knife Creek, 1–2.5 km above their confluence.

In the narrowest reach of the central VTTS there are four principal gorges, ultimately convergent, that are sheer, of similar depth, and floored within the tuff—from west to east, the River Lethe and the mid-valley, main, and Juhle forks of Knife Creek (fig. 74). The Juhle fork drains a glacier on Mount Griggs, the River Lethe drains glaciers on Mount Mageik, and the main fork integrates drainages from all five Knife Creek Glaciers. The mid-valley fork is now spring-fed where its gorge heads abruptly in the center of the valley; it had been for half a century the principal channel of Knife Creek before capture of its headwaters by what is now the main fork, sometime between 1951 and 1976. In the lower VTTS, there are shorter gorges, 10–20 m deep, along Moraine Creek and two minor tributaries of Knife Creek (see map of Hildreth and Fierstein, 2003).

Alluvial Plains, Channels, and Veneers

Extensive braided plains of alluvial gravel had been laid down by the Ukak and Savonoski Rivers and by Katmai River and its tributaries long before 1912, but the abrupt introduction of eruption products caused severe aggradation for years afterward. Aggradation of the broad Katmai River floodplain (fig. 132) and conversion of the former Katmai Bay to braided alluvium and mudflats were extensively illustrated by Griggs (1917, 1918f, 1922). Comparison of the 1908 U.S. Coast and Geodetic Survey Chart No. 8555 (Fenner, 1923) with

Figure 131. View northwest down gorge of River Lethe, ~25 m deep, in central VTTS, ~4.5 km upstream from Three Forks. Upper half of exposed ignimbrite is nonwelded to weakly sintered where jointed; inner gorge is sintered to partially welded, with base not exposed. Large oxidized zone of fumarolically indurated tuff on far right wall. Brown mudflow deposit overlies thin white layer of dacite fallout on right rim. Mount Katolinat in cloud in background. Geologist at lower left.

Maynard's topographic map of 1917 (Griggs, 1922) suggests that, even though the Katmai River estuary and former anchorage had been severely infilled, any general advance of the shoreline had been modest. By 1951, the coastal sandspit and sandbars had changed little since 1917, but tidal mudflats had apparently advanced ~1 km seaward (USGS 1:63,360-scale topographic map Mt. Katmai A–3). The gravel plain of the lower Ukak River likewise advanced ~1.5 km at the expense of the Iliuk Arm of Naknek Lake between the mapping of 1917 and 1951 (USGS 1:63,360-scale topographic map Mt. Katmai C–5). Morphological changes at the mouths of the Ukak and Katmai Rivers appear to have been modest since 1951, but no quantitative observations have been maintained.

In the VTTS, there were also a number of rapid posteruption alluvial responses that are interesting enough to record. A shallow pond ~400 m across formed in Novarupta basin west of the dome, impounded by the subtle east-dipping monocline north of Falling Mountain that marks the west rim of the vent structure. Still there in 1919, the pond drained due west toward the River Lethe along a shallow gorge that remains little changed today. The pond was later covered by a fan of alluvial pumice (fig. 18) that grew outward from the mouth of a gulch draining several ravines on the vent-wall scarp ~1 km southeast of Novarupta dome.

In the uppermost Lethe arm, a dark-colored composite alluvial fan (fig. 26) that extends2 km northward from the compactional bench at the head of the valley had already formed by 1919. Fed by three streams that cut through the ignimbrite bench, the confluent fans are rich in clasts of andesite-dacite lavas from Mount Mageik. The fans scoured away the stratified dacite fall deposits and rest directly on the ignimbrite. Farther down the Lethe arm, the fall deposits are widely intact on the valley floor, except close to the River Lethe itself, which has incised a gorge 2–8 m deep into the ignimbrite.

Likewise, in uppermost Knife Creek an alluvial fan in front of Glaciers 1 and 2 rapidly extended ~2 km northward atop the ignimbrite, scouring away most of the dacite fallout. Pre-1912 clasts of lava (from Trident) are present here as well, but, compared to the Lethe fans, a much larger share of the alluvial gravels here are 1912 eruption products reworked off the glaciers.

Figure 132. Braided Katmai River floodplain viewed south-southeastward from slope of Trident Volcano. Coastline of Shelikof Strait at Katmai Bay is 23 km from camera, and Kodiak Island lies an additional 55 km beyond it. At right, lower gorge of Mageik Creek opens onto floodplain to join Katmai River. Barrier Range at upper left. Ash-covered glacier tongue at lower left separates Trident lavas from brown Jurassic siltstone overlain by dark-gray lavas of Mount Katmai on far wall of gorge. Near-white Novarupta pumice-fall deposits and derivative pumiceous screes mantle countryside far and wide. Floodplain was initial route taken by Griggs and his companions in approaching Mount Katmai in 1915–16. Before severe aggradation and complex braiding of floodplain in response to 1912 eruption, it had also provided principal trail from Savonoski to Katmai village.

Downstream from the confluence of the Griggs fork, Knife Creek rapidly fashioned a shallow braided alluvial channel 150–300 m wide, floored by sintered ignimbrite and walled principally by the stratified dacite fall deposits. For ~4 km below the confluence, its general configuration has changed little since 1919, as shown by National Geographic Society photographs. Just northwest of the nose of Broken Mountain, however, Knife Creek spread out across the central VTTS into a braided fan, 1.7 km wide and 3.5 km^2 in area, that lasted more than four decades. The fan funnelled distally into the mid-valley gorge of Knife Creek and split off a minor left distributary to the River Lethe, as documented on the 1951 USGS 1:63,360-scale topographic map Mt. Katmai B-4. The fan gravels are 1–7 m thick and rest directly on indurated ignimbrite, from which the fallout had already been stripped in the central VTTS by 1919 (Allen and Zies, 1923). Sometime between 1951 and our first visit in 1976, however, the fan was abandoned and the principal course of Knife Creek was captured by what is today the main fork, parallel but farther northeast (fig. 74).

The terminal Ukak lobe of the ignimbrite below Three Forks (fig. 29) records a complex fluvial history of erosion and sedimentation. The narrow corridor of the Ukak River (fig. 30), which runs 2 km downstream from Three Forks, was rapidly stripped of its thin ignimbrite cover, exhuming a steep pre-1912 channel cut in Jurassic shale. Another pre-1912 channel along the east side of the corridor has been only partly cleared of ignimbrite by incision of present-day Moraine Creek (see map of Hildreth and Fierstein, 2003). After emplacement of mudflow deposits atop the Ukak lobe ignimbrite, the exhumed Ukak River channel spread a thin sheet of gravel across both of them. Subsequently, as the river cut deeper into the ignimbrite barrier at the lower end of the corridor, at least three sets of terrace gravels were inset before gorge incision reached the base of the 1912 deposits, establishing a drainage configuration that has now persisted for decades.

Moraine Creek, which bends northeast at the toe of the steep corridor and becomes the marginal stream along the southeast edge of the Ukak lobe, was the site of various tuff-water interactions during and after ignimbrite emplacement—phreatic explosions, gravel stringers and insets between ignimbrite flow units, and ignimbrite remobilization and drainaway. Subsequently, two right-bank tributaries of Moraine Creek, temporarily blocked by the ignimbrite before incision of Moraine Creek, distributed sheets of coarse gravel atop and against the ignimbrite. These gravels were at least in part derived from nearby glacial till deposits. Because one of the tributaries was never overrun by the ash flows and the other only thinly so, it is unlikely that breakout of temporarily impounded ponds caused remobilization of the till. The severe seismicity that continued until at least 17 June suggests an alternative mechanism.

The pre-1912 channel of lowermost Windy Creek was buried by ~10 m of nonwelded ignimbrite (fig. 87) but is still marked by a gentle swale on the tuff surface that extends ~1.3 km northward to its former confluence with the River Lethe. Windy Creek never reoccupied its former channel but instead cut a gorge ~250 m long through the blockage and established a new course around the western margin of the ignimbrite, ~1 km west of its original one. The blockage temporarily impounded a lake that lasted at least through the summer of 1918 (see "Lakes Impounded" in chapter 10).

Aeolian Deposits

A century of notorious Katmai wind has relocated and sorted millions of tons of pumice and phenocryst sand from the 1912 eruptive products. Much of the crystal-dominant ashfall in the central and lower VTTS has been stripped from the ignimbrite surface and redeposited in part as dunes and sand sheets along the northeast margin of the valley and in the hilly morainal terrain northeast of Three Forks. There the aeolian deposits variously overlie the layered dacite fallout, thin ignimbrite, 1912 soils, or glacial deposits. Behind the Three Forks moraine (figs, 29, 64), longitudinal dunes 6–8 m high, now stabilized by willow growth, are separated by gravel-armored chutes with veneers of rippled sand. Some of the sand deposits expose as many as 70 pairs of coarse/medium sand laminae. One sand sheet is gradually encroaching on a swamp impounded by the ignimbrite. Aeolian sand sheets locally climb up and over kames and irregular hills of till, around which the sluggish ignimbrite had merely banked.

Around the site of Katmai village, a pre-1912 dune field has been augmented by as much as 3 m of post-1912 sand that overlies complete fallout sections (Layers A-H) as thick as 23 cm (fig. 21). Much of the sand spit that shelters the former estuary and anchorage (now mudflats), however, predates the eruption.

The pumice-fall deposits have been extensively stripped from ridgecrests by the region's severe and frequent winds, which are intensified by the arterial role of Katmai Pass, one of the lowest saddles along the volcanic axis. Wedges of reworked pumice have piled up below the walls enclosing the upper and central VTTS, as well as in the downwind fallout sector to the east and southeast—along Ikagluik Creek, Katmai River, and Soluka Creek. Nearly every gully that descends to the VTTS from the Buttress Range, Mount Griggs, and the valley's northeast wall has built a fan of pumiceous alluvium (Hildreth and Fierstein, 2003). From the fans, pumice then is frequently remobilized as slurries down the local drainages, so that floating clasts can still today be distributed tens of kilometers down the Naknek Lake system.

The central VTTS is a wind-abraded desert marked by sandy windrows, sculpted pumice-block ventifacts, and etched surfaces of sintered ignimbrite. Pioneering vegetation cannot survive the sandblast and is able to establish a toehold only in the most sheltered gullies. Despite a century of wind scour, a few dry summer days will loosen ashy substrates sufficiently to permit intense duststorms in and near the VTTS. Windstorms still occasionally loft white vitric ash from the barren surface to elevations of many kilometers. Such high-altitude plumes of remobilized 1912 dust have been the cause of several spurious eruption reports (Hildreth and others, 2000), and such misleading reports still recur almost annually.

Behavior of Ash-Covered Glaciers

We earlier discussed some effects of the pumice-fall mantle on the Knife Creek Glaciers and the glaciers on the north slope of Mount Mageik, as well as local evidence for ignimbrite interactions with the glacial termini (see "Emplacement over and against ice" in chapter 6). Here we summarize more broadly the behavior of glaciers affected by the 1912 eruption.

Mount Mageik has a substantial icecap that extends radially into nine glaciers, of which two terminated in 1912 on the floor of the Lethe arm at elevations lower than 2,100 ft (640 m). Termini of all nine have retreated 500–1,200 m (in plan view) since 1951 and a few as far as 1,700 m since 1912 (Hildreth and others, 2000). The generality of 20th century recession here is no doubt dominated by climatic control of the ice budget and not significantly influenced by the mantle of 1912 fallout. The fall deposits thinned across Mount Mageik from ~50 cm on its northeast to ~5 cm on its southwest, and nearly all of it was removed from the glaciers within a few decades.

Mount Griggs, highest peak in the Katmai region, has six active glaciers, which are thin, narrow, and distributed radially around its summit. Elevations of their termini during the 20th century were in the range 3,000–3,800 ft (915-1,160 m), except for the one south-facing glacier (5,000 ft; 1,525 m). All were thickly mantled with 1912 fall deposits, but all were too high to be reached by ash flows. Today only the lower third or so of each glacier is still covered with tephra, and this has generally been disrupted chaotically by ice movements. The upper reach of each glacier is steep, active, crevassed, and now free of its 1912 pumice cover. All six may have thinned a bit, but comparison of 1951 and 1987 aerial photographs shows that the terminal positions of three had retreated, one had advanced, and two had changed little. The lack of systematic glacial behaviour on Mount Griggs is likely to reflect a combination of climatic and edifice effects, with little influence attributable to the fallout (Hildreth and others, 2002).

Snowy Mountain, 25–32 km east of Novarupta, is blanketed by an icefield from which a dozen glaciers extend radially. Comparison of aerial photographs taken in 1951 and 1984 shows that during that 33-year interval, termini of nine glaciers retreated, two were stationary, and the southwesternmost advanced ~150 m (Hildreth and others, 2001). The blanket of crystal- and pumice-rich 1912 fallout had a cumulative thickness of 1–2 m on the Snowy Mountain edifice, thinning to the north and east but thickening southwest to more than 3 m at the head of Katmai River. There, the southwestern glacier from Snowy Mountain (called Princess Glacier by Griggs, 1922, p. 131) is a major source of Katmai River, as is nearby Noisy Mountain Glacier, and both advanced ~150 m between 1951 and 1984. Ice motion and surface erosion stripped the tephra from most of the Snowy Mountain glaciers decades ago, but those to the southwest and west still retain thick fallout blankets, especially atop their distal reaches. The ash-free glaciers to the east have all retreated, while the ash-covered ones to the west and southwest of Snowy Mountain have advanced or stagnated (Hildreth and others, 2001). Where thick enough and enduring for decades, the pale gray to tan tephra mantle evidently retards ablation.

The Trident group, closest to Novarupta, received a thicker blanket of 1912 fallout than any other edifice in the Katmai cluster. In addition, Knife Creek Glaciers 1, 2, and 3 on the north side of Trident (fig. 133) were covered distally by thin ignimbrite, and the all-rhyolite ash flows (Package 1) ran all the way up Glacier 1 to feather out in the saddles between Trident summits. Glaciers 1 and 2, which each issue from multiple cirques on the north face of Trident, have been largely cleared of 1912 ejecta along steep upper reaches ~800 m long, by a combination of flowage and new ice accumulation. Medial and distal parts of both remain covered by as much as 12 m of 1912 material, supplemented locally by more recent avalanche rubble (Hildreth and Fierstein, 2003). Comparison of 1951 and 1987 aerial photographs shows that Glacier 1 advanced 250 m and Glacier 2 at least 300 m during that interval (Hildreth and others, 2003a). Along with nearby Glaciers 3 and 4, which head on Mount Katmai, both have thus overrun the edge of the 1912 ignimbrite sheet in the upper Knife Creek arm. In contrast, glaciers on the south side of Trident, where supraglacial fall deposits are 4–5 m thick, have thinned and wasted, though changes in terminal positions have been inconsistent (Hildreth and others, 2003a).

Behavior of glaciers on Mount Katmai involves the complication that some (not all) were partly beheaded by caldera collapse (fig. 134). Nonetheless, all of the dozen radial extracaldera glaciers around Mount Katmai were blanketed with several meters of Novarupta fall deposits. Some aspects of Knife Creek Glaciers 3, 4, and 5 were discussed earlier in the context of their interactions with 1912 ignimbrite (see "Emplacement over and against ice" in chapter 6). Glacier 3, although still partly fed from Peak 6128, lost at least half its catchment area to collapse of the summit. The glacier has since thinned and disrupted its thick ejecta blanket, but its upper reach remains active and crevassed; although its lower half looks hummocky and stagnant, its terminus actually advanced ~225 m between 1951 and 1987. Glacier 4 is fed from the extensive North Katmai Icefield, which was partly beheaded but remains robust, not shrinking. The vigorously lobate terminus of Glacier 4 advanced ~150 m between 1951 and 1987, and a National Geographic Society photograph taken in 1919 (fig. 128) shows that Glacier 4 had already advanced 500 m before 1951, for a total of 650 m since 1919. Knife Creek Glacier 5 is supplied from an isolated cirque 2.5 km northeast of the caldera rim and was thus not affected by caldera formation. Between 1919 and 1951, its terminus advanced ~1,300 m, but since 1951 its distal 700-m reach has thinned and stagnated, although its terminal position has changed little.

Figure 133. Ash-mantled snout of Knife Creek Glacier 1, 30–40 m high. Pyroclastic density currents crossed this glacier in 1912, and Novarupta fall deposits draped it; both have been attenuated and disarranged in a century of ice movement. At lower right, intact pale-gray dacite fall deposits many meters thick are capped by fine-grained Layer H and by a veneer of white pumiceous alluvium but are otherwise undisturbed.

Figure 134. Katmai caldera, 4 km across at rim, viewed south-southwestward from helicopter in 1999. Several extracaldera radial glaciers persist in spite of loss of mountain's summit in 1912. Icefield at upper right mantles slope of twin peaks 6128 and 6200, and since 1920s it has extended a narrow intracaldera tongue down west wall. Slump-block bench (at center) on southwest wall progressively accumulated an intracaldera glacier that now reaches lake level. South-rim rhyodacite lavas form crest and crag above that glacier. Thick dacite lava flow caps north-ridge cleaver in right foreground.

Two substantial glaciers west and south of the caldera exhibit inconsistent behavior. The Wishbone Glacier, fed from the Katmai-Trident saddle, divides into two tongues that extend several kilometers southward into narrow gorges. The southeast tongue advanced 200 m between 1951 and 1989, while the southwest tongue thinned and receded slightly, although it has a small western lateral lobe that advanced 110 m (Hildreth and others, 2003a). Due south of the caldera rim, Metrokin Glacier was partly beheaded but continues being supplied by the west-rim ice shelf below Peaks 6128 and 6200 (Hildreth and Fierstein, 2003). Its canyon-confined terminus retreated ~600 m between 1951 and 1989 and another 400 m between 1989 and 2001. This is the glacier that shed the Katmai River pumiceous debris flow (fig. 120), so its insulating blanket of fallout was substantially less than on other glaciers of Mount Katmai.

The steep southeast slope of Mount Katmai supports three isolated glaciers, all beheaded by collapse and today respectively 1.8, 2.1, and 1.2 km long (Hildreth and Fierstein, 2003). The northern one fills the valley below the east notch in the caldera rim. Termini of all three have retreated more than 1 km since 1951.

The East Katmai Icefield (Hildreth and Fierstein, 2003) is largely fed by three ice tongues on the steep east slope of the postcaldera summit of Mount Katmai (Peak 6715). Only its upper parts have been cleared of 1912 ejecta, but the whole icefield is actively crevassed. It contributes northward to Ikagluik Glacier, the terminus of which has moved little since 1951, and contributes southward to Noisy Mountain Glacier, which has advanced ~150 m since 1951.

The North Katmai Icefield (Hildreth and Fierstein, 2003) is also robustly active, feeding icefalls that descend toward Glacier 4 and toward Ikagluik Creek. The icefield is partly supplied by a series of small cirque glaciers on the steep north rim of the caldera. Although these were deprived of contributions from higher on the foundered edifice, each remains actively crevassed and shrinkage is limited to notches on the caldera rim.

Katmai Caldera Lake and Intracaldera Glacier Development

Caldera collapse left a sheer jagged rim, which was initially capped by ice cliffs of the beheaded glaciers in four sectors, 90°–150°, 190°–260°, 280°–350°, and 0°–10°, thus fringing ~60 percent of the caldera's upper margin (Griggs, 1922; photos on p. 175, 178). Because much of the ice had been fed from higher on the vanished edifice, the ice cliffs wasted over a period of decades, thinning and receding 50–800 m from the rim and leaving rubbly deposits of ablation debris—largely consisting of what had been 1912 supraglacial ejecta. On the southwest side, however, Peak 6200, which stands 275 m higher than the rim, has continued to sustain an ice shelf (fig. 134). Much of the shelf has receded from the rim, but a narrow ice tongue has extended from its north end into a rim scallop and partway down the west wall toward the lake (figs. 103, 134).

Complex collapse left kilometer-wide slump blocks that rose to inward-sloping benches 300–400 m above the caldera floor at its north and southwest margins. The benches had accumulated snow patches by 1917 (Griggs, 1922, p. 172) and still only modest snowfields by 1923 (figs. 12, 17 of Fenner, 1930). By Hubbard's first trip to the rim (1929), the snowfields had turned to ice and he reported a bergschrund (Hubbard, 1932), but the earliest record of extensively crevassed glaciers on the benches is a set of 1951 aerial photographs. As observed in 1953–54 by Muller and Coulter (1957a), the southwest intracaldera glacier terminated in ice cliffs 50–80 m above the lakeshore, the north glacier mantled only the upper half of its bench, and the narrow ice tongue on the west wall reached halfway down to the lake. In 1976, the western edge of the southwest glacier had barely reached the lake, but later it wasted back. Although all three intracaldera glaciers have in the past calved icebergs into the lake, a set of 1987 aerial photographs shows all of them terminating atop steep rocky slopes just shy of entering the lake. By 1999, however, small tongues of both slump-block glaciers had reached the lakeshore (figs. 134, 135). The steep western ice tongue had surged in 1976 and did so again in 2001, twice temporarily reaching the lakeshore but subsequently wasting back upslope. Regional snowline was estimated by Muller and Coulter (1957a) to be about 1,675 m (5,500 ft), which is higher than both iceclad benches and higher even than some segments of the caldera rim. Snow that accumulates directly on the benches is clearly augmented by drift and avalanching from the rim.

The lake itself has a complex century-long filling history. Although a turquoise lake was 10–15 m deep during the expeditions of 1916–19 (fig. 136), it had disappeared by 1923 when Fenner and Yori descended to a caldera-floor mudflat (Fenner, 1930). Fenner described numerous steam columns rising from the floor and the fringing talus; fumarolic incrustations and the odor of H_2S; boiling mudpots as wide as 8 m that produced gigantic bursting mud bubbles; and a mud geyser (fig. 137) that erupted at short intervals in jets as high as 60 m with loud detonations. He also showed that Horseshoe Island was a dome of platy to blocky dacite lava ~350 m in diameter with ~20 m of visible relief (fig. 138). It had a steep-walled central lagoon, ~75 m wide, >20 m deep, and open to the northwest (figs. 15, 16 of Fenner, 1930), that had been explosively reamed through the dome; Fenner described stratified deposits dipping radially from the dome that may have been related to such an event. He suggested but could not confirm that Jurassic sedimentary rocks cropped out near the base of the caldera wall. While this is not unreasonable in view of the elevations and extracaldera distribution of Naknek Formation exposures (Riehle and others, 1993), such rocks have not been identified in the breccias deposited explosively on the caldera rim. The question remains unresolved because the base of the wall now lies beneath at least 200 m of lake water.

By 1929, Horseshoe Island was fully concealed by the rising lake (Hubbard, 1932), implying a filling rate of at least 3–4 m/yr after 1923. When it was 10–15 m deep in 1917, Maynard had surveyed the elevation of the lake level to be ~1,003 m (Griggs, 1922), but by 1951 it was ~1,188 m (Motyka, 1977), yielding an average filling rate of 5.4 m/yr for 34 years. For the short interval 1951–53, Muller and Coulter (1957a) also estimated a change of ≥5 m/yr. In 1974, Motyka (1977) surveyed lake level at 1,235 m (and 1.2 m higher in 1975), giving a reduced rate of ~2 m/yr (47 m in 23 years) for the interval 1951–74. Returning in 1977, however, Motyka (1978) found the level to have risen to 1,242 m, implying an increase in rate to ~3 m/yr for 1975–77. We are not aware of any precise measurements since 1977, but observations on our many trips to the caldera rim (1976–2001) suggest that additional deepening of ~10 m had taken place.

From an inflatable raft on the lake in 1977, Motyka (1978) measured a depth no greater than 200 m, an apparent discrepancy of ~50 m with the 1917 survey: [(1,242–1,003 m)+(10–15 m) = 249–254 m depth by 1977]. The discrepancy could reflect large uncertainties in one or both surveys or as much as 50 m of lake-floor sedimentation in 60 years. Whether adequate to account for the discrepancy or not, sediment accumulation is still substantial in the 21st century, as shown by continuing rockfall from unstable walls, several muddy meltwater waterfalls, and aeolian and sheetwash transport of 1912 ejecta and glacial ablation debris across the rim.

According to pilots quoted by Muller and Coulter (1957a), the lake had not frozen over in winter as late as 1953. Motyka (1977) reported that it did freeze in 1967 but not in 1975, suggesting variations in heat flux through the lake floor. Winter and spring freezing has become common in recent decades, however. Since 1953, summer observers have witnessed calved icebergs circulating around the lake and melting within hours. A yellow-green upwelling in the northeastern part of the lake has been observed intermittently for decades, roughly over the site of the mud geyser photographed in 1923 by Fenner (1930). In 1975, Motyka (1977, 1978) paddled through the upwelling, reporting a flux of small sulfur particles, bubbles, and the odor of H_2S. In 1977, he measured a lake-bottom temperature of 20.5°C, which declined to 12°C only 10 m above the floor, then remained nearly constant at 8.8°C at depths between 150 m and 10 m. Motyka noted that this represented a substantial temperature increase from the 5.5°C recordings he had made at depths of 10–60 m only 2 years earlier.

Figure 135. North and northeast wall of Katmai caldera viewed from southeast rim. Slump-block bench at foot of north wall progressively accumulated an intracaldera glacier that now reaches lake level. Hydrothermally altered wall below east rim notch (right) is one of several such internal zones that supplied acid-altered ejecta during caldera collapse (see fig. 106). Northwest-dipping lavas of true summit (peak 6715 on central skyline) are cut by vertical dike complex; southeast-dipping strata above east notch (at right) are part of same edifice, gutted in 1912. Pedestal on skyline rim above glacier is head of north-ridge dacite lava flow prominent in figure 134.

A

Figure 136. Katmai caldera: North bench and Horseshoe Island. Views northward from south rim showing intracaldera lake and slump-block bench on north wall: *A,* 1917 and *B,* 1989. Lake was ~10 m deep in 1917, >200 m in 1989. North bench had thin snow shelf in 1917, but no glacier was seen before 1929. By 1989, however, the glacier was >50 m thick and locally flowed into lake. Crest at right is true summit (Peak 6715 at 2,047 m); lake level in 1917 was 3,290 ft (1,003 m), and snow shelf on north bench was at 4,800–4,900 ft (1,463–1,493 m). Volcanic strata above glacier dip northwest away from dike complex (upper right) at core of eviscerated edifice. Horseshoe Island was a dome of platy to blocky dacite lava ~350 m in diameter that extruded before 1916 and has been wholly submerged since 1929 at latest. When the lake temporarily disappeared in 1923, the dome had ~20 m of visible relief. It also had a steep-walled central lagoon, open to the northwest, which was ~75 m wide, >20 m deep, and is presumed to have been reamed explosively (see fig. 138). For several views of caldera floor when lake-free in 1923, see Fenner (1930).

B

Figure 137. Mud geyser with its own ejecta ring on northeastern floor of Katmai caldera when Yori and Fenner descended west wall in July 1923 (Fenner, 1930). Caldera lake, ~10 m deep during all previous ascents to rim (1916–1919), had drained by 1923, but by next known ascent in 1929, lake was tens of meters deep and floor was submerged for good. Note numerous fumaroles in talus at base of north wall, which is now draped by intracaldera glacier (figs. 134, 135). Fenner wrote that geyser threw mud as high as 60 m in frequent roaring explosions.

Figure 138. View eastward of Katmai caldera floor mudflats from partway down west wall in July 1923 (Fenner, 1930), the only time the caldera was ever seen without a lake. At center is Horseshoe Island dacite dome, ~350 m in diameter with ~20 m of visible relief; its steep-walled central lagoon, probably reamed out explosively, was ~75 m wide and >20 m deep. See figure 136 for 1917 view of dome as island in lake.

Two water samples taken at a depth of 60 m in 1975 and analyzed at the USGS in Menlo Park contained (in mg/L) 1,350–1,750 Cl, 1,200–1,250 SO_4, 590–760 Na, 120–140 SiO_2, and 12–14 B; their pH was 2.5–3.0 as measured in the field or 2.05 as calculated in the lab (Motyka, 1977). Another sample taken in the summer of 1989 had a similar composition (Cameron and Larson, 1992). The acidity suggests dissolution of gaseous CO_2 in the lake water and little or no bicarbonate ion. Such chloride-sulfate waters are typical of volcanic crater lakes and are consistent with meteoric meltwater mixed with a substantial input of fluids from drowned fumaroles.

Photographic Comparisons: Then and Now

We made intermittent efforts over the years (1982–2001) to photograph features in and near the VTTS for which the National Geographic Society had provided images from 1917–19. Paired images are distributed throughout the text as appropriate to the subject matter, but they are all cited, abstracted, and highlighted here.

Mount Katmai Beheaded (fig. 100A, B).—Images of postcaldera Mount Katmai taken in 1915 and 1989 from the Katmai River floodplain show little change in its southern flank and caldera-rim profile.

Katmai Caldera: North Bench and Horseshoe Island (fig. 136A, B).—Views northward from south rim showing intracaldera lake and slump-block bench on north wall of caldera, 1917 and 1989. Lake was ~10 m deep in 1917, more than 200 m in 1989. North bench had thin snow shelf, but no glacier was seen before 1929. Crest at right is true summit of present-day Mount Katmai, which is cut by a dike complex (upper right) at the core of the eviscerated edifice. Horseshoe Island was a 20-m-high dome of platy to blocky dacite lava ~350 m in diameter that has been wholly submerged since before 1929.

Katmai Canyon Landslide Deposit (fig. 123A, B).—A chaotic 0.08-km^3 slide mass broke loose from the south wall of Mount Katmai on the night of 6/7 June after deposition of Novarupta fallout Layer A and part of Layer B. Covering ~1.2 km^2 and as thick as 125 m, the shattered andesitic debris blocked the narrow gorge of Katmai Canyon and impounded a 5-km-long lake, which breached by 1915. Views from 1917 and 1998.

Novarupta and its Ejecta Ring (fig. 55A, B).—Views northeastward of the 380-m-wide lava dome and its asymmetrical ring of Episode III ejecta, 1919 and 1982. Above the dome-encircling moat, the ejecta ring ranges in height from ~7 m in the southwest to 225 m in the northeast, where it broadens into the 1-km-wide northeast-elongate mound called the Turtle. Configuration of the dome, ring, faults, and fissures has changed little since 1919. The dome steamed vigorously for decades, as did several fumarolic fissures along the lower ring, but emissions were modest atop the Turtle and are weak and wispy today. At lower right of 1919 image, a fallen block of shattered dacite of Falling Mountain crumbled in place to form a 12-m-high hummock that has sometimes been mistaken for a domelet or "volcanette."

Extensional Faults Atop the Turtle (fig. 99A, B).—Views toward northwest along southern set of northwest-striking faults near crest of the Turtle, ~1 km northeast of Novarupta dome, 1919 and 1982. Taken from near intersection with northeast-striking fault set (fig. 97). Lower scarp represents ~3 m displacement, upper scarp ~20 m. Degradation of scarps reflects 63 years of sloughing induced by wind and snowmelt.

Griggs Fork of Knife Creek (fig. 128A, B).—Views eastward from north-bank rim at lower end of Griggs fork up its 3-km-long braided floodplain, 1919 and 1982. At head of valley in center distance, Glacier 5 has advanced 1,300 m. Glacier 4 at right has advanced 650 m. Both glaciers remain heavily tephra-covered a century after the eruption. Cutbanks consist of ashy fines-rich ponded sediments overlain by terrace gravels and phreatic surge beds.

Overview of Fissure Lake (fig. 84A, B).—Views westward along trough occupied by Fissure Lake (Griggs, 1922, p. 212, 244) during first decade after eruption, 1917 and 1982. Trough was excavated by phreatic explosions along hingeline of compactional bench at southwest edge of VTTS. Glacier occupied present-day site of West Mageik Lake.

Floor of Fissure Lake (fig. 94A, B).—Views westward along 100-m-wide trough of Fissure Lake, 1917 and 1982. Ignimbrite scarp at right is ~20 m high and was locally still steaming in 1917; that at left is more than 30 m high. Sediment from the glacier (center distance, ~1 km from the camera) was already by 1919 filling the lake, which was largely silted in by 1923. Later withdrawal of the glacier terminus created present-day West Mageik Lake (not in view), which drains north (right) to the River Lethe.

Lethe Arm of VTTS (fig. 108A, B).—Views northwest and north across Lethe arm of upper VTTS to Buttress Range (at left), 1917 and 1982. Distance from bench at northwest toe of Mount Cerberus to steaming bench across valley at left is 4 km; fissure control of many (not all) fumarolic emissions in upper valley is conspicuous. From just west of Mount Cerberus, 1982 image shows River Lethe and 14-km-long reach of VTTS, fumarole-free.

Horses of a U.S. Geological Survey pack train departing the Valley of Ten Thousand Smokes after leaving C.N. Fenner and Charles Yori and a well-supplied base camp at the foot of Mount Cerberus on 23 June 1923. The horses brought Fenner and Yori to the area from Kanatak via Kejulik and Yori Passes for two months of fieldwork. No other geologists are known to have ridden to the area on horseback. Novarupta dome and surrounding fumaroles are seen in upper right. Photo by Charles Yori.

Chapter 12

The Katmai Volcano Cluster

An extraordinary aspect of the greater Katmai magmatic system is the clustering of andesite-dacite volcanoes close to the 1912 eruption site (figs. 5, 6). Along the volcanic chain here, crater-to-crater spacing between adjacent, mostly contiguous, edifices is typically 5 km or less, one of the world's most tightly spaced lines of arc stratovolcanoes. Discrete eruptive edifices far outnumber the stratocones and lava domes formally named. Mount Martin is a relatively simple cone and Mount Griggs a fairly symmetrical cone, but both overlie ruggedly eroded remnants of older edifices. Both Mount Katmai and Snowy Mountain consist of pairs of contiguous cones, and Mount Mageik and Trident Volcano each have four overlapping cones. In addition, five dacitic lava domes form an array around the Trident edifice. There are thus no fewer than 21 separate volcanic vents in the Katmai cluster, of which 16 lie within 11 km of Novarupta and 7 are within 8 km (Hildreth and Fierstein, 2003). Numerous K-Ar and $^{40}Ar/^{39}Ar$ ages for most of these volcanoes were presented by Hildreth and others (2003) and are summarized in table 1. Martin, Mageik, Trident, Katmai, Griggs, and Snowy have all had Holocene eruptions, but not one of the 21 pre-1912 vents ever produced high-silica rhyolite—the dominant product of the 1912 eruption. Mount Katmai has produced modest amounts of rhyodacite (68-72% SiO_2) and basalt, but all the rest have produced exclusively andesite-dacite. Coincidence of such an anomalous clustering of intermediate-composition vents with the anomalous site of rhyolitic magma generation, each a rarity on the Alaska Peninsula, induces inference of a relationship. The unusual degree of magmatic focussing, particularly beneath the Katmai-Trident cluster, may have provided the thermal environment necessary for intracrustal production and storage of phenocryst-poor high-silica rhyolite magma.

Here we provide overviews of the morphology, composition, and eruptive history of each of the principal volcanoes of the Katmai cluster (table 1).

Alagogshak Volcano

Oldest edifice in the Katmai cluster, Alagogshak lies on the rangecrest 15 km southwest of Katmai Pass (Hildreth and Fierstein, 2003) and is centered only 3 km southwest of Mount Martin (figs. 5, 26). Originating by 680±20 ka, the glacially dissected volcano produced 14±4 km^3 of andesite-dacite eruptive products during several episodes in the middle and late Pleistocene (Hildreth and others, 1999). From a central vent marked by hydrothermal alteration and remnants of a cratered fragmental cone on the drainage divide, glacially incised stacks of lava flows dip radially and extend 6–10 km in most directions. The basal flow in the northwest sector, resting on Jurassic basement, gave a K-Ar age of 680±20 ka. In the southwest sector, intracanyon lava flows yielded ages of 394±46 ka and 389±71 ka, and to the south, the top flow of a stack of seven andesite lavas gave an age of 104±10 ka.

The youngest unit at Alagogshak is an andesitic lava flow (61% SiO_2) that descended steeply southeastward from the rangecrest and spread into upper Alagogshak Creek, where its 100-m-thick distal lobe now forms a glacially scoured plateau. Originating at an ice-clad knob, which is probably a flank-vent dome 1.5 km east of the crater, the lava flow gave an age of 43±8 ka. No evidence is known for subsequent activity until construction of the nearby cone of Mount Martin in the Holocene. The medium-K calcalkaline eruptive suite (57–66% SiO_2) of Alagogshak Volcano is compositionally varied, probably reflecting independent evolution of different magma batches supplied intermittently over a lifetime of more than 600,000 years, much longer than that of other volcanoes in the Katmai cluster.

Mount Martin

Youngest stratocone in the Katmai cluster, Mount Martin consists of a 2-km-wide fragmental cone (fig. 26) and ten overlapping coulees of blocky dacite (fig. 5), each lobe 75–100 m thick, that descend northwest for 10 km (Hildreth and Fierstein, 2000, 2003). Although it exceeds 1,860 m in elevation, the cone itself has local relief of only ~500 m, owing to its construction upon a high ridge of Jurassic basement rocks. Of the 7-km^3 present-day volume estimated for Mount Martin, the cone makes up only 5 percent and the 31-km^2 lava-flow field about 95 percent. Despite a cone-encircling collar of small active glaciers, erosion of the cone and coulees is insignificant, indicating that Mount Martin is a Holocene volcano in its entirety. Because the topmost flow distally is overlain by soil and tephra layers as old as 6 ka (Fierstein, 2007), the package of lava flows must have been emplaced in the early Holocene, conceivably in an effusive episode only years or decades long. The cone has been marked by a steam plume persistent throughout the century since the edifice was

Table 1. Physical data for Katmai cluster volcanoes.

[Present volumes for Mount Katmai are calculated for edifices prior to 1912 caldera collapse. Volume erupted from Southwest Katmai includes ~25 km^3 for edifice, 5±2 km^3 for off-edifice dacite pyroclastics, and 12±5 km^3 for 22.5-ka rhyolite pyroclastics]

Volcano/edifice	SiO_2 range of products	Age range of activity	Present area (km^2)	Volume today (km^3)	Volume erupted	Notes
Alagogshak Volcano	57–66	680–43 ka	20	5	14±4	Central-vent stratocone plus outliers
Mount Martin	58–64	Holocene	33	7	8	Vigorous fumaroles in crater atop cone
Mount Mageik	56–68	Since 93 ka	80	20	30	Four overlapping subedifices, each with its own vent
Southwest Summit	56–68	93–80 ka	32	9	17	
Central Summit	58–66	80–60 ka	10	4	6	
North Summit	60–64	60–10 ka	13	1.5	2	
East Summit	60–64	Holocene	25	5	5.5	
Trident Volcano	53–72	Since 143 ka		~10	22±3	Four contiguous cones and several domes
East Trident	58–65	142±15 ka	7–8	2.3	6–9	
Trident I	53–64	101–58 ka	14–15	4	7–10	Pyroclastic-flow remnants 57–72% SiO_2
West Trident	55–64	44±12 ka	8–9	2.4	3–4	
Trident domes	54–65	late Pleistocene	5	0.7	1	Six domes include Mount Cerberus and Falling Mountain
Southwest Trident	57–65	1953–1974	8.5	0.65	0.7	Cone and lava apron built since 1953
Mount Katmai	51–72	Since 100 ka		43		Two overlapping cones, both cut by 1912 caldera
Northeast Katmai	52–66	100–50 ka	25	10	15±3	Early basaltic cone followed by andesite-dacite lavas
Southwest Katmai	51–72	60 ka to 1912	61	20	42	Horseshoe Island dome concentric with pre-1912 crater
Novarupta	51–78	1912 only	3	13	13	New flank vent drained magma from beneath Mt. Katmai
Snowy Mountain	55–64	200 ka to Holocene		7		Two contiguous cones; two vents 4 km apart, largely ice covered
Northeast Snowy	61–64	180 ka to Holocene	20	5	8±2	Dome plugs amphitheater of Holocene sector collapse
Southwest Snowy	55–62	200–90 ka	13	2	5±2	Glacially dissected aretes and windows through ice
Mount Griggs	51–63	292 ka to Holocene	60	20–25	35±5	Sector collapse refilled by Holocene inner cone

first observed in 1913. The plume coalesces from as many as 20 fumaroles, rich in CO_2, H_2S, and SO_2, that are precipitating sulfur in the talus northwest of a shallow lake on the floor of its steep-walled 300-m-wide crater (fig. 139). Scoriaceous and massive glassy blocks of the cone are andesitic (59–61% SiO_2), whereas the stairstep sequence of coulees is dacitic (62.5–64.2% SiO_2). Eruptive products yield a tight compositional array that is similar to the andesite-dacite suite erupted at Novarupta in 1912 but is poorer in Sr and marginally more potassic (Hildreth and Fierstein, 2000). Mount Martin is essentially a Holocene successor to Alagogshak Volcano, their craters being only 3 km apart (figs. 26, 140). Long persistent earthquake activity and recurrent small swarms beneath and near Mount Martin were punctuated by a strong swarm in January 2006; Alaska Volcano Observatory located 860 events, mostly M_L 0–1.5 and largely at depths of 1–4 km, but they were accompanied by no other evidence of volcanic unrest (Dixon and Power, 2009).

Mount Mageik

An ice-capped compound andesite-dacite stratovolcano, Mount Mageik is one of the most extensive (80 km²) and voluminous (30 km³) edifices in the Katmai cluster (fig. 140). Each of its four summits is a discrete eruptive vent (Hildreth and others, 2000), and each is the source of numerous lava flows (56–68% SiO_2). At least 57 lava flows are exposed, many of them 50–200 m thick distally (owing to ice-contact emplacement). Three of the four overlapping centers are severely eroded glacially, but the East Summit is a Holocene edifice, from which a dozen leveed lava flows (60–64% SiO_2) have descended toward Katmai Pass (figs. 5, 28), where blocky terminal lobes are 60–150 m thick. Ice-filled craters mark three of the summits, but the Central Summit is capped by an ice-clad dacite dome. An ice-free 350-m-wide phreatic crater between the East and Central Summits contains an acid lake (pH ~1) and many superheated fumarolic jets. Holocene debris avalanches have broken loose from three different sites on the south slopes of Mount Mageik, the youngest triggered by seismicity attending collapse of Katmai caldera in 1912 (Hildreth and others, 2000).

The four crowded cones of Mount Mageik epitomize the close clustering of the Katmai volcanoes. Physical overlap and relative erosion show the Southwest Summit cone to be the oldest, followed in order by the Central, North, and East Summit cones. The Southwest Summit subedifice is larger, longer lived, and compositionally more varied than its companions. Basal lavas of the Southwest Summit, resting on Jurassic basement rocks along Martin and Alagogshak Creeks, respectively, yield K-Ar ages of 93±8 ka and 89±8 ka. The lowest lava flow from the Central Summit in the southeast sector gave 71±11 ka, and a 150-m-thick andesite lava from the

Figure 139. Crater floor of Mount Martin, where numerous fumaroles rich in sulfur gases and CO_2 combine to produce prominent plume, which has persisted above cone since first reported in 1913 (see photos in Griggs, 1922).

North Summit, which rests on basement rocks at the northwest toe of the mountain, gave 59±11 ka. No attempt was made to date the Holocene lavas of the East Summit, but tephras overlying them suggest that some could be as young as 2,500 years (Fierstein, 2007). Judging by their relative degrees of degradation, the dozen Holocene lava flows represent several discrete episodes (Hildreth and others, 2000). The postglacial eruptive volume is ~5–6 km^3, to which dacitic fallout and a small andesitic pyroclastic-flow deposit contribute no more than 0.1 km^3.

All products of Mount Mageik are plagioclase-rich pyroxene-dacites and andesites that form a calcic medium-K arc suite (Hildreth and Fierstein, 2000; Hildreth and others, 2000). We estimate eruptive volumes of 17, 6, 2, and 5.5 km^3 for the Southwest, Central, North, and East Summit edifices, respectively, yielding a total of ~30 km^3.

Mount Mageik's fumarolic plume is often conspicuous, but none of the sporadic reports of 20th century tephra eruptions (for example, Jaggar, 1927) are plausible. Configuration of the crater has not changed since it was first photographed in 1923 (Fenner, 1930; Hildreth and others, 2000), and the only tephra deposits younger than 2 ka on the flanks of Mount Mageik are the Novarupta pumice falls of 1912 and Trident ash of 1953–74.

Trident Volcano and its Domes

Like Mount Mageik, Trident consists of four contiguous andesite-dacite cones (fig. 141), only one of which was constructed in the Holocene. That cone, Southwest Trident, erupted from 1953 to 1974 but was largely built in the 7-year interval 1953–60, during which four lava flows and most of the fragmental cone were emplaced. The three Pleistocene edifices young from east to west, and all three are severely eroded glacially (Hildreth and others, 2003a).

East Trident, a small andesite-dacite cone (Peak 6010; 58–65% SiO2) almost contiguous with Mount Katmai, is gutted by five glacial cirques, which expose a hydrothermally altered core and stacks of 10–20 radially dipping lava flows and breccia sheets. Most are silicic andesites 10–30 m thick, but 100-m-thick lavas that support the southwest ridge

Figure 140. Aerial view of ice-clad Mount Mageik toward south-southwest. Each of its four summits is an independent eruptive vent for andesite-dacite lavas as old as 93 ka and as young as middle Holocene (Hildreth and others, 2000). At right are Mount Martin and Alagogshak Volcano. At lower left is Mount Cerberus dacite dome, guarding entrance to Katmai Pass. At lower right, 1912 ignimbrite banks against toe of Mount Mageik and displays well-developed "high sand mark" bench that rises ~50 m above valley floor.

and cap the northeast ridge are dacites (63–65% SiO2). The northeast ridge dacite, which extends nearly to the summit (indicating that most of East Trident is older) yields a K-Ar age of 142±15 ka. An analytically indistinguishable age of 143±8 ka for a basal andesite lava low on the north-northwest arête suggests, however, that activity at the small East Trident cone was short lived. Peak 5700, central prong of the "tri-dent" as viewed from the southeast (the aspect that inspired the volcano's name; Griggs, 1922, p. 98–9), is not a separate cone but simply the ruggedly eroded southwest flank of East Trident. Reconstruction suggests that the cone may once have exceeded 6,600 ft in elevation, which may have once made it the highest of the Trident summits.

Trident I is a discrete andesitic cone (Peak 6115; 53–63% SiO_2) that forms the central and highest edifice (1,864 m) of the Trident group. Its lavas bank against East Trident at their mutual saddle. The core of the edifice is eviscerated by a north-face cirque, the sheer headwall of which reveals ~200 m of crudely stratified andesitic breccia overlain by a 150-m-thick near-summit lava (63% SiO_2), which yields a K-Ar age of 101±12 ka. This dacite lava may be a dome remnant that overlies crater fill and dome breccia on its north side. The southwest slope consists of a mafic (53–55% SiO_2) pyroclastic complex that gives an age of 58±15 ka. Activity at Trident I is thus loosely bracketed between approximately 142 and 58 ka. Although the edifice is glacially ravaged, its lower southeast flank has a field of sulfur-depositing fumaroles. Another large fumarole on the southwest ridge of Trident I became the vent for the Southwest Trident cone of 1953–74.

West Trident (Peak 5605), smallest and youngest of the prehistoric Trident stratovolcanoes, is the closest cone in the Katmai cluster to Novarupta. Its lavas are heavily mantled by 1912 fallout. Those exposed are silicic andesite to dacite (58–64.4% SiO_2), and many are unusually rich in chilled enclaves (1 to 20 cm) of phenocryst-poor mafic andesite (54–58% SiO_2). The summit lava (61% SiO_2) is >125 m thick and may be a dome remnant. Summit-derived lavas dip east against Trident I at their mutual saddle. An andesite lava flow 30-40 m thick that dips 20° down the north ridgeline yields a K-Ar age of 44±12 ka. Although largely ice-free today, West Trident is everywhere glacially modified. Some lava flows bank against Falling Mountain dome (70±8 ka), and, just east

Figure 141. Trident volcanic group viewed northwestward from lower Mageik Creek near its confluence with Katmai River. Four separate eruptive vents are visible. Two peaks at right are parts of East Trident; high peak at center is Trident I; and at left, black cone of Southwest Trident (1953–74) largely obscures snowclad cone of West Trident. Trident cluster is as old as 143 ka (Hildreth and others, 2003a, b). At far left is Observation Mountain. At right, black bench of glaciated andesite from Mount Katmai overlies brown bench of Jurassic siltstone along canyon of Mageik Creek. Pale-gray pumice-fall deposits from Novarupta mantle most slopes.

of there, the distal parts of adjacent lavas were destroyed on 6 June 1912 by collapse into the Novarupta vent, thereby providing the andesitic lithics among the 1912 ejecta.

As many as eight lava domes of pyroxene dacite (62–65% SiO_2), all glacially scoured, were emplaced within and adjacent to the West Trident and Trident I edifices. The two largest domes, 425-m-high Falling Mountain and 365-m-high Mount Cerberus (each 0.3–0.4 km^3), are compositionally similar to the smaller (unnamed) domes, the volumes of which are in the range 0.015–0.12 km^3. Like West Trident, the domes contain chilled enclaves (1 to 20 cm) of phenocryst-poor andesite (54–58% SiO_2). Falling Mountain and Mount Cerberus, which frame the entrance to Katmai Pass (figs. 5, 26), appear to be little modified by ice in gross morphology and were once suspected of being fairly young (Hildreth, 1983). Repeated search, however, turned up few remnants of glassy carapace, and late Pleistocene K-Ar ages were later measured for both domes: 70±8 ka for Falling Mountain and 114±46 ka for Mount Cerberus (Hildreth and others, 2003b). The superficiality of glacial erosion may reflect their compact profiles and locations close to the drainage divide. Attempts to date other late Pleistocene domes along the southwest periphery of the Trident group failed.

Glacial ice and the blanket of 1912 fallout obscure so much of Pleistocene Trident that counting lava flows is an uncertain exercise; we estimate roughly 40 for the eastern edifice, 25 for the central, and 10 for the western. Eruptive volumes, as reconstructed (table 1), were roughly 8 km^3 each for East Trident and Trident I, only 3–4 km^3 for West Trident, a total of 1 km^3 for the several domes, and 0.7 km^3 for the 20th century eruption of Southwest Trident, giving a total eruptive volume for the Trident group of 22±3 km^3.

Southwest Trident (Eruption of 1953-1974)

Beginning in February 1953, a new andesitic edifice (0.7 km^3) was built at the southwest margin of the Trident group (Snyder, 1954; Ray, 1967). Although sometimes referred to informally as "New Trident," we have called it **Southwest Trident**, in anticipation of the day it ceases to be Trident's youngest component. During two decades of sporadic activity (vulcanian and effusive), a new 3-km^2 composite cone was constructed of block-and-ash deposits, scoria, agglutinate, stubby lava lobes, and the intercalated proximal parts of the main lava flows that spread as an apron around the cone (fig. 142). The cone grew to an elevation of 1,515 m (GPS

Figure 142. Southwest Trident in July 1979 (5 years after cessation of 1953–74 eruptive activity), viewed eastward from glacier on Mount Mageik. Fragmental cone, asymmetrically built on slope of glaciated Trident lavas, rises 750 m above Katmai Pass in foreground; its crater is 350 m wide and still weakly fumarolic. Leveed lava flow at left erupted in 1959–60, and overlapping flows at right in 1953 and 1958. Petrology of andesite-dacite eruptive products was reported by Coombs and others (2000) and geology by Hildreth and others (2003a).

measurement reported by Coombs and others, 2000) on the former site of a 100-m-wide fumarolic pit at an elevation of about 1,175 m on the steep southwest flank of Trident I. Although relief on its south side exceeds 700 m, the slope-draping new cone has a central thickness of only 340 m and a volume of about 0.3 km^3 (Hildreth and others, 2000, 2003a). At successive stages of cone construction, four blocky leveed lava flows effused from its central vent, in 1953, 1957, 1958, and during the winter of 1959–60. Each flow is 25–60 m thick and 2.5–4 km long, and altogether they add about 0.35 km^3 to the eruptive volume. The cone's summit is today marked by a shallow crater 350 m wide (Hildreth and others, 2000, 2003a) that was the site of several small ephemeral plugs, which were emplaced after the final lava flow and were repeatedly destroyed by intermittent explosive activity (1960–74).

Black ash clouds rose 6–9 km at least 10 times between 1953 and 1974 and perhaps to 12 km once or twice. Several times during the first month of activity, light ashfall dusted areas as far as 30–50 km from the vent, in all sectors. By far the most voluminous fallout appears to have resulted from the initial outburst of 15 February 1953 (Snyder, 1954). A single nongraded scoria-fall layer (5–17 cm thick) deposited during that event is preserved at a few protected sites as far away as Mount Katmai and upper Knife Creek. Sieve data for bulk samples of this layer yield median and maximum particle sizes of 6.5 mm and 10 cm, respectively, in the saddle 1 km north of the vent and 2.1 mm and 2 cm in the saddle 7 km northeast of the vent—between the twin western summits of Mount Katmai. Thin sheets of finer ash that fell during the many subsequent events have been almost entirely removed or reworked by wind and runoff. Abundant ballistic blocks, variously breadcrusted, densely vitrophyric, or scoriaceous, are scattered as far as 3 km from the vent and are products of many discrete explosive episodes (none of which were closely observed) distributed over two decades. Liberal estimates of total fallout yield no more than 0.05 km^3, contributing less than 10 percent of the total eruptive volume of 0.7±0.1 km^3.

The period of most frequent observation was from February to September 1953, principally by military reconnaissance aircraft during the early months (Snyder, 1954) and by a U.S. Geological Survey party that camped at Knife Creek during the summer (Muller and others, 1954). When the vent was first seen through the cloud layer on the fourth day of activity (February 18, 1953), an effusive lava flow (then already 250 m wide) was upwelling centrally and spreading radially (fig. 15 of Hildreth and others, 2003a). Although a fumarolic pit as deep as 40 m was conspicuous at the impending vent site on aerial photographs taken in 1951 (and had probably been further excavated by the explosive outburst of February 15), any such crater was soon filled and buried by the effusive lava, which continued to extrude and spread slowly throughout the seven months of intermittent observation in 1953. At different times, lava lobes emerged laterally through the chilled carapace at the foot of the pile, or the pile itself "expanded like a balloon" and extruded lobes by overflow from the vent, or small slumps and slide masses detached from the steep flow margins (Snyder, 1954). By June 1953, the main southerly tongue of lava, ultimately 4.2 km long, had advanced only 1,250 m from the vent. Snyder (1954) estimated the volume of fallout and lava produced by June 17, 1953, to have been between 0.23 and 0.3 km^3, roughly a third of the eventual output. During the summer, steady steaming and continued spreading of the lava was punctuated sporadically by steam bursts or occasional "smoke columns" (Muller and others, 1954) that rose 1–3 km and dusted various proximal sectors with minor additional ashfall.

Observations after September 1953 were sporadic and few. A general chronology of major events was compiled by Decker (1963) and augmented by Ray (1967), largely from intermittent National Park Service reports. The 1953 lava flow may not have attained its final dimensions until early 1954 or later. It appears that there were no observations at all during eruption and outflow of the lava flows of 1957, 1958, and 1959–60, merely aerial snapshots taken in the summer seasons following emplacement of each. Time of emplacement of the lava flow attributed to the winter of 1959–60 is the least well known, because no photographs are known to have been taken between September 1958 and August 1960. The 1958 lava flow partly overran the 1953 flow and impounded a small lake on upper Mageik Creek that soon filled in with pumiceous alluvium, becoming a mudflat ("Ferruginous Flat") now marked by numerous iron-precipitating warm springs (Keith and others, 1992).

Growth of the fragmental cone only began after much or all of the 1953 lava flow had been emplaced. It accumulated progressively during the later 1950s, as shown by emergence of the successive lava flows at different levels of the fragmental edifice. National Park Service photographs show that the cone had attained nearly its full height by 1960, but explosive showers of blocks continued to augment the cone until 1974. In addition to the four main lava flows, cone construction included emplacement of several stubby lava lobes limited to its proximal southwest slope. The southeast side of the cone completely buried a 1-km^2 cirque glacier, producing no recognized effects on eruptive behavior or edifice structure, though it is possible that enhanced steaming could have contributed to the stronger fumarolic emission and alteration on that side of the cone. Explosive ejections of tephra, some involving blowout of plugs and at least one spine, took place from 1960 to 1974, but volumetrically significant eruptions ended by 1963. Numerous sulfurous fumaroles, superheated in the 1960s but below or at the boiling point today, persist on upper parts of the cone. Dark-gray bouldery debris flows reworked from the pyroclastic deposits have built a proximal fan and thin distal sheets (1 to 4 m thick) that cap stream terraces for 3 km downstream along Mageik Creek. Some debris flows resulted from the initial February 1953 fallout over snow, others in later years by avalanching of rubble from the steep slopes of the cone.

The lava flows and ejecta are olivine-poor, two-pyroxene andesite and dacite that span a compositional continuum from 56 to 65.5 percent SiO_2. All products are plagioclase-rich and contain 33 to 45 percent phenocrysts, many of them complexly zoned, resorbed, and overgrown (Ray, 1967; Kosco, 1981; Coombs and others, 2000). The least evolved bulk material identified is the initial scoria fall of February 1953 (56–57 %) SiO_2). Andesitic enclaves are abundant, relatively mafic (56–59 % SiO_2), contain the same phenocryst species as the host material, and fall mostly in the size range from 1 to 30 cm (Coombs and others, 2000). No evidence of contamination by leftover 1912 magma (such as jacketed quartz phenocrysts) has been found, and small but consistent compositional differences (Hildreth and Fierstein, 2000) indicate that the andesite-dacite magmas erupted in 1912 and in 1953–74 were different batches. For the Southwest Trident batch, Coombs and others (2000) presented experimental and analytical evidence for mixing between resident dacite magma stored at 890°C at a depth of about 3 km and a newly arrived batch of 1,000°C andesite magma. They interpreted diffusion profiles in phenocrysts to suggest that extensive mixing could have produced the linear compositional array within about a month before the eruption began.

Mount Katmai

Centered 10 km east of Novarupta and adjacent to Trident, Mount Katmai is a compound stratovolcano that consisted of two contiguous cones, both beheaded by caldera collapse in 1912, as discussed above in chapter 8. On the caldera walls are exposed separate foci of hydrothermal alteration and abutting stacks of lavas and ejecta that dip against each other from discrete Northeast Katmai and Southwest Katmai centers (figs. 5, 103, 135). The caldera has an area of 4 km^2 at the floor and 8.8 km^2 at the rim (Hildreth, 1991; Hildreth and Fierstein, 2000). Its internal volume is today about 4 km^3, and the foundered superstructure amounted to an additional 1.1 to 1.5 km^3, giving a collapsed volume of about 5.5 km^3. The Horseshoe Island dacite dome (figs. 136, 138) and numerous fumaroles on the caldera floor (fig. 137; Fenner, 1930) were by 1929 covered by rising lakewater, now deeper than 200 m and more than 1 km^3 in volume.

Eruptive products of the Katmai edifice (Hildreth and Fierstein, 2000) range from basalt to rhyolite (51.6–72.3% SiO_2), an array wider than that of any other stratovolcano in the cluster (table 1) and surpassed only by the zoned pyroclastic eruption of 1912 (50.4–77.8% SiO_2). Andesitic lava flows are the dominant material of both Katmai cones, and products more mafic than 55 percent SiO_2 are limited. Silicic lavas (63–72% SiO_2) are common, dominating the stratigraphically youngest products of both cones. No lava domes are recognized in the caldera walls, though some of the thick silicic lava flows truncated by the walls could have issued from summit domes obliterated by the 1912 collapse. The most silicic materials identified in the Northeast Katmai edifice are dacite lava flows (66% SiO_2) as thick as 60 m that cap the north rim (figs. 134, 136). The Southwest Katmai cone likewise has many dacites but additionally produced several rhyodacitic to rhyolitic lava flows (68–72% SiO_2) that armor its upper southern slopes and cap much of its south to southeast rim (fig. 143). The most evolved of the south-rim lavas yielded

Figure 143. Thick glassy rhyodacite lava capping southeast rim of Katmai caldera, draped by several meters of Novarupta fall deposits. Geologist stands 600 m above lake level. Beyond east rim notch (center) is Northeast Katmai edifice (see figs. 5, 135, 144).

a $^{40}Ar/^{39}Ar$ age of 22.5±1.6 ka, but a zoned dacite-andesite scoria fall and several rhyodacite lavas overlie it (Hildreth and Fierstein, 2003).

The highest point surviving on the caldera rim belongs to the oldest component of Mount Katmai, a 1-km-thick dike-laced pile of rubbly mafic lava flows (52–55% SiO_2) that makes up the northeast wall of the caldera (fig. 135) and dips steeply outward (fig. 144). The core of the Northeast Katmai edifice consists entirely of this structurally simple mafic cone, the base of which is exposed only distally, not within the caldera. A basaltic lava midway through the cone sequence (where cut by the caldera rim) gave a $^{40}Ar/^{39}Ar$ age of 89±25 ka. An andesite lava flow resting (with apparent conformity) on the pile of mafic lavas at the northeastern toe of the cone gave a K-Ar age of 70±13 ka. Several ridge-capping andesite-dacite lavas are stratigraphically younger still.

Caldera-wall stratigraphy shows that the two Katmai cones overlap in age but that the southwest edifice is in large part younger. A basal andesite lava flow of the Southwest Katmai cone, resting on basement at Katmai River, gave a K-Ar age of 39±12 ka, and a dacite lava flow (also basal or nearly so) west of the caldera gave 47±13 ka. The Southwest Katmai cone erupted at least twice postglacially (Hildreth and Fierstein, 2000, 2003), as well as producing the Horseshoe Island dacite dome on the caldera floor after the 1912 collapse.

Along and near Katmai River, several lava flows relatively low in the Mount Katmai pile are 50–150 m thick, pervasively glassy and columnar, and display entablatures of slender curving columns, many of which are subhorizontal and locally form rosettes. Such features, which occur in both silicic and mafic andesites of Southwest Katmai and even in a few basaltic lavas of Northeast Katmai, suggest ponding of many early Katmai lavas against ice near paleocanyon floors. Despite persistence of active glaciers on both cones, the stacks of generally thinner andesite-dacite lavas higher on the compound edifice rarely display such pervasive ice-contact features.

Off-edifice pyroclastic deposits erupted at Mount Katmai (Fierstein, 2007) include the following: (1) Valley-floor remnants of a late Pleistocene plinian pumice-fall deposit and associated ignimbrite are chemically (71–72% SiO_2) and mineralogically similar to the hornblende-bearing rhyolite lava flow on the south rim, $^{40}Ar/^{39}Ar$-dated at 22.5±1.6 ka. Organic material at the base of the plinian deposit gave an age of 19,240±70 ^{14}C yr BP, equivalent to a calendar age of 22.8 ka (Hildreth and Fierstein, 2000). (2) In upper Knife Creek, sintered remnants of a dacitic scoria-flow deposit (Hildreth and others, 2002), 4–10 m thick, are compositionally similar (64–66% SiO_2) to an agglutinated scoria-fall deposit that drapes the west rim of the caldera (figs. 5, 134, 145) and ponded inside a paleocrater (Hildreth and Fierstein, 2003). (3) Debris flows and hyperconcentrated sand deposits rich in dacite pumice are as thick as 25–30 m where exposed in the lower VTTS (Hildreth and others, 2000). These deposits and those described in (2) have not been dated directly but were

Figure 144. Aerial view of eastern flank of Northeast Katmai edifice (fig. 5) looking northwestward. Foreground relief ~400 m. Steeply dipping lavas and red scoria layers are largely mafic, in contrast to products of Southwest Katmai edifice, which range from andesite to rhyolite. For intracaldera view of steep inner wall of these crags, see figure 135. Beyond knife-edged caldera rim rises andesitic cone of Mount Griggs, 15 km from camera.

emplaced during end-Pleistocene deglaciation, probably 16–8 ka (Fierstein, 2007). (4) Remnants of an undated quartz-bearing ignimbrite (76.6% SiO_2) as thick as 100 m cap a divide ~6 km northeast of the caldera rim (Hildreth and Fierstein, 2003). Although this ignimbrite is not confirmed to have issued from Mount Katmai, the only center known to have produced Quaternary products more silicic than dacite, no alternative source has been identified anywhere in the district.

Snowy Mountain

Snowy Mountain consists of two contiguous edifices (Hildreth and others, 2001; Hildreth and Fierstein, 2003) that were supplied by long-lived central vents only 4 km apart. Because glacial ice still covers nearly 90 percent of both edifices, rock exposures are limited to narrow ice-bounded arêtes at higher elevations and ice-scoured lava-flow benches at lower elevations. Each cone appears to consist of only 12–15 andesite-dacite lava flows, apparently erupted as a few packages at widely spaced intervals. Many products of the southwest vent are olivine-bearing andesites (55–62% SiO_2), whereas those of the northeast vent are largely pyroxene dacites (62–64% SiO_2).

For Southwest Snowy, a basal andesite lava flow resting on basement rocks at the north toe of the cone gave a K-Ar age of 199±9 ka, and the summit-capping andesite gave 196±8 ka. However, another andesitic lava flow exposed through the ice on the southeast flank (probably also basal in its sector) yielded an age of 92±10 ka. For Northeast Snowy, the second flow of a basal stack of dacite lavas on the north apron gave an age of 171±8 ka, which is probably close to the inception age of that center. Sporadic activity apparently alternated from the paired vents during the first half of the eruptive lifetime of Snowy Mountain, but only the northeast cone remained active through the late Pleistocene into the Holocene. The only known Holocene eruption was emplacement of a dacite dome within the amphitheater of the northeast cone shortly after a sector collapse about 1.5 ka (Hildreth and others, 2001). Eruptive volumes are estimated to be 8±2 km^3 for the northeast edifice, 5±2 km^3 for the southwest, and 13±4 km^3 for the Snowy Mountain center as a whole. As much as half has been removed by glacial erosion.

Mount Griggs

Highest peak in Katmai National Park, Mount Griggs is a fumarolically active andesitic stratovolcano that stands alone 12 km behind the main volcanic chain (Hildreth and others, 2002). Symmetry of the apparently little-dissected cone (fig. 146) reflects postglacial effusions of andesitic lava that have healed and concealed older scars of glacial erosion. Mount Griggs consists of three main parts: (1) Remnants of a middle Pleistocene edifice as old as 292 ka, exposed in only a few windows; (2) a late Pleistocene cone consisting of at least 50 lava flows, well-preserved except for its southwest quadrant, which underwent postglacial sector collapse; and (3) a Holocene inner edifice that filled the collapse amphitheater by building a cone with nested inner craters, a 12-km^2 fan of at least 20 blocky andesite lava flows, and a volume of about 2.2 km^3. Bedded scoria falls and layers of poorly sorted phreatomagmatic ejecta are exposed on rims of the inner craters but rarely elsewhere. In contrast to the compound volcanic-front centers, Mount Griggs has maintained a stable central-vent position throughout its multi-stage history. Present volume is 20–25 km^3, but reconstruction of the eroded components of the edifice yields an estimate of 35±5 km^3 for the total volume erupted (Hildreth and others, 2002).

Eruptive products define a typically Ti-poor, medium-K calcalkaline arc suite, largely in the range 55–63 percent SiO_2; most units are andesitic, ranging from olivine-rich mafic andesite to two-pyroxene silicic andesite and dacite. Olivine-rich andesite makes up a larger fraction of Mount Griggs than of the other volcanoes in the Katmai cluster. All lavas are plagioclase-rich. Relative to products of the arc-front centers, those of Mount Griggs are generally enriched in Na, Rb, Sr, Ba, Al, and P, and consistently enriched in K, Zr, Hf, Ta, Th, and U (Hildreth and Fierstein, 2000; Hildreth and others, 2004). Fumarolic gases at Mount Griggs are richer in He than those at other Katmai volcanoes and have $^3He/^4He$ values as high as 7.7 Ra (Poreda and Craig, 1989; Sheppard and others, 1992). The magmatic plumbing system of Mount Griggs is independent of those beneath the main volcanic chain, apparently well down into the mantle. Consistent chemical differences from all the volcanic-front centers indicate that magma typical of Mount Griggs did not contribute to the 1912 eruption.

Mount Griggs is as old as 292±11 ka and thus predates inception of all the nearby volcanic-front centers except Alagogshak. The basal lava flow at the north toe of the volcano yields that age, and two windows of a dissected andesitic edifice yield K-Ar ages of 160±8 ka (crags on the mountain's west shoulder) and 133±25 ka (a southerly kipuka). Such a range of ages for widely spaced glaciated remnants, respectively 1,600 m, 500 m, and 1,300 m lower than the modern summit, signify that a large fraction of the long-lived middle Pleistocene edifice was eroded away before being covered by the late Pleistocene cone, which, in contrast, is only slightly eroded.

The stratigraphically lowest lava on the east side of the late Pleistocene cone ponded in an older glacial canyon and yields a K-Ar age of 90±7 ka. Another thick basal lava flow resting on basement rocks at the northeast edge of the cone gave 54±8 ka. Attempts were made to date three additional lava flows from other parts of the late Pleistocene cone. To the north, a thick distal andesite flow that rests on basement rocks gave an age of 15±18 ka. On the west slope, the topmost andesite flow beneath the debris-avalanche deposit that resulted from the Holocene sector collapse gave 21±11 ka. Adjacent to the latter flow, the lone dacitic lava flow exposed at Mount Griggs (also overlain by the debris-avalanche deposit) failed to yield measurable radiogenic Ar despite repeated extractions, apparently because it is very young,

Figure 145. Southwestern inner wall of Katmai caldera from helicopter. Bench-forming rim cliff is massive to layered dacite agglutinate, here as thick as 60 m, that dips modestly inboard where it filled a precaldera crater. Stack of andesitic lavas on lower wall dips ~25°SW (opposite to the agglutinate) and is part of Southwest Katmai edifice (fig. 5) that held the crater. Lithologically described and mapped by Hildreth and Fierstein (2003), the agglutinate also drapes ridge at upper right to elevations as much as 225 m higher than rim ledge here. Narrow glacier tongue at right extends from rim bench nearly to lakeshore (see fig. 134).

Figure 146. Mount Griggs andesitic stratovolcano, highest peak in Katmai cluster at 2,332 m; aerial view northward across upper Knife Creek arm of VTTS. Exposed products range in age from 292 ka to Holocene (Hildreth and others, 2002). Pumice-covered leveed lavas of southwest sector (at left) and summit crater are postglacial. Mount Juhle at far left; Griggs fork and snout of Glacier 4 at far right.

possibly early Holocene. The dacite is also overlapped locally by the fan of leveed Holocene lavas (55–62% SiO_2) that cover a fourth of the volcano's surface. Flows of the fan retain rugged near-primary surfaces (fig. 146).

Other Volcanoes Behind the Volcanic Front

In addition to Mount Griggs, there are several additional rear-arc volcanoes, isolated or clustered, well behind the closely spaced, single-file chain of volcanic-front centers (fig. 6). These include the Savonoski River cluster, the Saddlehorn Creek cluster, and the Gertrude Creek cone (all on fig. 6*A*), for all of which radioisotopic ages have been determined (Hildreth and others, 2004). Most are of Pleistocene age, and all are small and either monogenetic or short-lived, except Mount Griggs. Analytical data and geologic descriptions are given for all by Hildreth and others (2004), and supplementary electronic files accompanying that article illustrate several of them.

Present-Day Magma Distribution?

Here we review what can be inferred about magma distribution beneath the Katmai cluster today, and we summarize what is known about the sizes and compositions of prehistoric magma batches in the cluster.

Gravity and seismic data that identify a low-density low-velocity region centered near Katmai Pass were summarized by Ward and others (1991). A southeasterly gravity traverse that crossed the volcanic axis by way of Katmai Pass showed cross-axial width of the anomalous region to be about 15 km, but the data are insufficient to know how far it extends along the axis. Of the several seismic stations in the Katmai array, only the one in Katmai Pass (central to the gravity anomaly) consistently showed traveltime delays for deep local and regional earthquakes. The delays suggest great thickness for the low-velocity domain, which may well extend deeper than 20 km (Ward and others, 1991), thus involving most or all of the crust. Their seismic data also indicated significant attenuation of P and S waves, but they found no screening of S waves (in contrast to seismic reconnaissance studies 25 years earlier). To explain the data, they invoked present-day crustal magma storage, favoring scattered small magma bodies rather than a large chamber.

The pattern of seismicity recognized by Ward and others (1991) has persisted during the past two decades; the Alaska Volcano Observatory continues to locate 40–130 earthquakes each month along and near the volcanic axis of the Katmai cluster (Jolly and McNutt, 1999). Most of the earthquakes fall into three persistent clusters: (1) beneath the volcanic line from Martin to Mageik; (2) beneath Katmai Pass and the northwest slope of the Trident group; and (3) beneath Mount Katmai, especially its northern slopes. In a fourth area, north of Snowy Mountain, the seismicity is less abundant and more scattered. Nearly all the earthquakes in all four areas are smaller than magnitude 2.5, and only a few events have been in the range *M* 3.0–4.5. Many are shallower than 5 km, and nearly all are shallower than 10 km. The dense shallow seismicity seems inconsistent with present-day upper-crustal storage of voluminous magma and is more likely to reflect fluid flow, volume changes, and hydraulic microfracturing in hydrothermal systems, along with slip-threshold reduction in altered rocks. If magma bodies other than small pods and dikes are really present in the seismically active, evidently brittle, top 10 km of the crust, they are more likely to underlie the gaps between the seismic clusters. If shallow chambers exist, they could lie beneath East Mageik or between Trident and Mount Katmai. Beneath the seismic clusters themselves, any magma bodies today are probably deeper than 10 km. Nonetheless, shallow components of the pre-1912 reservoir might once have occupied what are now seismically active domains that were effectively drained by the 1912 eruption, any residues having subsequently crystallized.

Vigorous fumaroles in the Katmai cluster are evidence that magma stored somewhere below is actively degassing, though this provides little or no control on depths, distribution, volume, or degree of crystallization of such magma. Superheated jets high on Griggs and around the crater lake of Mageik are steam-dominated but rich in CO_2 and H_2S and have magmatic He-isotope ratios 7–8 times the atmospheric value. Boiling-temperature fumaroles on the southeast flank of Trident I have similarly magmatic signatures and have discharged vigorously since at least 1916 when first recorded (photo on p. 99 of Griggs, 1922). Though not yet sampled directly, many fumarolic jets in the crater of Mount Martin are precipitating abundant sulfur, are probably superheated, and combine to discharge the biggest and (olfactorily) H_2S-richest plume in the Katmai district. USGS flights through the Martin plume have measured high levels of CO_2 and SO_2 (Doukas and McGee, 2007).

From the point of view of resolving the magmatic plumbing puzzle for the Katmai cluster, there are too many "smoking guns." The vigorous fumaroles on Martin, Mageik, Trident I, and Griggs have apparently been continuously active with little change in style or configuration since they were first photographed in the period 1913–17 (Griggs, 1922), but there is no evidence that any of these volcanoes produced any ejecta in 1912 or erupted at any time since. In contrast, fumaroles at the three vents that did erupt in the 20th century have been in decline.

Those on the Southwest Trident cone were superheated during the 1953–74 eruptive episode and were still depositing sulfur and remained at or above the boiling point during our sampling in 1979, but they have since declined both thermally and in H_2S output (R.B. Symonds, oral commun., 1998). The many fumaroles on the Katmai caldera floor (Fenner, 1930), although drowned under rising lakewater by 1929, continued for decades to discharge beneath the lake. Yellowish-green upwellings and failure of the lake to freeze over fully in winter attested for many years to their survival and to concomitant thermal flux through the caldera floor (Muller and Coulter, 1957a; Motyka, 1977). Our own visual observations (1976–2003) suggest, however, that these chemical and thermal fluxes weakened appreciably during the 1990s. In the 1912 vent area, within 1.5 km of the lava dome, numerous high-temperature fumaroles (Allen and Zies, 1923) declined gradually, passing through an acid-argillic alteration phase and leaving today only odorless wisps of near-neutral wet steam that discharge weakly at 20°C–90°C (Keith, 1991b). Declines in temperature and magmatic-gas flux at the three 20th-century vents (in contrast to sustained discharge at Trident I, Griggs, Mageik, and Martin) may signify that Mount Katmai, Southwest Trident, and Novarupta are no longer linked by fracture networks to deep magma reservoirs. If so, does this reflect self-sealing of fluid-filled fractures, cooling and crystallization of magma conduits, or nearly complete removal of the magma formerly under each of them? For Novarupta, the rapid fumarolic decline supports the likelihood that no part of the voluminous magma reservoir directly underlay the 1912 vent, bolstering the case for transport by dike or sill from reservoir components elsewhere (Hildreth, 1987; Wallmann and others, 1990).

Although modulated by passage through upper-crustal hydrothermal systems, the magmatic gases discharging at so many sites indicate that cooling magma is distributed widely and discretely at indeterminate depths beneath the Katmai cluster. Depletion or exhaustion of local upper-crustal magma bodies by eruption need have little impact on coexisting independent components of deep-crustal magma storage systems distributed beneath the volcanic line.

Typically Small Magma Batches

Eighteen vents distributed along the 25-km volcanic line from Mount Martin to Mount Katmai have built the chain of cones and domes by sporadic eruption of many small batches, mostly of pyroxene-plagioclase dacite and andesite rich in phenocrysts. Volumes of lava domes here are in the range 0.01 to 0.3 km^3, most lava flows 0.01 to 0.4 km^3, and quasi-conformable stacks of compositionally similar lava flows generally in the range 0.1 to 1 km^3. The whole 1953–74 eruptive episode that built Southwest Trident amounted to only 0.7 km^3. The most voluminous individual lava flows of East Mageik are smaller than 0.4 km^3, but sets of morphologically and compositionally similar ones may have erupted in short order from magma batches as large as 2 km^3. As an extreme possibility, the ejecta cone and ten Holocene lava flows making up 7-km^3 Mount Martin are similar enough that they might have been derived from an effectively monogenetic magma batch. Most exceptional of all are the two major plinian eruptions, of 1912 and of late Pleistocene rhyodacite at Mount Katmai. Each released on the order of 10 km^3 of magma, but they involved the most silicic magmas ever erupted in the Katmai cluster, providing the exceptions that illustrate the rule—small phenocryst-rich andesite-dacite batches rather than large magma chambers undergoing strong internal differentiation. Accordingly, the two large plinian events produced the only phenocryst-poor eruptive products in the later history of the Katmai cluster. All other late Pleistocene and Holocene products of Martin, Mageik, Trident, Katmai, and Griggs Volcanoes contain at least 20 percent phenocrysts (plagioclase >> pyroxenes > Fe-Ti oxides ≥ olivine), and many have 30–50 percent.

Widespread incidence of chilled magmatic enclaves (varied but generally of mafic to silicic andesite) in most of the andesitic and dacitic lava flows of Mount Katmai and Trident is further evidence of repeated mingling of separate magma batches, supporting the idea of a continually changing plexus of dikes, sills, and modest irregular or podiform magma bodies. Such enclaves are conspicuously absent in the voluminous rhyolite of Layer A (which initiated the 1912 zoned sequence), as they typically are in early-emplaced parts of many major pyroclastic deposits erupted from thick zoned chambers.

Intermediate lava flows here characteristically have several populations of complexly zoned phenocrysts (Coombs and others, 2000), suggesting disorderly histories of mingling and changing physical conditions. Many flows are internally heterogeneous (in the 1 to 2 percent SiO_2 range) rather than having the homogeneity or simple zonation that might be expected from a large chamber. If large reservoirs of intermediate magma do exist here, they probably lie in the mid-to-deep crust and sporadically yield ascent batches, typically smaller than 1 km^3, as dikes that erupt, stall, or augment an upper-crustal plexus. Only occasionally would the ascent flux be sufficient that parts of such a plexus would enlarge into shallow chambers big enough to undergo sustained systematic zonation and evolution of silicic roof zones capable of voluminous plinian eruptions and attendant caldera collapse.

Baked Mountain Hut, the only fixed shelter in the Valley of Ten Thousand Smokes (VTTS), stands 4 km northwest of Novarupta. Built in 1965 by the Geophysical Institute of the University of Alaska as a research station, the facility lacks water and power, but through the years it has accommodated scores of academic and U.S. Geological Survey scientists and countless visiting backpackers. Windblown sand and pumice have scoured the walls and piled up against the upvalley (right) side of the hut. View is northeastward across the Knife Creek arm of the VTTS toward Mount Griggs. Photo by Wes Hildreth in August 1976.

Chapter 13

1912 Magma Compositions and Preeruptive Storage

The ~13 km^3 of magma released in 1912 consisted of ~4.5 km^3 of dacite and ~1 km^3 of andesite (both having 25–42 weight percent phenocrysts) and 7–8 km^3 of crystal-poor high-silica rhyolite with only 0.5–3 weight percent phenocrysts (fig. 147). A large compositional gap separates the slightly zoned rhyolite from the strongly and continuously zoned andesite-dacite (figs. 59, 148). A second apparent compositional gap suggested by our first 30 pumice analyses (Hildreth, 1983)—separating black andesite and light grey dacite (fig. 4)—was an artifact of limited sampling and does not exist. The 1912 ejecta include neither rhyodacite nor phenocryst-rich rhyolite comparable to the products of many other large zoned ignimbrite eruptions. Homogeneous pumice containing 69–77 percent SiO_2 is absent.

1912 Bulk Compositions

Major Elements

Major element contents were determined by XRF for ~265 samples of 1912 eruptive products, including ~70 pumice blocks from the ignimbrite, 8 lava samples from Novarupta dome, 8 blocks from Phantom dome, and ~180 pumice clasts representing all fall units. Data are tabulated in appendix A, and several elements are plotted in figures 59 and 148. Apart from the very unusual coeruption of high-silica rhyolite, the 1912 intermediate suite represents an ordinary subalkaline, low-Ti, medium-K, calcalkaline arc array. The narrow array (fig. 148) is strongly calcalkaline by the criterion of Miyashiro (1974) but calcic by the alkali-lime index of Peacock (1931), and it straddles the low-Fe/medium-Fe boundary of Arculus (2003). Table 2 summarizes and compares the compositional ranges for all rhyolite, dacite, and andesite samples analyzed.

The phenocryst-rich andesite-dacite array forms a dense continuum, with no discernible compositional distinction between pumice clasts emplaced as fallout or ignimbrite. The conventional boundary between andesite and dacite at 63 percent SiO_2 happens to coincide roughly here with a change in pumice color from black to pale gray (fig. 4), but it should not be inferred that they represent different magma batches or that 63 percent SiO_2 represents any kind of compositional discontinuity. SiO_2 contents range continuously from 57.9 to 68.6 percent whereas Al_2O_3 drops from 17.1 to 14.6 percent. Across the array from andesite through dacite, K_2O rises from 1.2 to 2.2 percent and Na_2O from 3.3 to 4.3 percent; contrarily, FeO* declines from 7.4 to 4.2 percent, MgO from 4.6 to 1.5 percent, CaO from 7.8 to 3.9 percent, and TiO_2 from 0.75 to 0.51 percent. The minor components MnO and P_2O_5 likewise decline with SiO_2 enrichment within the low concentration ranges 0.15–0.09 percent and 0.22–0.09 percent, respectively. Although most black scoria clasts (fig. 4) fall within the array at 58–63 percent SiO_2, two that we took from the S bed (fig. 48) atop the Turtle were uniquely mafic among 1912 products in having only 54.6 and 50.4 percent SiO_2; only the second was noticeably enriched in olivine. In addition, three mafic enclaves within Novarupta dome rhyolite gave SiO_2 contents of 57.1, 56.5, and 55.5 percent, significantly outside the main array (fig. 148; table 2).

The phenocryst-poor, volumetrically dominant rhyolite of 1912 shows limited but real compositional variations. SiO_2 is in the range 76.5–77.8 percent, but only (slightly contaminated) Novarupta dome samples have less than 77.1 percent. Al_2O_3 ranges from 11.9 to 12.6 percent, and alkali data are tightly grouped at 4.0–4.4 percent Na_2O and 3.1–3.3 percent K_2O. Other major element data define quite limited ranges for the 1912 rhyolites: 0.13–0.23 percent TiO_2; 1.2–1.5 percent FeO*; 0.04–0.07 percent MnO; 0.09–0.19 percent MgO; 0.81–1.16 percent CaO; and 0.02–0.09 percent P_2O_5.

Major-element compositions of matrix glass from 1912 pumice clasts have been determined by electron microprobe by Avery (1992), Westrich and others (1991), Fierstein and Hildreth (1992), and Hammer and others (2002). For the phenocryst-poor rhyolite, glass composition is not much different from bulk composition. For the crystal-rich dacite pumice, glass composition is high-silica rhyolitic, but (despite a little overlap) it is generally distinguishable by microprobe from the slightly more evolved glass of the rhyolite pumice (fig. 149). For the crystal-rich andesite scoriae, glass compositions reported are far more varied, ranging from dacitic to rhyolitic (65–73% SiO_2). A summary of the abundant glass data follows, for rhyolite, dacite, and andesite, respectively, in weight percent for each oxide: SiO_2 77–79, 76–79, 65–74; TiO_2 0.08–0.21, 0.20–0.35, 0.6–0.8; Al_2O_3 11.9–12.2, 11.7–12.8, 13.5–15.5; FeO* 1.0–1.4, 1.3–1.85, 2.5–4.6; MgO 0.09–0.14, 0.2–0.28, 0.55–1.4; CaO 0.7–0.9, 0.8–1.4, 2.2–3.6; Na_2O 4.0–4.4, 4.0–4.1, 4.1–4.6; and K_2O 3.1–3.3, 2.8–3.2, 2.2–2.7.

Table 2. Compositional summary for 1912 eruptive products.

[The 10 major-element oxides are in weight percent; trace elements in parts per million; —, not analyzed . Chemical data are from Appendices A–C. Isotope data are in Hildreth, 1983, 1987; Reagan and others, 2003, Hildreth and others, 2004; and Turner and others, 2010.]

	Rhyolite	Dacite	Andesite	5 Mafic Samples
		Major elements		
SiO_2	76.5–77.8	63.0–68.6	57.9–63.0	50.4–57.1
TiO_2	0.13–0.23	0.51–0.71	0.67–0.75	0.69–0.81
Al_2O_3	11.9–12.6	14.6–16.2	16.1–17.15	14.0–17.9
FeO*	1.20–1.53	4.2–5.95	5.9–7.35	7.7–10.1
MnO	0.04–0.07	0.09–0.13	0.12–0.15	0.14–0.17
MgO	0.02–0.34	1.5–2.7	2.65–4.6	4.1–11.4
CaO	0.81–1.16	3.85–5.7	5.7–7.8	7.6–10.1
Na_2O	4.0–4.4	3.6–4.3	3.2–4.0	2.0–3.7
K_2O	3.10–3.28	1.56–2.18	1.17–1.57	0.50–0.92
P_2O_5	0.02–0.09	0.09–0.22	0.12–0.22	0.09–0.16
		Trace elements		
Rb	54–79	24–50	23–37	8–23
Cs	2.3–3.0	1.1–1.7	0.9–1.2	0.36–0.72
Sr	58–95	210–280	260–330	270–360
Ba	790–1025	370–800	330–580	155–340
Y	40–48	30–42	20–31	14–25
Zr	120–165	120–170	100–140	42–102
Hf	4.3–5.2	4.1–4.9	2.9–3.7	1.6–2.5
Nb	4.0–5.4	3.0–3.9	2.5–3.5	1–2
Ta	0.40–0.44	0.26–0.42	0.20–0.34	0.08–0.19
Th	4.8–5.4	2.8–3.5	1.1–3.1	1.0–1.9
U	2.1–2.5	1.3–2.0	0.5–1.5	0.4–0.8
La	14–21	8–16	6–15	5–7.5
Ce	32–38	22–26	19–22	11–16
Nd	18–21	13–15.5	11.5–14	7.5–11
Sm	5.0–5.9	3.8–4.4	3.3–4.0	2.3–3.1
Eu	0.71–0.86	0.97–1.14	0.94–1.07	0.8–0.9
Gd	5.2–6.1	4.0–4.7	3.5–4.3	2.4–3.5
Tb	0.94–1.11	0.72–0.84	0.62–0.74	0.43–0.56
Yb	3.8–4.9	2.7–3.9	2.3–3.0	1.6–2.5
Lu	0.62–0.72	0.44–0.51	0.37–0.45	0.24–0.38
Sc	7–9	17–23	23–28	29–46
V	1–12	44–120	120–175	—
Cr	0–2	4–18	18–55	28–230
Co	0.5–0.7	10–14	17–19	23–54
Ni	0–3	1–12	4–32	11–64
Cu	1–3	6–31	13–36	—

Table 2. Compositional summary for 1912 eruptive products.—Continued

	Rhyolite	Dacite	Andesite	5 Mafic Samples
		(Trace elements)		
Zn	30–44	47–90	64–113	69–84
Ga	12–16	16–28	17–29	—
Li	31–56	17–26	16–21	—
Be	1.5–2.0	0.5–0.8	0.6	—
B	31–58	28–41	<4	—
Mo	2.9–2.9	1.6–1.7	1.4–1.5	—
Sn	1.8–2.0	0.6–1.0	0.9–1.8	—
Sb	0.9–1.2	0.5–1.2	0.3–0.7	0.1–0.3
W	0.86–0.98	0.40–0.46	0.36–0.50	—
Pb	10–14	6–13	5–7	2.4
		Ratios		
FeO*/MgO	7–19	2.2–2.9	1.6–2.2	0.9–2.0
K_2O/Na_2O	0.71–0.80	0.38–0.56	0.35–0.43	0.24–0.31
K/Ba	25–32	20–35	21–34	22–27
K/Rb	340–505	320–530	320–530	320–480
Rb/Cs	22–28	20–23	20–22	24–35
Rb/Sr	0.90–0.94	0.11–0.17	0.08–0.11	0.03–0.07
Ba/Sr	13–14	1.8–3	1.3–1.8	0.6–1.0
Ba/La	47–55	42–52	37–48	37–52
Ba/Nb	190–200	130–160	120–160	144
Ba/Zr	5–6.5	2.6–4.5	2.6–4.4	2.8–3.3
Zr/Hf	27.5–29	33–34.5	33–34.5	35–42
Sr/Nd	3.0–3.2	13–20	19–25	32–36
Ce/Yb	8.3–8.9	7.5–8.6	7.5–8.5	5.6–8.1
La/Sm	3.0–3.9	2.6–3.6	2.3–3.4	2.1–2.5
Sm/Yb	1.1–1.3	1.2–1.4	1.3–1.5	1.55
Eu/Eu*	0.4–0.5	0.7–0.8	0.8–0.9	0.8–0.9
Th/U	2.05–2.11	2.09–2.17	2.15–2.21	2.2–2.6
$^{87}Sr/^{86}Sr$	0.70368	0.70354	0.70352	0.70338
	0.70374	0.70362	0.70374	0.70348
$^{143}Nd/^{144}Nd$	0.512987	0.512967	0.512995	0.512952
	0.512996	0.513010	—	0.513035
$^{206}Pb/^{204}Pb$	18.85	18.88	—	18.88
$^{207}Pb/^{204}Pb$	15.57	15.57	—	15.58
$^{208}Pb/^{204}Pb$	38.375	38.38	—	38.39

Trace elements

Trace element contents of the 1912 magmas were listed by Hildreth (1983, 1987) and Turner and others (2010), and abundant additional data are tabulated in appendices B and C and summarized in table 2. Some trace-element data for matrix-glass separates from 1912 rhyolite, dacite, and andesite pumices were given by Lowenstern (1993). Across the andesite-dacite continuum, the following ranges (all in ppm) have been determined for individual pumice clasts: 1.1–1.7 Cs, 23–50 Rb, 210–330 Sr, 330–800 Ba, 100–170 Zr, 2.5–3.9 Nb, 6–16 La, 2.3–3.9 Yb, 5–13 Pb, and 1.1–3.5 Th. The 1912 rhyolites are more enriched than the dacites in these and most other incompatible elements (table 2; figs. 150, 151, 152) with the exceptions of Sr (58–95 ppm) and Zr (120–165 ppm), the relative declines of which are attributable to fractionation of abundant plagioclase and traces of zircon. Trivalent rare earth elements (REE) are enriched progressively from andesite to dacite to rhyolite (fig. 153), but their normalized patterns remain nearly parallel across the whole compositional range, while Eu deficiency increases toward rhyolite (table 2; fig. 10 of Hildreth, 1983, Turner and others, 2010). Ni and Cr are quite depleted, 1–32 ppm and 4–55 ppm, respectively, in the dacite-andesite array and are, of course, still more impoverished in the low-MgO rhyolites.

Incompatible trace-element patterns as displayed on primitive-mantle-normalized diagrams (Hildreth and others, 2004; Turner and others, 2010) are typical of those from many magmatic-arc suites, showing spikes in fluid-mobile Cs, Ba, Sr, Pb, and U and troughs in less mobile Ti, Nb, Ta, and Th (fig. 152). Accordingly, the elevated values (table 2) of such ratios as Ba/La, Sr/Nd, Ba/Nb, Ba/Th, U/Th, and Cs/Rb are characteristic of the geochemical signature of the slab-derived fluid flux beneath arc volcanoes globally. For the Katmai region in particular, the across-arc waning of such geochemical subduction indicators was investigated in detail by Hildreth and others (2004), who compared compositions of volcanic-front magmas (including Novarupta and Mount Katmai) with those of several rear-arc volcanoes nearby.

Affinities to Nearby Volcanoes

In the course of mapping all eruptive components of the Katmai cluster (Hildreth and Fierstein, 2003), we assembled compositional data for more than 700 samples in an effort to

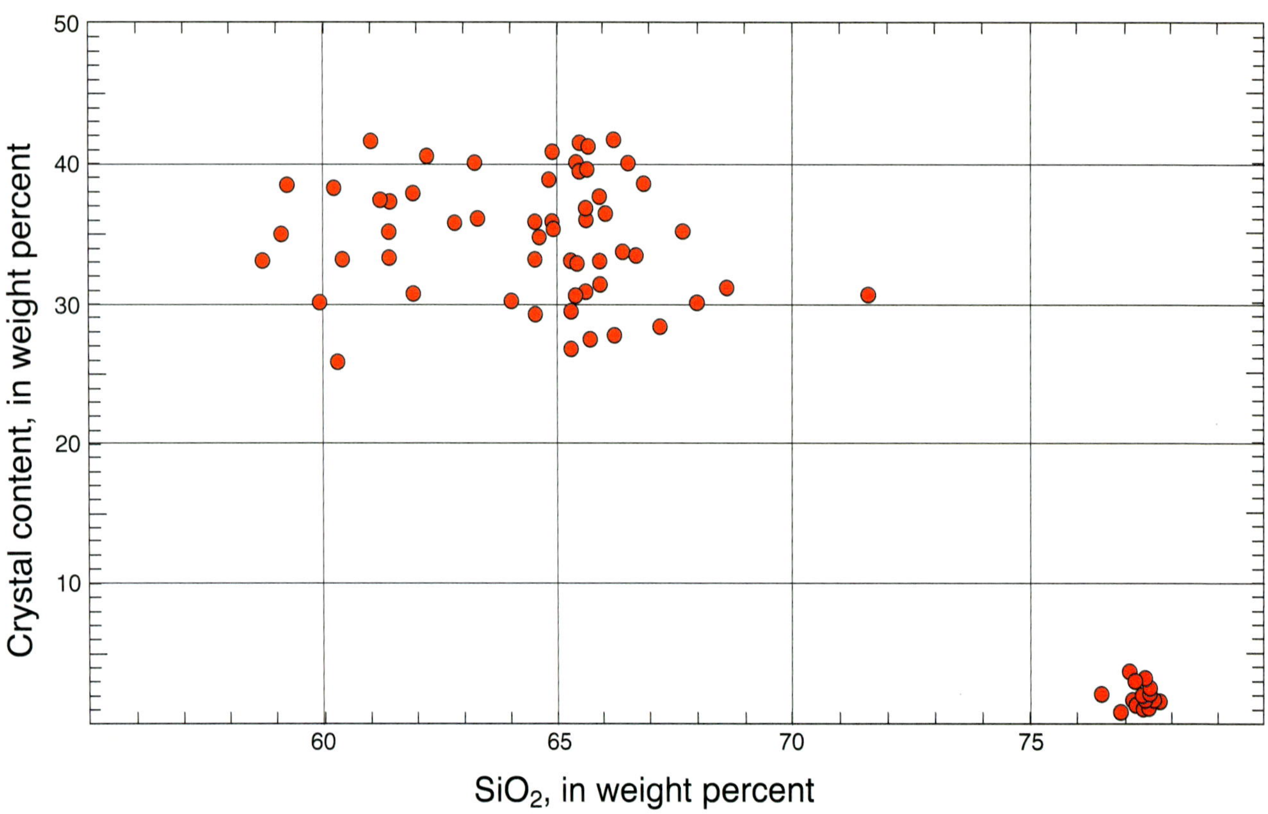

Figure 147. Weight percent crystals versus whole-pumice weight percent SiO_2 for ~70 blocks of 1912 pumice and scoria. After cleaning of their surfaces, individual blocks were halved—one fraction for chemical analysis, the other for mineral separation in heavy liquid. Phenocryst-poor rhyolite contrasts with unsystematically phenocryst-rich andesite-dacite continuum.

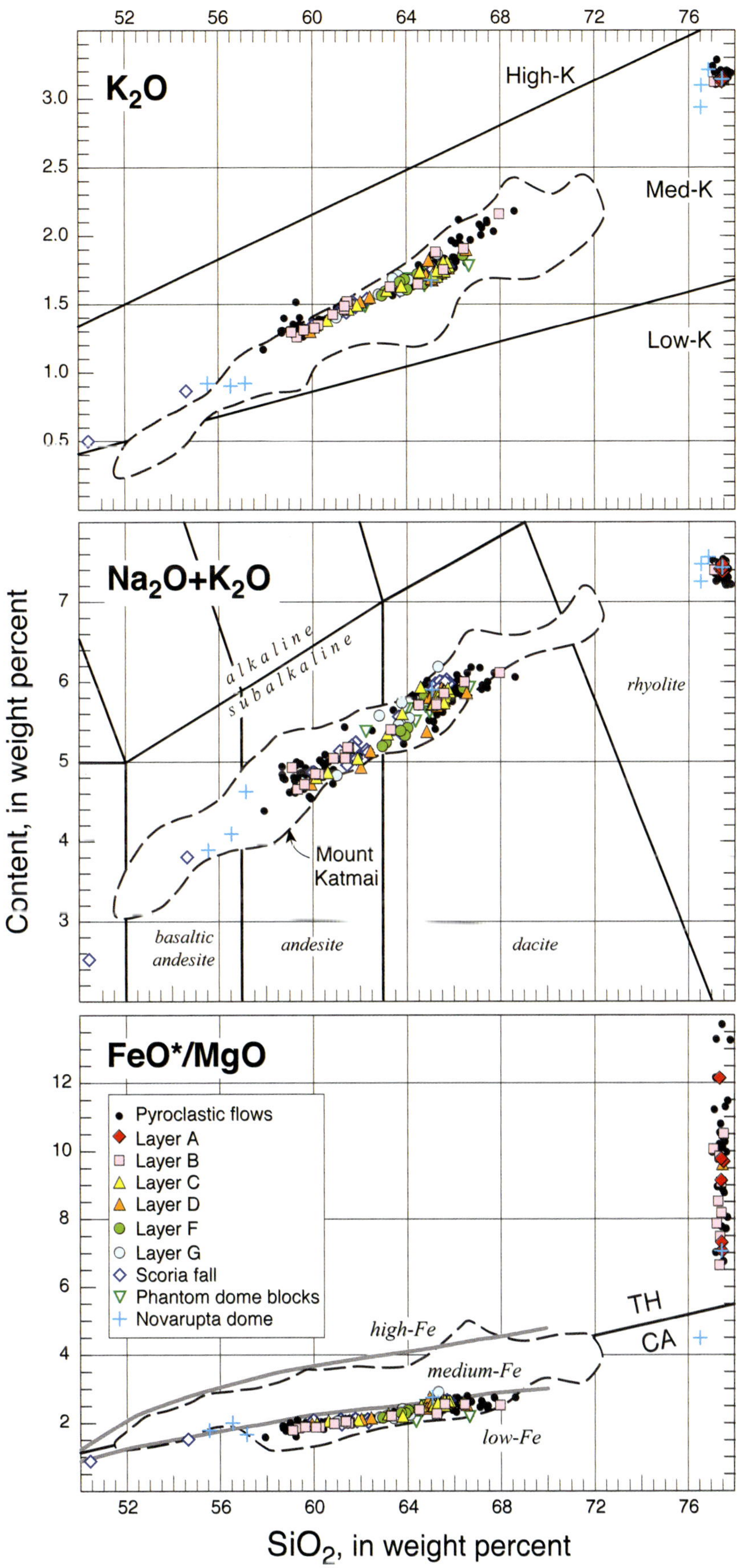

Figure 148. Compositional variations versus SiO_2 content (all in weight percent) for ~265 samples of 1912 Novarupta pumice and dome lava, subdivided as indicated in inset key. Principal features are continuity of andesite-dacite array, wide compositional gap from dacite to rhyolite, slight contamination of dome samples, and five (rare) mafic outliers. Outlined enclosures are fields for ~150 samples from Mount Katmai, illustrating its compositional affinity with Novarupta. Alkaline/subalkaline boundary and named fields follow LeBas and others (1986). Tholeiitic/calcalkaline (TH/CA) boundary after Miyashiro (1974). Low-, medium-, and high-Fe fields after Arculus (2003). Data in appendix A.

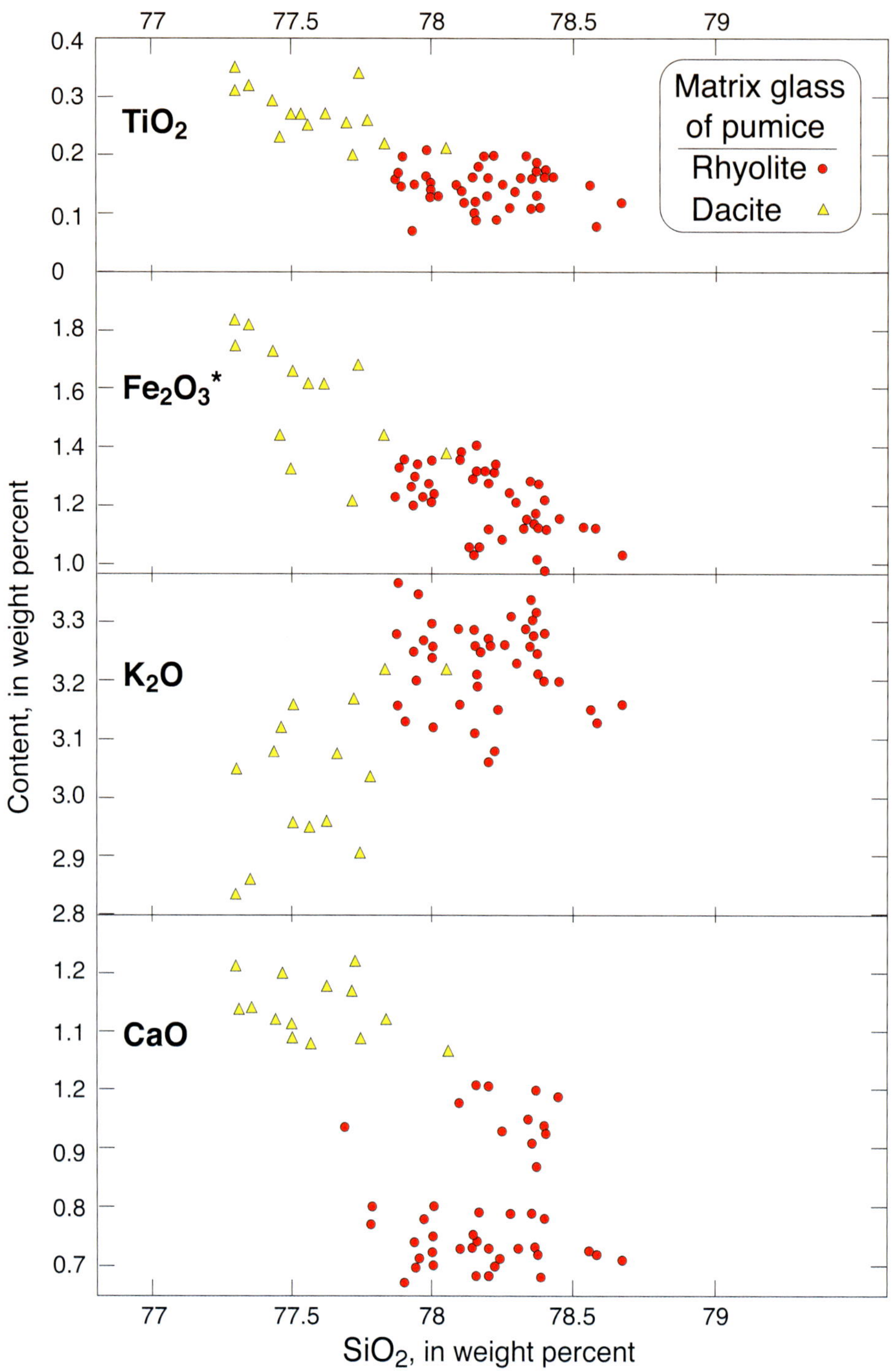

Figure 149. Matrix glass compositions for three rhyolitic and two dacitic pumice clasts, visually homogeneous, determined by electron microprobe by Avery (1992). SiO_2, TiO_2, K_2O, CaO, and total iron as Fe_2O_3, all reported in weight percent after normalization of analytical totals to 100 percent. In spite of wide compositional gap (69–77 percent SiO_2) between whole-pumice data for dacite and rhyolite, glass compositions are virtually continuous. A more extensive data set for pumice clasts and ash shards would extend the rhyolite range above 79 percent SiO_2 and the zoned dacite range below 77 percent SiO_2 (table 3b of Fierstein and Hildreth, 1992).

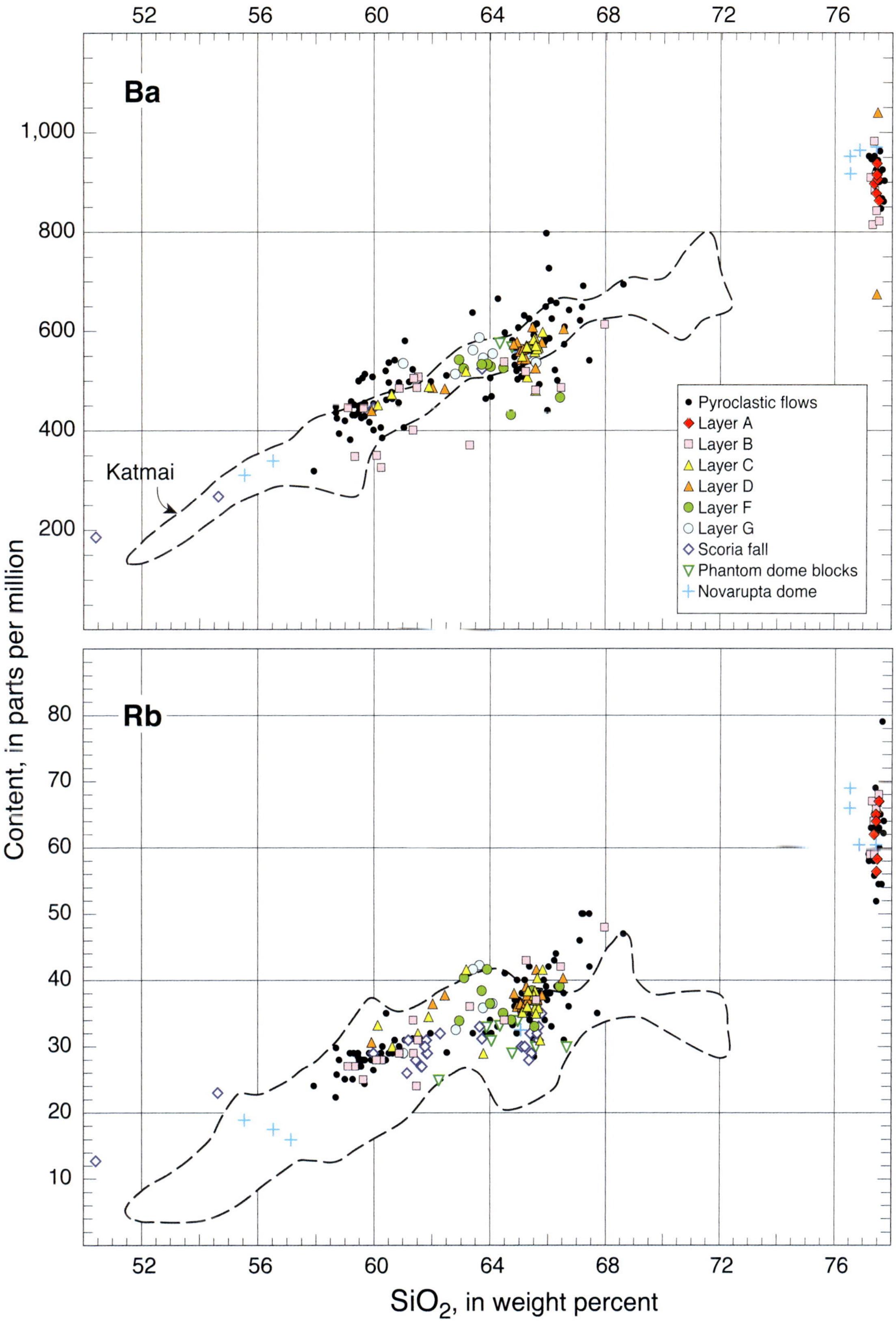

Figure 150. Variation of Ba and Rb (in ppm) versus weight percent SiO_2 for ~265 samples of 1912 Novarupta pumice and dome lava, subdivided as indicated in inset key. Outlined enclosures are fields for ~150 samples from Mount Katmai.

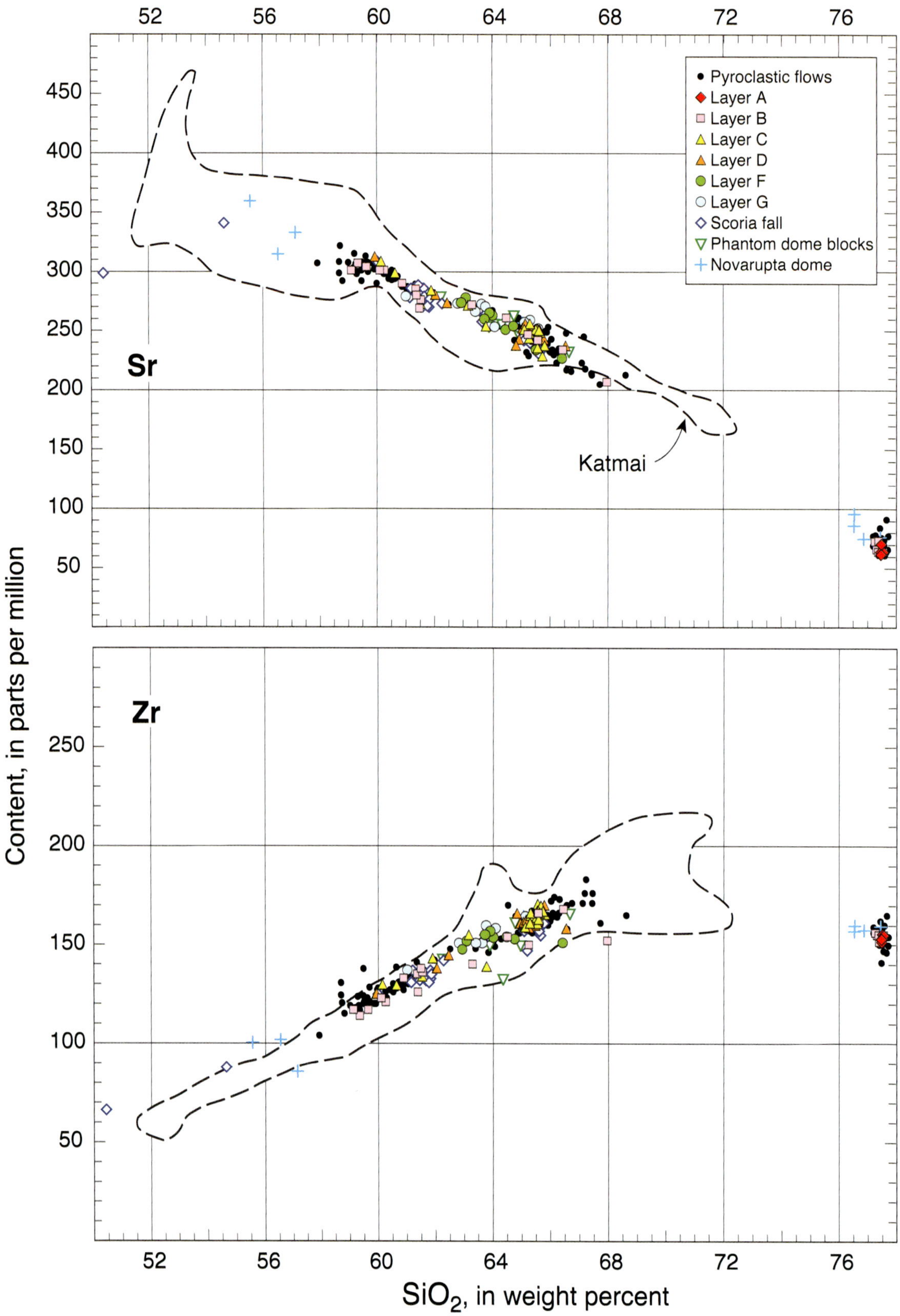

Figure 151. Variation of Sr and Zr (in ppm) versus weight percent SiO_2 for ~265 samples of 1912 Novarupta pumice and dome lava, subdivided as indicated in inset key. Outlined enclosures are fields for ~150 samples from Mount Katmai. Note pronounced effect of zircon fractionation on Zr content of rhyolites.

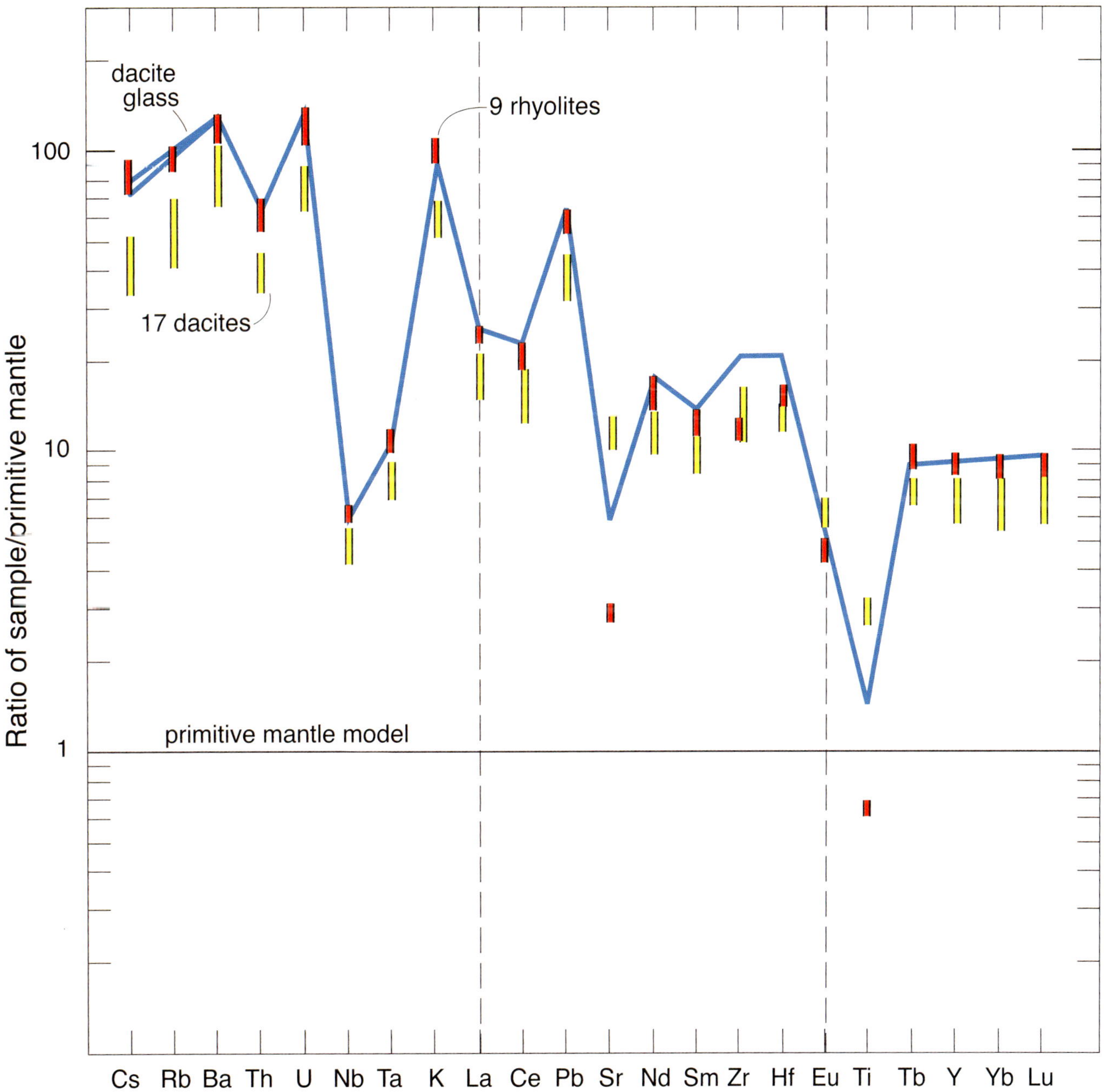

Figure 152. Incompatible trace-element patterns for 1912 dacite and rhyolite, normalized to average primitive mantle composition of Sun and McDonough (1989). Red bars represent nine rhyolite pumice clasts and yellow bars 17 dacite pumice clasts. Bold blue line represents two glass separates from crystal-rich dacite pumice clasts; within analytical errors, their plots coincide except for Cs and Rb. Diagram illustrates that melt extracted from bulk dacite is similar to bulk rhyolite, except that rhyolites contain lower Sr, Zr, Eu, and Ti, inferentially as a result of further fractionation of plagioclase, zircon, and Fe-Ti oxides.

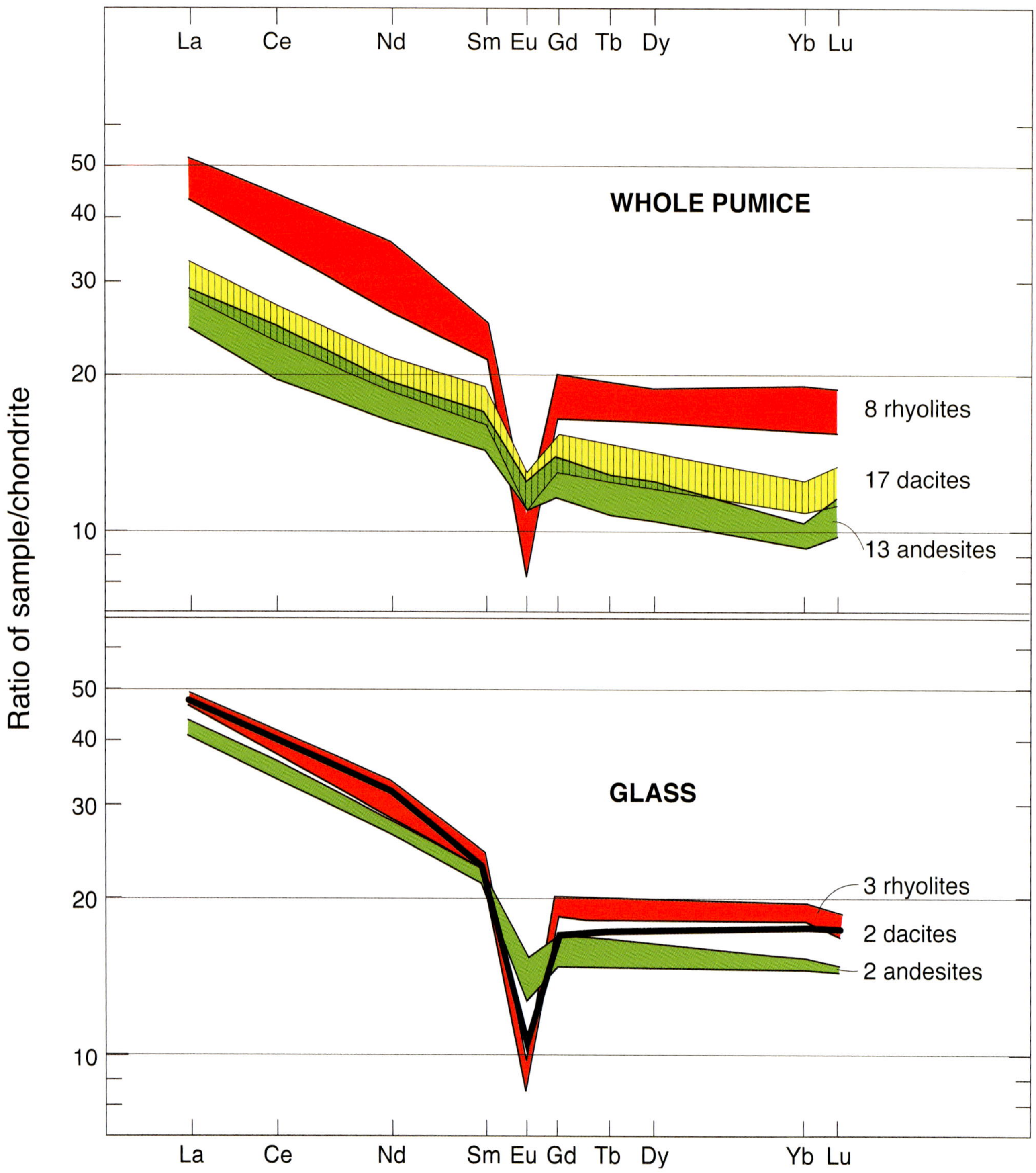

Figure 153. Chondrite-normalized rare-earth-element abundances for 1912 pumice blocks and for high-purity matrix-glass separates thereof, arranged by ionic radius. For whole-pumice blocks, dacite field (lined) partly overlaps that for andesites. For glass separates, two dacite samples are, within analytical errors, virtually identical and plotted as a single bold line.

assess magmatic affinities of the 1912 products with those of the several stratocones and domes nearby. Novarupta lies at the toe of the Trident complex, and the 1912 vent cuts across one of its peripheral lava domes (Falling Mountain). But it also lies close to Mageik and Griggs Volcanoes and, of course, Mount Katmai collapsed nearby during the 1912 eruption. All of these, and Mount Martin and Snowy Mountain as well, consist predominantly of andesite-dacite, and all have produced Holocene eruptions. Not one has erupted any late Quaternary high-silica rhyolite.

The analysis of Hildreth and Fierstein (2000) showed that all products of Mount Griggs and all Pleistocene products of Trident and Mageik are unambiguously distinguishable geochemically from the andesite-dacite continuum of 1912. Some Holocene dacites of Mount Mageik are similar to 1912 dacites, but many others are relatively enriched in K, Fe, and Zr. The Southwest Trident lavas and ejecta of 1953–74 are similar in many respects to the 1912 andesite-dacite array, but they are consistently lower in K and the andesites are lower in Fe. Geochemical affinity of the 1912 suite is closest to the products of Mount Katmai (figs. 59, 148, 150, 151), in particular to the dacite-rhyodacite array erupted from its southwest cone in the interval ~23–10 ka. Moreover, Southwest Katmai is the only Quaternary vent in the Katmai cluster to have erupted both basalt and rhyodacite (51.6–72.3% SiO_2) in addition to its andesite-dacite array. Its caldera floor, of course, also provided the vent site for the Horseshoe Island dacite in the aftermath of caldera collapse synchronous with the eruptions at Novarupta. It thus appears most plausible that the andesite-dacite magma released at Novarupta came from beneath Mount Katmai, even though its transport path may have led under parts of the older Trident edifice. What was the preeruptive storage site of the 1912 rhyolite, and whether it was conjoint with the andesite-dacite or not, are questions addressed at length below.

Volatile Compositions

Preeruptive water contents were estimated by Fourier transform infrared spectroscopy (FTIR) and ion-probe analysis of melt (glass) inclusions (MI) trapped in phenocrysts, presumably during crystal growth in the magma reservoir prior to syneruptive decompression (Westrich and others, 1991; Lowenstern, 1993; Wallace, 2005). For MI in quartz and orthopyroxene phenocrysts from 1912 rhyolite, the three studies measured ranges of 3.8±0.7, 3.0–4.7, and 4.2–4.5 weight percent H_2O, respectively. Lowenstern (1993) found overlapping ranges and no consistent differences in water contents of MI in 1912 rhyolite emplaced as fallout, ignimbrite, or lava dome. For MI in plagioclase from 1912 dacite pumice, Westrich and others (1991) measured 2.3±0.6 weight percent H_2O by ion probe and Lowenstern (1993) 2.2–3.1 weight percent H_2O by FTIR. If such values accurately represent pre-eruptive water content of the dacite melt, the dissolved water content of the crystal-rich dacite magma would have been ≤2 weight percent. For 1912 andesite pumice, nearly all MI examined (in several laboratories) have been found to be partly crystallized and/or fractured; nonetheless, Westrich and others (1991) reported 1.0±0.1 weight percent H_2O in eight bubble-bearing MI in andesite plagioclase, although it remains unresolved whether such low water contents might reflect gas loss during preeruptive magma ascent or posteruptive modification of MI.

CO_2 content of MI in 1912 rhyolite quartz has been determined only by Wallace (2005), who found a low range of 20–50 ppm by FTIR spectroscopy (using a CO_2-purged system with a very low detection limit). Lowenstern (1993) reported CO_2 contents to be below his 50 ppm detection limit in MI from both dacite and rhyolite. The combined H_2O–CO_2 data, in conjunction with experimentally determined vapor-saturation isobars (fig. 8 of Wallace, 2005; Newman and Lowenstern, 2002), suggest preeruptive storage of the rhyolite magma at 100–130 MPa (1–1.3 kb), consistent with the phase-equilibrium experiments of Coombs and Gardner (2001). Lowenstern (1993) independently made a strong case for preeruptive gas saturation (in all 1912 magmas), based on the presence of Cu-rich bubbles in MI, the low CO_2 and S contents of MI, and the need for partitioning into a fluid phase to account for marked decreases of S and Cu concentrations and suppression of Cl enrichment from andesite to dacite to rhyolite.

Halogen contents of MI were determined by microbeam analysis by Westrich and others (1991), who reported 1,930±120 ppm Cl and 630±210 ppm F for MI from rhyolite pumice; 1,790±220 ppm Cl and 640±220 ppm F for MI from dacite pumice; and 1,690±180 ppm Cl and 530±200 ppm F for MI from andesite pumice. They also emphasized that large fractions of the magmatic Cl and F were retained by the matrix glass of the ejecta, leaving them available for release by leaching or devitrification during the subsequent decade of acid-gas fumarolic activity. Palais and Sigurdsson (1989) also determined halogen contents of MI (by electron microprobe) from several samples of 1912 dacite (only), obtaining ranges of 1,710–2,220 ppm Cl and 220–420 ppm F.

Sulfur contents of MI in 1912 dacite, as measured by electron microprobe by Palais and Sigurdsson (1989), range widely, 48–359 ppm S. By the same method, Westrich and others (1991) obtained the following concentrations for MI in 1912 ejecta: <65 ppm S for rhyolite, 120±40 ppm S for dacite, and 170±60 ppm S for andesite. All 1912 ejecta contain pyrrhotite ($Fe_{1-x}S$), implying magmatic sulfur saturation across the whole compositional range.

Mineralogy of 1912 Products

Mineral separations from individual pumice blocks have yielded 0.5–3 weight percent crystals in 1912 rhyolite and a remarkably wide range of 25–42 weight percent crystals in the dacite and andesite (fig. 147). All ejecta contain plagioclase, orthopyroxene, titanomagnetite, ilmenite, apatite, and pyrrhotite, in that order of abundance. In addition, euhedral quartz is only slightly less abundant than plagioclase in the rhyolite, and clinopyroxene is comparable in abundance to orthopyroxene in the dacite and andesite (Hildreth, 1983). Rare zircon occurs in the rhyolite as tiny inclusions in other crystals, and inclusions of Fe-Cu sulfide (cubanite-like "intermediate solid solution") occurs within titanomagnetite in some andesitic scoriae (see table 5 of Lowenstern, 1993). Sparse olivine accompanies the abundant pyroxenes in eight andesites and one dacite—about half the scoria blocks from which minerals were separated.

Rare euhedral hornblende has been observed in mineral separates from nine pumice blocks and in the rhyolitic lava of the Novarupta dome. All the hornblende-bearing pumice blocks were taken from Layer B or from Packages 3 to 7 of the main ignimbrite, thus from compositionally mixed parts of the eruptive sequence. Four of the pumice blocks turned out to have bulk compositions (66.9–68.6% SiO_2) at the high-silica end of the andesite-dacite continuum, but two were rhyolites (77.1–77.4% SiO_2), two were lower-silica dacites (63.2–65.9% SiO_2), and one an andesite (59.1% SiO_2). Hornblende was not found in rhyolite pumice from Layer A.

Abundances of mineral phases in pumice are hard to estimate optically and seldom satisfactorily precise even in mineral separates. The following data represent the combined efforts of Hildreth (1983) and Avery (1992). In 1912 rhyolites, plagioclase is euhedral, generally smaller than 1 mm, and makes up 0.5–2 weight percent of the rock, while dipyramidal quartz is 0.2–0.8 mm across and typically somewhat less abundant (0.3–1 weight percent) than plagioclase. Orthopyroxene prisms are as long as 0.5 mm and roughly 0.2–0.5 weight percent in abundance. Clinopyroxene is absent in most rhyolite samples, but traces (often broken and probably transferred from coerupted intermediate magmas) were observed in mineral separates from the Novarupta lava dome and from a few rhyolitic pumices taken from Layers B, C, and D, but not from Layer A. Euhedral unrimmed hornblende found in Novarupta and in two rhyolite pumice blocks is stubby, 0.1–0.4 mm long, and (where present at all) is less than 0.1 weight percent in abundance. Titanomagnetite is generally 0.1–0.3 mm across and quite sparse in the rhyolite, ≤0.2 weight percent. Ilmenite is of similar size but at least an order of magnitude sparser than titanomagnetite. Apatite, sulfide, and zircon inclusions are tiny (≤0.1 mm), rare in the rhyolites, and have not been observed as free crystals.

In 1912 dacite and andesite, quartz is absent or very rare and presumed xenocrystic; plagioclase is by far the dominant phenocryst, constituting 20–35 weight percent of the rock. Plagioclase ranges in length from ~0.1 mm to 2.5 mm in dacites and rarely up to 4 mm in andesites. Orthopyroxene prisms are 0.1-1 mm long and make up 1.5–4 weight percent of dacites and 1–3 weight percent of andesites. Clinopyroxene prisms are typically 0.2–1 mm long and make up 1–3 weight percent of dacites and 4–8 weight percent of andesites. In mineral separates, opx:cpx ratios are estimated to be as great as 2:1 in some dacites, 1:1 in others, and as small as 1:3 in many andesites. Hornblende prisms are euhedral, unrimmed, and typically 0.1–0.4 mm but rarely ~1 mm long; absent from most pumice blocks examined, they are as abundant as 0.6 volume percent in a few (Avery, 1992). Fe-Ti oxides are as big as 0.5 mm in dacites and 0.3 mm in andesites, but they also occur as tiny inclusions (commonly ≤0.1 mm) in other phenocrysts (except quartz). Ilmenite is very sparse in andesites but may be half as abundant as titanomagnetite in some dacites, as estimated by microprobe in heavy-mineral concentrates (Hildreth, 1983). Olivine is rare, ≤0.5 mm, identified only in andesite scoriae (and one dacite pumice), and may or may not have reaction rims. Sulfides and apatite are widespread as tiny inclusions (≤0.1 mm), the former most common in pyroxenes and oxides, the latter as needles in plagioclase.

Compositions of most mineral phases were summarized by Hildreth (1983) and Avery (1992), who conducted extensive analytical work by electron microprobe. Plagioclase in 1912 rhyolite is unzoned or very weakly so and limited to the oligoclase range An_{25-29}. In dacites and andesites, most plagioclase is optically zoned and euhedral, though many crystals are broken. Plagioclase with sieve texture (myriad interconnected glass inclusions) is uncommon in dacites but abundant in andesites. Individual isolated melt (glass) inclusions (MI) are typically 20–80 µm across in both andesites and dacites. Zoning is generally normal in dacites, but internal dissolution discontinuities are common and zoning reversals present in many plagioclase grains in andesites. Avery (1992) illustrated and documented various styles of oscillatory, discontinuous, and irregular zoning, weak and strong, in andesitic and dacitic plagioclases. In general, plagioclase core compositions are An_{36-82} in dacites and An_{48-89} in andesites, whereas rims are An_{35-53} in dacites and An_{36-83} in andesites (Hildreth, 1983; Avery, 1992). A sparse subpopulation of highly calcic plagioclase—discrete grains almost unzoned in the range An_{87-94}—was found principally in andesitic scoriae (Hildreth, 1983) but also as rare contaminants in Novarupta dome rhyolite. No grains or overgrown cores of oligoclase transferred from 1912 rhyolite have been identified in 1912 andesite or dacite (or in Trident products erupted nearby in 1953–74).

Orthopyroxenes are virtually unzoned in rhyolites at En_{49-50} and have low contents of CaO (~1.1%) and Al_2O_3 (~0.3%). Across the dacite-andesite continuum, orthopyroxenes are modestly zoned but show no compositional overlap with those in the rhyolites. Core compositions are En_{61-65} in dacites and En_{62-73} in andesites, whereas rims are En_{59-63} in dacites and En_{59-74} in andesites. CaO content ranges from 1.2 to 1.9 percent ($Wo_{2.0-3.7}$) and Al_2O_3 widely from 0.6 to 2.6 percent.

Clinopyroxenes are unzoned or weakly zoned in Ca-Mg-Fe but fairly heterogeneous in minor elements. Compositions cluster tightly in the augite field at Wo_{40-45} for both andesites and dacites. Clinopyroxene cores are En_{40-43} in dacites and En_{40-45} in andesites, whereas rim compositions are En_{39-42} in dacites and En_{41-46} in andesites. Al_2O_3 contents of the clinopyroxenes are more varied, 1–2 percent in dacites and 1.4–4.4 percent in andesites; thin reversely zoned (Mg-richer) rims in some clinopyroxenes in andesite scoriae have Al_2O_3 contents as great as 5.0–6.7 percent. In an ion-probe reconnaissance, Shearer and others (1991) found such Mg+Al-enriched pyroxene rims (25–75 µm thick) to be depleted in REE relative to their cores, suggesting late interaction with a more primitive magma batch. Such mafic recharge not long before the eruption is also consistent with the An-rich rims of many plagioclase crystals (and the sparse unzoned An_{90} grains) in the 1912 andesite.

Olivines, found sparsely in about half the andesite scoria blocks separated, are in the compositional range $Fo_{73\text{-}82}$ and contain 0.09-0.18 percent CaO.

Quartz in the rhyolite is euhedral and shows subtle (if any) cathodoluminescence zoning and no significant zoning in Ti content (Payne and Eichelberger, 2009), providing temperature estimates consistent with the range determined from coexisting Fe-Ti-oxide geothermometry (Hildreth, 1983). The quartz dipyramids contain abundant melt inclusions, mostly 20–150 µm in diameter and exclusively high-silica rhyolite in composition (Clocchiatti, 1972, 1975; Westrich and others, 1991; Lowenstern, 1993).

Hornblendes from dacite pumice blocks analyzed by Avery (1992) and Hammer and others (2002) are magnesian and calcic, low in alkalies and alumina, and only weakly zoned. Even in the glassy dome lava of Novarupta, as in the pumice clasts, hornblende prisms lack reaction rims. Nonetheless, experimental work (Hammer and others, 2002) indicates that the sparse hornblende found in a few 1912 rhyolite samples is a magmatic contaminant transferred from the commingled dacite. By weight, MgO/(MgO+FeO*) in the hornblendes is 0.49–0.54, and by cations per formula unit, the range of Mg/Fe is 1.75–2.1. The following ranges were measured by electron microprobe: 10.8–11.2 percent CaO; 1.45–1.7 percent Na_2O; 0.23–0.38 percent K_2O; 1.3–1.6 percent TiO_2; and 5.5–7.0 percent Al_2O_3.

Iron-titanium oxide minerals were analyzed by Hildreth (1983), Coombs and Gardner (2001), and Hammer and others (2002) by electron microprobe, emphasizing coexisting titanomagnetite-ilmenite pairs. In titanomagnetites, X_{usp} is 0.29–0.31 in rhyolites, 0.25–0.29 in dacites, and 0.23–0.27 in andesites. In ilmenites, X_{ilm} is 0.85–0.975 in rhyolites, 0.76–0.81 in dacites, and 0.72–0.75 in andesites. Minor elements range widely in both phases. In the titanomagnetites, as MgO increases from 0.7 to 3.1 percent (from rhyolite to andesite), Al_2O_3 increases from 1.0 to 3.5 percent, and MnO declines from 0.8 to 0.35 percent. Likewise, in the ilmenites, as MgO increases from 1.6 to 3.7 percent, Al_2O_3 increases from 0.1 to 0.5 percent and MnO declines from 1.1 to 0.35 percent. Because of such compositional ranges and because coeruption of rhyolite, dacite, and andesite has introduced mixed populations of oxide minerals into some samples, all pairs were screened for equilibrium by the Mg/Mn partitioning test of Bacon and Hirschmann (1988), and touching pairs were preferred (but not always found). Using the QUILF algorithm of Andersen and others (1993), equilibration values of temperature and oxygen fugacity (fO_2) were then calculated for numerous preferred pairs, values inferred to represent magmatic conditions just prior to eruptive quenching. For 1912 rhyolites, the method yields 800ºC–857ºC and log fO_2 values of –13.5 to –12.0. For dacites, the ranges are 848ºC–950ºC and –11.7 to –10.0. And for andesites, they are 952ºC–990ºC and –9.2 to –10.1.

As emphasized by Hildreth (1983), such data imply a thermal continuum across the pre-eruptive magmatic spectrum in spite of the wide bulk-compositional gap between the rhyolite and the dacite-andesite continuum. It should be noted, however, that only four rhyolitic pumice samples have given oxide temperatures between 835ºC and 857ºC (whereas ~20 pairs yield 800ºC–833ºC). Conceivably, the four could represent rhyolite magma that had been heated in proximity to the intermediate magmas during or before eruption, but we have no independent evidence supporting that suggestion. Support for the higher temperature range for some 1912 rhyolite pumices comes from the work of Clocchiatti (1972), who measured homogenization temperatures of 850±10ºC (108 determinations) for melt inclusions with vapor bubbles trapped inside their quartz phenocrysts.

Contact heating may also have taken place in the rhyolite magma that was later extruded as the Novarupta dome, which has yielded pairs that give 833ºC and 840ºC. Several more equilibrium pairs in the dome rhyolite, however, gave values up in the dacite-andesite temperature range (Hildreth, 1983; Coombs and Gardner, 2001), suggesting contamination shortly before extrusion. A discrete dacite layer (65.1% SiO_2) within the rhyolite dome gave 915ºC. A sample from the Phantom dome block bed gave 935ºC and –10.35 fO_2, near the high ends of the ranges found for dacite plinian pumice.

Isotope Ratios

Oxygen isotope ratios were measured for plagioclase, quartz, orthopyroxene, and titanomagnetite separated from 1912 rhyolite, dacite, and andesite (fig. 154). Values of $\delta^{18}O_{plag}$ for rhyolites are +6.0–6.4, for dacites +5.8–6.0, and for andesites +5.6–5.9. The plagioclase data should represent close approximations to nonhydrated whole-rock values, and the slightly heavier values of the rhyolites are what is predicted for a normal fractionation sequence. Fractionations of ^{18}O among the coexisting phenocryst species (Hildreth, 1987) is consistent with the range of Fe-Ti-oxide temperatures (800–990ºC) determined for the 1912 andesite-to-rhyolite array (Hildreth, 1983; and section above titled “Mineralogy of 1912 Products”). Regardless of the extent to which crustal melting may have contributed to genesis of the rhyolites, the $\delta^{18}O$ data indicate that crustal rocks previously affected by hydrothermal alteration or surface weathering were insignificant contributors.

Nd isotope ratios for 1912 eruptive products and for the adjacent stratovolcanoes are within the narrow range found for the entire Alaska-Aleutian Quaternary arc. Values of $^{143}Nd/^{144}Nd$ for the whole arc range only from 0.51290 to 0.51310 (George and others, 2003; and many sources cited therein), while for products of the 1912 eruption, the range is

0.512952 to 0.513035 (table 2), a range that also brackets all the $^{143}Nd/^{144}Nd$ values measured for the entire Katmai volcano cluster (Hildreth, 1983; and authors' unpublished data). Because the small ranges determined for 1912 rhyolite, dacite, and andesite mutually overlap, no systematic fractionation-assimilation sequence is discernible among them on the basis of Nd isotopes. The only pre-Quaternary rocks exposed nearby for which Nd-isotope values have been determined (Lowenstern and Mahood, 1991) are Pliocene intrusions (more radiogenic than 1912 products, at 0.513067–0.513089) and Jurassic siltstone (less radiogenic, at 0.512725–0.512838), neither of them suitable assimilants to account for the observed range of 1912 isotopic diversity (fig. 1 of Lowenstern and Mahood, 1991).

Pb-isotope ratios for volcanic-front centers along the Alaska Peninsula have quite limited ranges (George and others, 2003): $^{206}Pb/^{204}Pb$ values determined extend from 18.73 to 18.88, $^{207}Pb/^{204}Pb$ from 15.52 to 15.58, and $^{208}Pb/^{204}Pb$ from 38.18 to 38.42. Only a few Pb-isotope ratios have been measured for the Katmai cluster. An andesitic enclave in the Novarupta dome gave values (18.88, 15.58, and 38.39) at the radiogenic ends of the ranges cited, and values for dacite and andesite of the 1953–74 eruption of Trident are very similar (George and others, 2003). A 1912 rhyolite pumice gave values (18.85, 15.57, and 38.375) trivially less radiogenic than those for dacite pumice (18.88, 15.57, and 38.38) or the Novarupta andesite (Reagan and others, 2003).

Sr-isotope ratios measured for 1912 eruptive products range fairly widely, but (like $^{143}Nd/^{144}Nd$) the ratios for rhyolite, dacite, and andesite exhibit wholesale overlap (table 2; fig. 155). Rhyolite pumice clasts (n=3) yield $^{87}Sr/^{86}Sr$ values of 0.70368 to 0.70374, and rhyolite lava (n=2) from the Novarupta dome 0.70372. Dacite pumices (n=10) are in the range 0.70354 to 0.70362, and andesite scoriae (n=17) 0.70352 to 0.70374. Two scoriae atop the Turtle, more mafic than the continuous array, gave less radiogenic values: 0.70344 (54.6% SiO_2) and 0.70338 (50.4% SiO_2, olivine accumulative); and a mafic enclave (55.2% SiO_2) in Novarupta rhyolite gave 0.70348 (George and others, 2003). For comparison, 15 samples from Mount Katmai range from 0.70335 to 0.70372, and 10 from Trident only from 0.70344 to 0.70359. The 1912 rhyolite and andesite thus have in common that they extend to Sr-isotope ratios more radiogenic than lavas of the adjacent edifices.

Because the rhyolite contains only a third as much Sr (58–95 ppm) as the andesite-dacite array (210–330 ppm), it should have been more susceptible to contamination (by either magma mixing or assimilation), but its Sr is no more radiogenic than that in much of the andesite. We again emphasize that the pale-gray dacite and black andesite (fig. 4) are simply convenient nametags along a phenocryst-rich compositional continuum, that the andesite-dacite distinction at 63 percent SiO_2 is an arbitrary convention, and that they were coerupted throughout much of the 1912 sequence. The mafic outliers

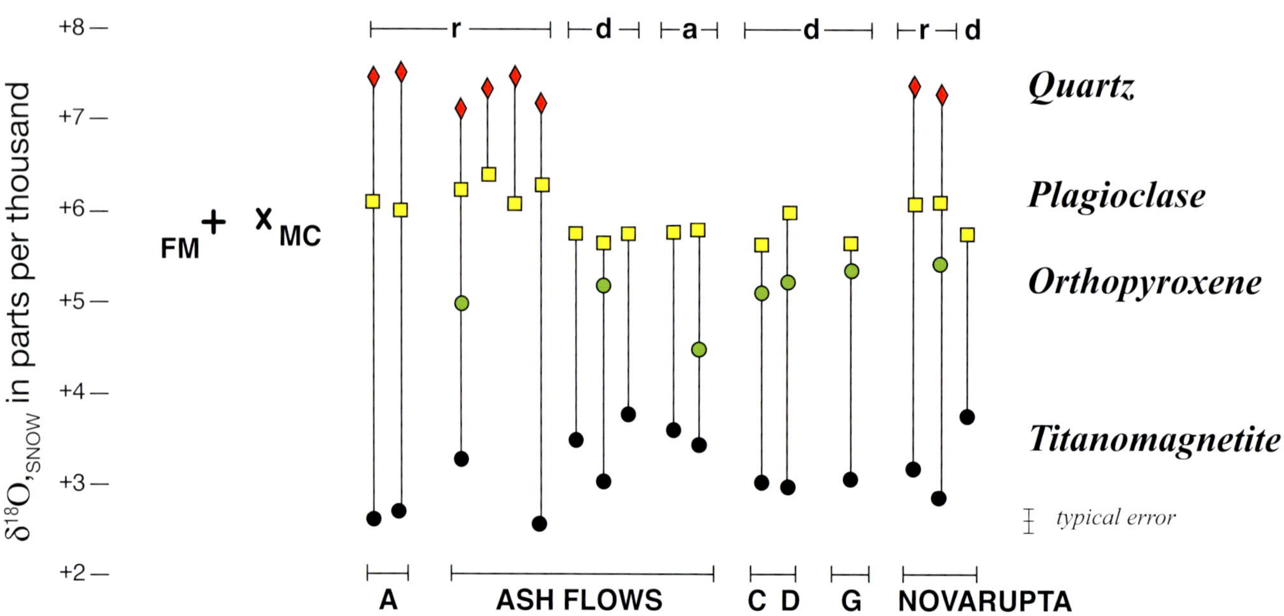

Figure 154. Values of $\delta^{18}O$ for coexisting quartz, plagioclase, orthopyroxene, and titanomagnetite separated from Novarupta lava and from individual pumice blocks of 1912 ignimbrite and fallout Layers A, C, D, and G. Symbols at top: r, rhyolite; d, dacite; a, andesite. Isotopic determinations for Falling Mountain (FM) and Mount Cerberus (MC) were for whole-rock crystalline dacite. Analyses by Hildreth in USGS laboratory of J.R. O'Neil at Menlo Park, California, in 1980. Values are deviations in parts per thousand from that of standard mean ocean water (SMOW).

have the least radiogenic Sr (0.70338–0.70348), but, within the andesite-dacite continuum, ratios for andesites (0.70352 to 0.70374) wholly encompass those for dacites (0.70354 to 0.70362). There was therefore clearly no systematic assimilation process, neither within the continuum nor relating the continuum to the rhyolite (0.70368 to 0.70374).

Some have tried in the past to model Sr-isotope variation in 1912 products by assimilation of the exposed Jurassic sandstone-siltstone of the Naknek Formation, which is subhorizontal here and extends from the surface to a depth of ~1,500 m (Riehle and others, 1993; Detterman and others, 1996). We see this as a pointless exercise because the magma reservoir was probably at least twice that deep and likely to have been enclosed by earlier Mesozoic sedimentary strata or by Paleozoic basement rocks (see fig. 13 of Hildreth and Fierstein, 2000). Small but varied and unsystematic contamination by feldspar-rich sedimentary rocks comparable to Naknek arkoses (267–705 ppm Sr; 0.7042–0.7048) might nonetheless help account for some of the $^{87}Sr/^{86}Sr$ variations in the 1912 magmas; but so could partial melts of arc-intrusive rocks at any depth.

Numerous Miocene and Pliocene tonalite-granodiorite plutons exposed in the Katmai area are reasonable candidates to provide crustal assimilants; about 30 samples of such local granitoids give ranges of 59–67 percent SiO_2 and 280–650 ppm Sr (authors' unpublished data), but the only $^{87}Sr/^{86}Sr$ value yet determined for any of them is 0.703288 (Lowenstern and Mahood, 1991) for one porphyritic pluton, far less radiogenic than any 1912 product. A sill of high-silica rhyolite (0.70427; 238 ppm Sr; 3.1 Ma) adjacent to that pluton exemplifies a local shallow intrusion presumed to be enriched in crustal melt contributions. A mixing line between the two intrusions (both intruding Naknek Formation strata near Mount Mageik; Hildreth and Fierstein, 2003) passes through high-Sr members of the 1912 andesite-dacite array (fig. 155), illustrating potential modes of introducing isotopic variability antecedent to continuing crystal-melt fractionation. Many other potential modes can be envisaged, entailing unexposed pre-Mesozoic basement or the (isotopically unstudied) granitoid plutons nearby, some of which are as old as 20 Ma (Hildreth and Fierstein, 2003).

Another general process promoting Sr-isotopic variation within the 1912 suite would be upper-crustal coalescence of many crystal-rich magma batches, each with a unique history of assimilation and fractionation during ascent through 30 km of peninsular crust. For example, products of the 1953–74 eruption of Trident, which was concluded by Coombs and others (2000) to have involved advanced mixing of a new batch of andesite with a shallow body of resident dacite, yield a range of $^{87}Sr/^{86}Sr$ values—from 0.703455 at 56 percent SiO_2 to 0.70352 at 57 percent SiO_2 through to 0.70357 at 64.2 percent SiO_2. Intrabatch variations at the 0.0001 level are potentially generated at any depth, deep to shallow, by interaction with Paleozoic or Mesozoic crustal rocks, and then further complicated by blending of magma batches.

U and Th isotopes have been measured for ~33 samples of 1912 eruptive products by Reagan and others (2003) and Turner and others (2010). Values of ($^{230}Th/^{232}Th$) are consistently high at 1.35 to 1.46, and ($^{238}U/^{232}Th$) ranges from 1.28 to 1.51, yielding a modest range in ($^{230}Th/^{238}U$) values from 0.95 to 1.05. Parentheses indicate activity ratios, not elemental ratios. On a U-Th equiline diagram (fig. 156), the three most mafic samples have 3–5 percent excess ^{230}Th, while most of the andesite samples plot close to the equiline, and most dacite samples plot to the right of it at 1–3 percent excess ^{238}U. The 1912 rhyolites have the highest values (1.43–1.46) of ($^{230}Th/^{232}Th$) and plot not far from the equiline—slightly to the left of it with Reagan's data, slightly to the right with Turner's data.

Any process that fractionates U and Th from each other introduces radioactive disequilibrium, thus requiring either ingrowth or decay of ^{230}Th to restore secular equilibrium, processes that take about 350,000 years. Their extreme incompatibility in mantle minerals governs the conventional presumption that mantle processes fractionate U from Th only trivially. In crustal rocks, however, zircon and apatite are widespread and strongly fractionate U and Th relative to host melts. Katmai dacites contain apatite but apparently lack zircon, whereas both the 1912 rhyolite and the 22.8-ka plinian rhyodacite from Mount Katmai clearly fractionated zircon (fig. 8 of Hildreth and Fierstein, 2000).

Fractionation of zircon (with Th/U typically lower than the melt) is likely to have contributed to U-Th isotopic variations in the 1912 rhyolite. The effects of zircon fractionation are obvious in the suppression of Zr and Hf enrichment in the rhyolites (fig. 151) and the reduced Zr/Hf ratios relative to the dacites (table 2). Such effects were plotted in figure 6 of Hildreth (1987) before zircon was identified in 1912 rhyolites and were at the time attributed to residual source-rock zircon. A consequence of zircon subtraction (whether refractory or fractionated) is, of course, to shift melts with excess ^{238}U instantaneously to the left, back toward or across the equiline, independent of any secular ingrowth of ^{230}Th (fig. 156).

Both U-series studies illustrated a 40-k.y. pseudoisochron that passes through many of the mafic, andesitic, and dacitic samples (fig. 156), but it is not clear what this might actually represent. Three andesite-dacite samples from the unrelated 1953–74 mixed eruption of Trident also fall along that pseudoisochron, as do the dacites of Falling Mountain and Mount Cerberus (Reagan and others, 2003), which are 2–3 times older than 40 ka (table 1). Far more likely than an expression of any time relationship, the trend probably represents a crude mixing array that reflects crustal assimilation at various depths and blending of successive magma batches; an open-system interpretation is strengthened by the fact that the zircon-free andesite-dacite array crosses the equiline from ^{230}Th excesses to ^{238}U excesses.

Moreover, Mount Katmai had erupted voluminous rhyolite and rhyodacite starting around 22.8 ka (Hildreth and Fierstein, 2003), which is hard to reconcile with steady U-Th maturation of its andesite-dacite magma reservoir since 40 ka.

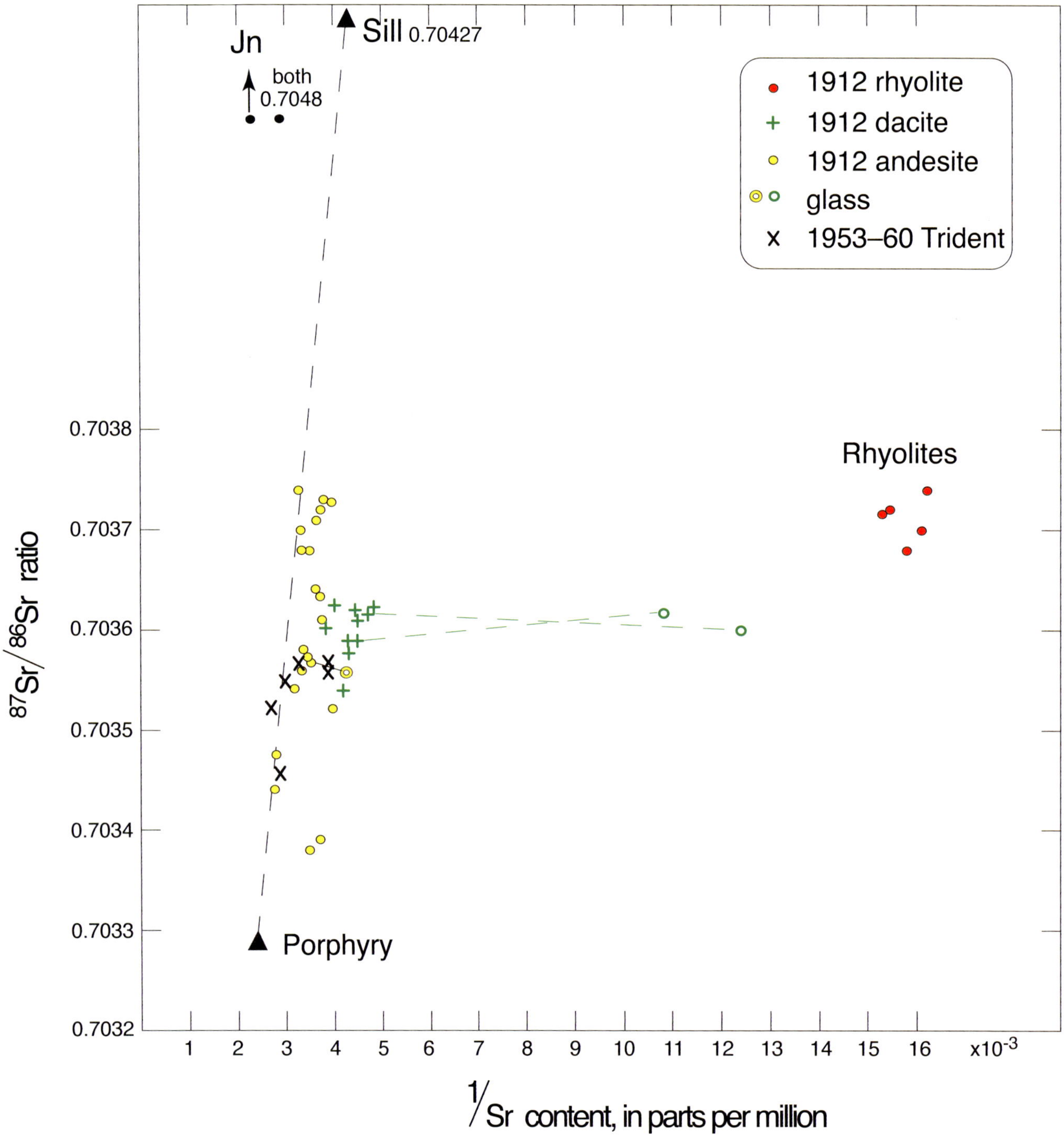

Figure 155. Sr-isotope ratios plotted against reciprocal of Sr concentration (ppm^{-1}), a convention that transforms a mixing hyperbola into a straight line. Line connects data (s) for Pinnacle Porphyry (63–65% SiO_2) and high-silica rhyolite sill (77% SiO_2), both near West Mageik Lake and both ~3.1 Ma (Lowenstern and Mahood, 1991; Hildreth and Fierstein, 2003). Symbols are identified in inset for 1912 ejecta and for 1953–60 lava flows of Southwest Trident. Tie-lines connect data for three whole-pumice samples and their matrix glasses, illustrating fractionation trend from andesite-dacite continuum toward rhyolites. Percent SiO_2 of the three pumice-glass pairs, respectively, are: sample K-75, 59.3, 66.8; sample K-45, 64.6, 75.2; and sample K-1490, 67.7, 76.0. It is not implied that these particular Pliocene intrusions provided assimilants to 1912 or Trident magmas, but, instead, that comparable Tertiary intrusive rocks are more likely candidates than the shallowly exposed Naknek Formation (Jn).

The 22.5-ka rhyolite lava flow at Mount Katmai has lower U/Th and lower (^{230}Th/^{232}Th) than 1912 rhyolites and plots farther left on an equiline diagram (Reagan and others, 2003), with significant excess ^{230}Th: (^{238}U/^{230}Th) = 0.973. Products of a voluminous plinian eruption that accompanied or directly preceded the 22.5-ka rhyolite lava had clearly undergone zircon fractionation (fig. 8b of Hildreth and Fierstein, 2000), thus reducing its U/Th ratio, as likewise inferred for the 1912 rhyolite.

The U-Th isotope data thus reinforce the inference from the Sr-isotope results that the 1912 compositional range was not produced by closed-system fractionation. The array of isotope values from excess ^{230}Th to excess ^{238}U, combined with a wide range of (^{226}Ra/^{230}Th)—(1.0–2.5; Turner and others, 2010)—suggests accumulation and blending of many different magma batches with ranges in crustal depth of batch coalescence, residence time, and extent of intracrustal assimilation. Because basalts are rare, primitive compositions absent, and andesites dominant, we assume that most batches arose from deep-crustal MASH zones that blended varied crustal melts with differentiates of varied basalts, as outlined by Hildreth and Moorbath (1988) and modelled by Annen and others (2006).

The distinctively higher (^{230}Th/^{232}Th) range of the 1912 rhyolites (relative to the andesite-dacite array) and their position not far from the equiline can be interpreted in various ways: (1) The conventional interpretation would infer a far longer residence time (100,000–350,000 years) for the rhyolite magma than for the coerupted andesite and dacite, a scenario that we find implausible for the 140-k.y.-old, recurrently active, predominantly intermediate, Katmai-Trident volcanic group. (2) Zircon fractionation from the 1912 rhyolite itself may have shifted the melt from originally greater ^{238}U excess back toward the equiline. (3) Rhyolitic partial melt contributions from felsic

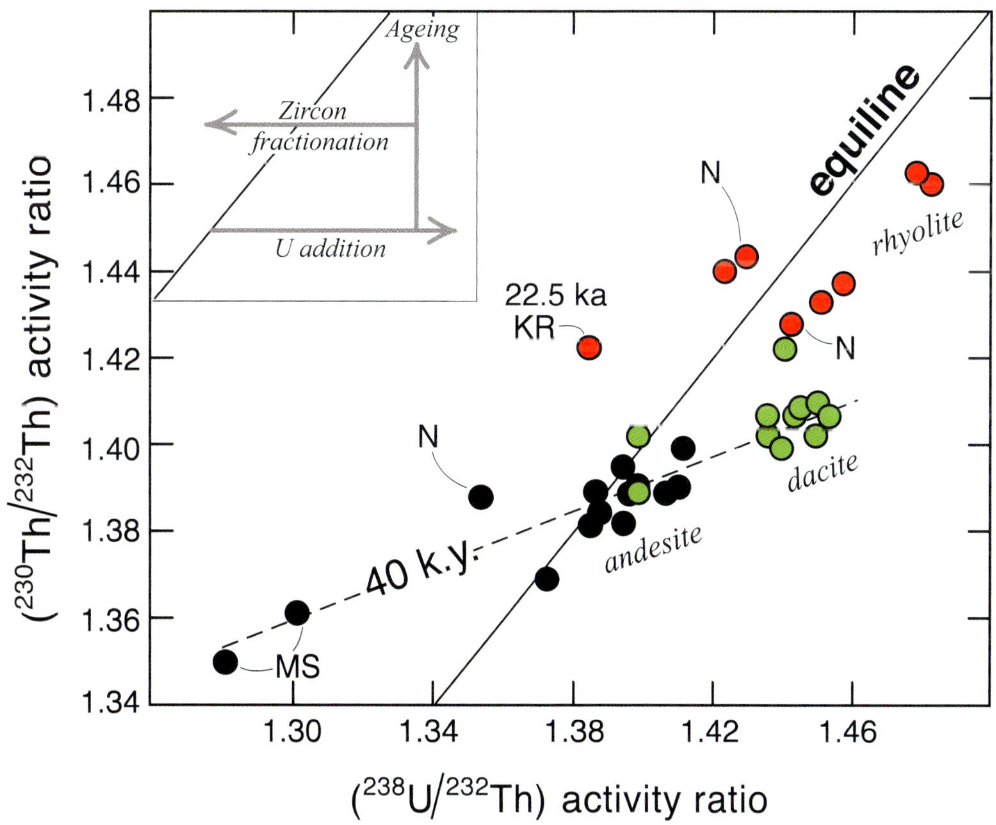

Figure 156. U-Th equiline diagram, combining Novarupta-Katmai data of Reagan and others (2003) and Turner and others (2010) for 33 samples. Equiline represents secular equilibrium. Parentheses indicate activity ratios, not elemental ratios. Distinguished by colors are rhyolite (red), dacite (green), and andesite (black). N, Novarupta dome lava; all others are 1912 pumice and scoria, except KR, Mount Katmai 22.5-ka rhyolite lava on south rim of caldera. MS, 1912 mafic samples (as distinguished in table 2 from andesite-dacite continuum). Dashed 40-k.y. pseudoisochron and all three rhyolites left of equiline are from Reagan and others (2003). Inset schematically shows trends expected to result from instantaneous addition of U, ingrowth of ^{230}Th by ageing, and either zircon fractionation or its retention in residue of crustal partial-melt contributions.

arc plutons old enough to be at secular equilibrium might have modified the U-Th budget of the dacite-derived rhyolite melt with which it blended. (4) Had such a remelting pluton been of late Pleistocene age, refractory zircon might have shifted the rhyolitic partial melt extract toward the equiline, away from probable bulk-rock excess ^{238}U. (5) The felsic plutonic rock remelted might have actually been an early differentiate of Mount Katmai andesite, compositionally akin to 1912 magmas, young enough to retain some degree of ($^{230}Th/^{238}U$) disequilibrium, and subject to the same shift promoted by refractory zircon. In view of the hundreds of late Pleistocene and Holocene eruptions that built Katmai and Trident, not one of which had previously produced high-silica rhyolite, we consider option (1) the least plausible, option (5) the most consistent with the many other compositional constraints, and options (2), (3), and (4) all plausible contributors. We return to this topic as part of a comprehensive summation in the section below titled "Origin of the 1912 Rhyolite."

Experimental Petrology

Phase-equilibrium experiments were undertaken on natural samples of 1912 rhyolite by Coombs and Gardner (2001) and of 1912 dacite and andesite pumice by Hammer and others (2002). The rhyolite study was conducted under water-saturated conditions, in accord with the evidence of Lowenstern (1993) for preeruptive gas saturation, and within a temperature range bracketing those indicated by Fe-Ti-oxide pairs (800ºC–850ºC; Hildreth, 1983). It was found that the natural phase assemblage was stable at 780ºC–860ºC at H_2O-saturated pressures less than 100 MPa (1 kb). The best compositional matches for natural plagioclase and orthopyroxene were found at 70–100 MPa for 800ºC and 40–50 MPa at 840ºC. Quartz was found in only 4 of 24 experimental runs, and because quartz stability increases with pressure, it has been suggested that the low pressures are underestimates. Nonetheless, FTIR determinations of 4.0–4.5 weight percent H_2O and 20–50 ppm CO_2 in melt inclusions within 1912 quartz, combined with experimentally calibrated mixed-vapor-saturation isobars (fig. 8 of Wallace, 2005; Newman and Lowenstern, 2002) yield entrapment pressures of 100–130 MPa, in agreement with the upper end of the H_2O-saturated estimates of Coombs and Gardner (2001). Assuming densities of 2,550–2,700 kg/m^3 for the local sedimentary section, these results converge on a probable depth range of 3–5 km.

The 1912 rhyolite was thus stored in a shallow reservoir where it crystallized its sparse assemblage of unzoned euhedral phenocrysts, which are nowhere found to contain higher pressure melt inclusions. Amphibole is absent in 1912 rhyolite pumice, but at water-saturated pressures of 100-150 MPa Coombs and Gardner (2001) found it to be the liquidus phase, thus sustaining the evidence against a higher pressure crystallization history for the rhyolite. No antecrysts have been found in 1912 rhyolite pumice, and xenocrysts are limited to rare clinopyroxene, typically broken, transferred from coerupted dacite or andesite. The sparsity and euhedral habit of the unzoned phenocrysts, especially the gem-quality quartz dipyramids, suggest weak undercooling, slow growth, and stable shallow storage of the rhyolite magma. Thermal continuity from rhyolite to dacite (as indicated by Fe-Ti-oxide pairs) and the need for a heat source to keep the rhyolite crystal-poor suggest preeruptive contiguity.

Hammer and others (2002) experimentally investigated 1912 pumice blocks from the bulk-compositional extremes bracketing the andesite-dacite continuum, a dacite (68% SiO_2) that gave Fe-Ti-oxide temperatures of 848ºC–876ºC and an andesite (58.7% SiO_2) that gave 952ºC–963ºC. The dacite was one of the rare pumice clasts found to contain small amounts of euhedral unrimmed hornblende. Matching of modes, melt compositions, and mineral-rim compositions of experimental charges and natural samples was best achieved under water-saturated conditions at 930ºC–960ºC and 75–100 MPa for the andesite and 850ºC–880ºC and only 25–50 MPa for the dacite.

Lowenstern's (1993) determination of 2.8±0.5 weight percent H_2O in melt inclusions in 1912 dacite phenocrysts is consistent with melt entrapment at the low-temperature, high-pressure end of these ranges for water-saturated dacite.

CO_2 contents of melt inclusions (MI) in 1912 andesite and dacite have never been measured accurately, however, owing in part to crystallization and breakage of inclusions in the andesite and low CO_2 in the dacite. Nonetheless, global evidence suggests that CO_2 and SO_2 should contribute to any hydrous vapor phase coexisting with such intermediate arc magmas, and Lowenstern (1993) demonstrated that 1912 andesite and dacite did indeed contain a free gas phase before eruption. Hammer and others (2002) thus undertook numerous additional runs with a CO_2-H_2O mixed fluid phase, adjusting the mole fraction of water in the fluid ($X_{H_2O}^{fl}$) between 0.34 and 0.85, supplementing their abundant data from water-saturated experiments. As they discuss in detail, reasonable intermediate values of $X_{H_2O}^{fl}$ yield preferred storage depths between 4 and 5 km for the dacite and 4.5–5.7 km for the andesite, consistent with overlapping to slightly shallower estimates for the rhyolite (fig. 157).

Hornblende similar in composition to the trace amphiboles in the dacite pumice block selected for experimental work was produced experimentally only at slightly higher P_{H_2O} and in somewhat less evolved melt than under equilibrium conditions determined for the dacite sample itself (fig. 6 of Hammer and others, 2002). It was suggested that the natural hornblende could have been drawn up from a deeper level of the voluminous zoned dacite and added syneruptively to the highly evolved dacite in which it was found. Most 1912 dacite and virtually all 1912 andesite, however, lack amphibole. Hammer and others (2002) also showed that amphiboles grown experimentally in 1912 rhyolite and andesite are compositionally unlike any natural 1912 hornblendes (their fig. 8). The euhedral unrimmed hornblende found in Novarupta dome rhyolite has the composition of those in the natural and experimental dacites, not that of hornblende grown experimentally from the rhyolite. Moreover, plagioclase included in and attached to dome hornblende crystals are more calcic than those of 1912 rhyolite but within the compositional range of plagioclase in the dacite (Coombs, 2001). The hornblende was evidently transferred from dacite to rhyolite magma during the intraconduit mingling made manifest by the dacite layers in the rhyolite dome. Ascent and transfer of the amphibole from its higher P_{H_2O} site of origin into a small fraction of the erupting dacite and thence into the dome rhyolite must have required only a few days in order to avoid development of dehydrated reaction rims (Hammer and others, 2002). This suggests that Novarupta extruded less than a week, not months or years, after the plinian eruptions. Prompt extrusion of the dome rhyolite is likewise suggested by preservation within the dome of subordinate Fe-Ti-oxide pairs characteristic of the andesite-dacite contaminants that had failed to reequilibrate to the lower temperature of their new rhyolitic host.

Preeruptive Magma Storage and Syneruptive Transport

Here we summarize the data, constraints, and inferences that bear upon the distribution of magma storage at the onset of the 1912 eruption (fig. 158). Any distribution model needs to account for arrival of rhyolite, dacite, and andesite at Novarupta on 6 June, caldera collapse at Mount Katmai on 7 June, initiation of the plinian sequence by unadulterated rhyolite for several hours, and subsequent mingling of the confluent magmas for two more days.

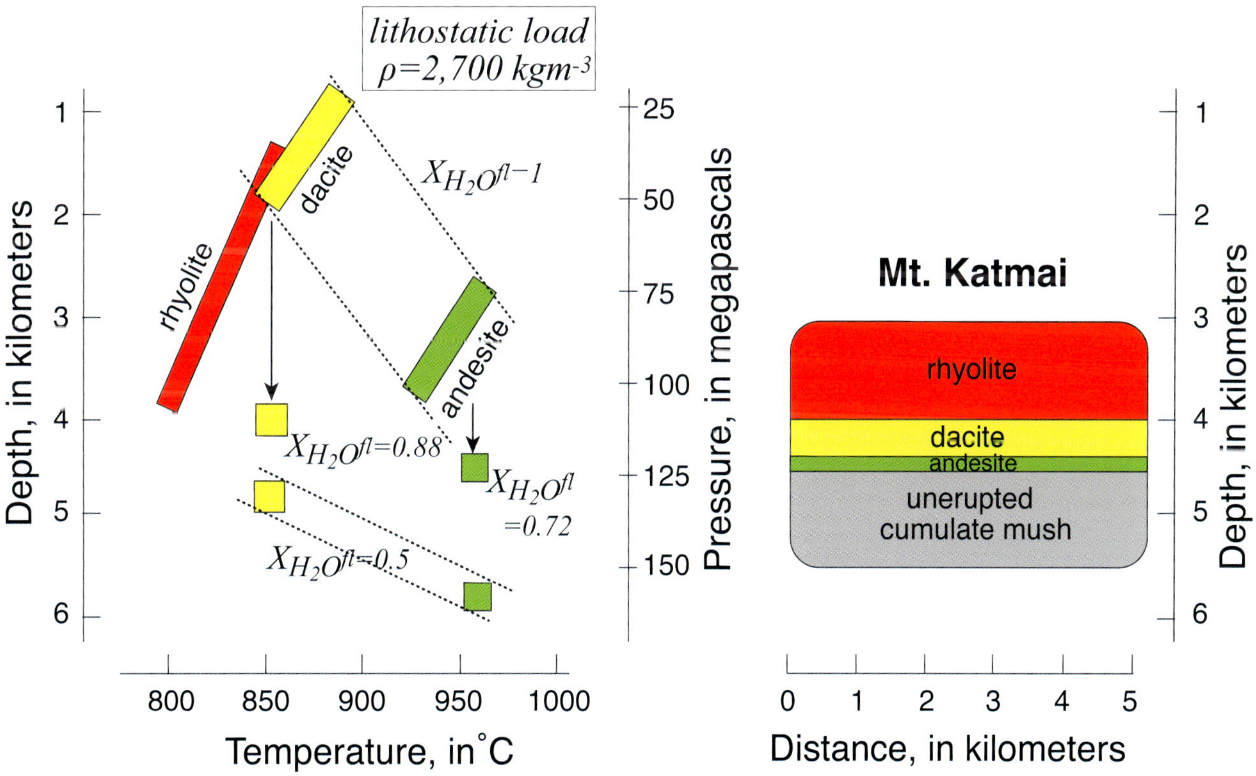

Figure 157. Summary of constraints on 1912 magma-storage conditions inferred from phase-equilibrium experiments of Coombs and Gardner (2001) and Hammer and others (2002), conducted on 1912 rhyolite and on end-member samples of the andesite-dacite continuum (ADC). Experimentally bracketed temperature ranges for all three compositions were in good agreement with those inferred from equilibration of FeTi-oxide pairs (Hildreth, 1983). Upper quadrilaterals represent P-T ranges determined under water-saturated conditions (X_{H_2O}fluid = 1). Hammer and others (2002) also made runs for andesite and dacite samples under fluid-saturated conditions with ranges of CO_2 and H_2O proportions in a mixed volatile phase. Lowenstern (1993) had earlier provided independent evidence for preeruptive vapor saturation in all three 1912 compositions; and Wallace (2005) subsequently showed melt inclusions in rhyolite quartz to contain 20–50 ppm CO_2 as well as 4.0–4.5 weight percent H_2O, indicating minimum entrapment pressures of 100–130 MPa. As yet uncertain CO_2 contents of andesite and dacite magma were probably greater. Accordingly, P-T estimates are illustrated here for X_{H_2O}fluid = 0.5—a probable minimum value for ADC. Also illustrated are P-T estimates for intermediate values of 0.88 for dacite and 0.72 for andesite, which correspond to chamber-model depths portrayed in cartoon at right, as proposed by Hildreth and Fierstein (2000).

Plausible Storage Sites of 1912 Magmas

In contrast to most earlier products of Mageik and Trident, the Holocene eruptive products of East Mageik, Southwest Trident, and Mount Martin are not easily distinguishable chemically from the intermediate suites of Novarupta and Katmai (Hildreth and Fierstein, 2000). Because so many Holocene products of Novarupta, Katmai, Trident, Mageik, and Martin are similar pyroxene-plagioclase andesites and dacites, it is useful to summarize here the volcanological evidence capable of supplementing the chemical distinctions as means to sharpen understanding of magma storage and transport in 1912.

From Novarupta, Mount Katmai is 10 km east, the Holocene vent of East Mageik is 8 km southwest, that of Mount Martin 15 km southwest, and the 1953–74 vent of Southwest Trident only 4 km south (fig. 5). There is no evidence (observational, stratigraphic, or morphologic) that Mount Martin has erupted historically, and there has been no obvious change in the configuration of its crater or fumarolic discharge since it was first photographed in 1913 (Griggs, 1922, p. 85). Separated from Novarupta by the massive Mageik volcano, Mount Martin's distance and its small but consistent differences in K and Sr contents seem sufficient to exclude its plumbing system as a plausible contributor to the 1912 eruption.

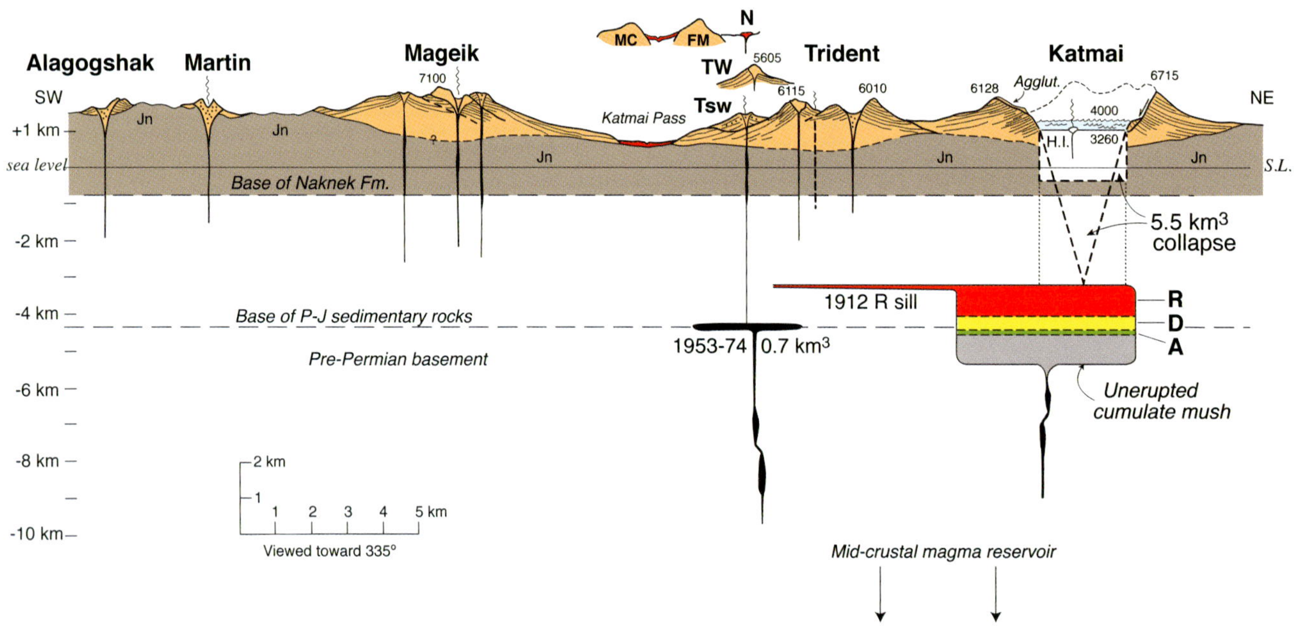

Figure 158. Katmai volcanic cluster magmatic plumbing system; schematic illustration drawn to scale along N65°E volcanic axis, looking N25°W with no vertical exaggeration. Elevations on profile are in feet (1 m = 3.28 feet). Wavy vertical lines indicate present-day fumarolic emissions. Jn, marine sedimentary rocks of Jurassic Naknek Formation. Novarupta (N), Falling Mountain (FM), Mount Cerberus (MC), and West Trident (TW) are not actually along section, but their relative positions (see fig. 5) are indicated above profile, respectively 4.5 km, 4 km, 4 km, and 1.5 km northwest of line of section. G indicates direction toward Mount Griggs, 12 km northwest of the profile. Precollapse profile of Mount Katmai is adapted from figure 102; volume displaced in 1912 was about 5.5 km^3 and surface subsidence was 1.2±0.1 km vertically. Displaced volume is modelled alternatively as conical funnel and cylindroid, each having top surface area (4 km^2) that of caldera floor. H.I., Horseshoe Island dacite dome; Agglut., remnant of pre-1912 agglutinated dacite fallout that mantles west rim (fig. 145) and Peak 6128 and thickens into a pre-1912 crater. Magma reservoir of 1912 is depicted (just before eruption) as a unitary chamber zoned from andesite (A) to dacite (D) to rhyolite (R) in the proportions (1/4.5/7.5) erupted. Because the magma volume erupted was 2.3 times greater than the volume collapsed, the 13-km^3 chamber is depicted as elongate along the volcanic axis, 1.3 km thick, 2 km across axis (into the profile), and 5 km along axis, thus extending well beyond caldera footprint. Rhyolite sill penetrates subhorizontal sedimentary strata, extending an additional 6 km toward Novarupta or Trident, whence the magma presumably diked its way toward surface, ultimately through the Naknek Formation (Jn). Chamber would have been larger to extent that additional magma failed to erupt. Depth of reservoir is constrained loosely by experimental work (fig. 157) but is represented arbitrarily at an upper-crustal discontinuity and beneath but (for clarity) clear of the caldera-collapse models. The 0.7 km^3 of magma that erupted at Southwest Trident (Tsw) in 1953–74 is shown stored as a circular sill 100 m thick; as a vertical dike, it would presumably have had an aspect ratio even thinner.

For more than 140 k.y., Trident has produced andesites and dacites that are clearly distinguished chemically from those of Novarupta, and our extensive fieldwork has turned up no record that its magmas ever evolved as far as rhyodacite or rhyolite. Moreover, except for Southwest Trident, all component cones and domes of Trident appear to be older than 44±12 ka (Hildreth and others, 2003a, b). The magmas that built Southwest Trident in 1953–74 are compositionally closer to the Novarupta andesite-dacite array than any previous products of Trident, but they appear to represent independent andesite and dacite batches uncontaminated by leftovers from the 1912 reservoir (Coombs and others, 2000; Hildreth and Fierstein, 2000). The small but significant differences in K, Fe, and Sr and—despite proximity to Novarupta—the lack of quartz xenocrysts or chemical evidence for rhyolite contamination of the 1953–74 magmas appear to exclude participation of any Trident precursors in 1912.

East Mageik (the Holocene component of Mount Mageik) is actually closer to the 1912 vent than is Mount Katmai, but it has not erupted historically and its dozen Holocene lava flows are limited to the range 60–64 percent SiO_2 (Hildreth and others, 2000a). Rich in plagioclase and pyroxenes, some of its lavas are compositionally so similar to the 1912 andesites and dacites that it would not be radical to entertain a 1912 scenario that involved magma stored beneath East Mageik. There is no evidence, however, for eruption or structural disturbance of Mount Mageik, in 1912 or subsequently. And many other Holocene dacites of Mount Mageik are richer in K, Fe, and Zr than the 1912 dacites (Hildreth and Fierstein, 2000).

Southwest Katmai is most plausible of the clustered centers to have had some or all of the 1912 magma stored beneath it before the off-axis eruption took place at Novarupta. Katmai caldera collapsed during the sequence of plinian eruptions at Novarupta, and the only postclimactic extrusions were at Novarupta and on the Katmai caldera floor. Of the entire Katmai cluster, only Southwest Katmai had previously erupted rhyodacite and rhyolite, and nearly all of its younger products have been dacite or rhyodacite (63–72% SiO_2). Although they are not identical, the 1912 andesite-dacite suite is compositionally closer to many products of Southwest Katmai than to those of any of the other centers (Hildreth and Fierstein, 2000).

A model envisaging storage of all the 1912 magma beneath Southwest Katmai needs, however, to account for the following observations: (1) The 5.0–5.5-km^3 caldera compensates for no more than 40–45 percent of the 13.5 km^3 of magma erupted. (2) The complex 60-hour eruptive sequence issued from a new vent that broke out through flat-lying Mesozoic sedimentary rocks at the foot of West Trident, 10 km from Mount Katmai. (3) No magma erupted at Mount Katmai in 1912 until the Horseshoe Island dacitic lava dome was emplaced on the caldera floor sometime following collapse of the edifice. (4) Although rhyolite dominated the 1912 eruption, only one true rhyolite (72.1% SiO_2; 22.5 ka) is exposed at Mount Katmai, and no rhyolitic lithics are recognized among the basaltic-to-rhyodacitic lithic ejecta thrown out by hydrothermal explosions during the caldera collapse (Hildreth and Fierstein, 1992). (5) Compositions of the numerous Katmai rhyodacite lavas do not exhibit the influence of zircon fractionation so conspicuous for the 1912 rhyolite array; only the voluminous 22.8-ka rhyolitic plinian pumice fall and ignimbrite (LPFI; 71–72.3% SiO_2) from Mount Katmai does show such zircon control (Hildreth and Fierstein, 2000). If the distinction principally reflects the influence of enhanced magmatic water content in promoting both zircon fractionation (by lowering viscosity) and plinian eruption, then it would not be implausible that the LPFI and 1912 rhyolite magmas were produced at different times in the same evolving reservoir. (6) Novarupta dome rhyolite is locally contaminated with streaks and blobs of contemporaneous andesite and dacite. If comparably commingled rhyolite were visible in Horseshoe Island dacite (66.4% SiO_2), Fenner (the only geologist who ever laid hands on it) would probably have noticed it, because magmatic assimilation was one of his foremost preoccupations (Fenner, 1923, 1926, 1930, 1950; Wright, 1952). Unless Fenner missed it, withdrawal of the (more buoyant) rhyolite layer must have been effectively complete from the part of the Katmai reservoir that later leaked dacite to the caldera floor. Nonetheless, analogously, postclimactic blocks of the ephemeral Phantom dome strewn around Novarupta basin are also dacite (62.2 to 66.7% SiO_2) that is not visibly contaminated by the rhyolite that earlier and later issued from the same vent. The evidence from Fenner's hurried 1923 visit to the Horseshoe Island dome thus remains tantalizingly inconclusive.

No Preeruptive Magma Storage Beneath Novarupta

In addition to the rapid decline of magmatic gas contributions and temperatures of fumaroles in Novarupta basin, several other lines of evidence support the inference that the 1912 magma reservoir did not directly underlie the vent at Novarupta. Falling Mountain dome adjacent to the 1912 vent (fig. 96) is not as young as previously thought; it yields a K–Ar age of 70±8 ka and consists of dacite compositionally similar to West Trident, thus providing no support for recent magma storage beneath the Novarupta depression. Lithic ejecta from Novarupta are limited to lithologies exposed at the surface, suggesting a shallowly flaring vent funnel (as elaborated in chapter 7) rather than subsidence into a subjacent chamber. Moreover, the 1912 vent region underwent no structural resurgence, visible uplift, or eruptive activity of any kind after the Novarupta lava dome plugged the Episode III conduit. Although the asymmetrical ejecta ring surrounding Novarupta is cut by numerous extensional faults (figs. 70, 97), they reflect spreading of the 225-m-high pile in response to compaction and welding, as described in chapter 7, above. Because the pumice-fall layers are locally downfaulted several meters but nowhere uplifted relative to continuously exposed unfaulted parts of the ring, the faulting provides no evidence for postclimactic, areally distributed, magma pressure beneath the 1912 vent depression. Causing no significant deformation of the basin floor, the Phantom and Novarupta lava domes simply squeezed up the same conduit to the same restricted point-source vent that had earlier fed the ejecta ring and plinian Layers C through G (fig. 96).

Data Bearing Upon Models of the 1912 Plumbing System

The 1912 ejecta themselves provide several constraints for envisaging a model of the 1912 plumbing system. The ~13 km^3 of magma released in 1912 consisted of ~4.5 km^3 of dacite and ~1 km^3 of andesite (both having 25–42 weight percent phenocrysts) and 7–8 km^3 of crystal-poor high-silica rhyolite (with only 0.5–3 weight percent phenocrysts) (fig. 147). Homogeneous pumice with 68.7–77.0 percent SiO_2 has not been found, so a broad compositional gap separates the slightly zoned rhyolite from the strongly and continuously zoned andesite-dacite (fig. 148). The only rhyodacites (68–72% SiO_2) identified in the whole region are some of the youngest eruptive products at Mount Katmai, but such compositions are absent in the 1912 suite.

In spite of the gap in bulk composition, Fe-Ti-oxide geothermometry demonstrates continuity in both magma temperature and oxygen fugacity (Hildreth, 1983) between the nearly aphyric rhyolite (800°C–857°C) and the extensively crystallized dacite (848°C–950°C)-to-andesite (952°C–990°C) continuum. Thermal and redox continuity suggests physical contiguity, supporting the inference that the andesite-dacite magma was the proximate heat source that kept the rhyolite nearly aphyric. The phenocryst-rich intermediate magma may have graded down into a mushy cumulate substrate that served not only as thermal buffer but, by compaction and upward percolation of its evolved interstitial melt, as a contributor of rhyolitic liquid (see Bacon and Druitt, 1988; Philpotts and others, 1998; Sisson and Bacon, 1999).

About 3 km^3 of rhyolite magma erupted before being joined by any coerupting andesite and dacite (figs. 34, 60), indicating that the conduit that fed the Novarupta vent tapped the rhyolitic part of the reservoir first. It took 3–4 hours before the first andesite-dacite arrived at the vent, as recorded by Layer B1.

Despite fluctuations and pulses in the proportions of andesite and dacite that coerupted during each of the subsequent episodes (fig. 60), no compositional or textural evidence has been found for more than one batch of (continuously zoned) intermediate magma. Ranges in phenocryst zoning overlap extensively for the andesites and dacites (Hildreth, 1983; Avery, 1992) irrespective of position in the eruptive sequence, supporting collinearity of bulk compositions (figs. 59, 148, 151) as evidence that all the andesite and dacite are comagmatic. One might be tempted to speculate (1) that some or all of the andesite could have represented a new batch that had just arrived from depth and "triggered" the 1912 outburst or (2) that the andesite and dacite magmas came from separate chambers, but neither idea is easy to reconcile with the bulk compositional and thermal continuity of the andesite-dacite suite nor with the modal and compositional overlap of their abundant phenocrysts. Fluctuations in the proportions of andesite and dacite that erupted together are more likely to reflect the dynamics of magma withdrawal from a single unitary chamber (Spera and others, 1986; Blake and Ivey, 1986) and of transport in a complex conduit system (Carrigan and others, 1992) rather than confluence from separate chambers. The unusually long horizontal component of 1912 magma transport may be implicated in producing the complicated plumbing that favored andesite-dacite fluctuations and permitted those magmas to bypass the small fraction (<0.1 percent) of the rhyolite that was detained and contaminated somewhere along the distal conduit system and left to extrude postclimactically as the terminal lava dome.

If two or more chambers had contributed to the 1912 eruption, it might be expected that lithic fragments of breached wall-rock septa between them (Curtis, 1968; Hildreth, 1983) would appear among the ejecta, but neither hornfelsed siltstone nor alternative candidates have been recognized. The suite of nonjuvenile lithic fragments in 1912 ignimbrite and fallout layers (Hildreth, 1987) is derived predominantly from the Jurassic marine sedimentary Naknek Formation, which extends from the surface to a depth of about 1,500 m (fig. 96). Sparse fragments of Falling Mountain dacites and Trident andesites, both truncated by the 1912 vent, are also present, but no lithics from the roughly 3,500 m of Mesozoic strata that regionally underlie the subhorizontal Naknek Formation (Detterman and others, 1996) have been recognized.

Sparsity of rhyolitic pumice blocks and lapilli in Layers G, F, D, and all but the lowermost part of C suggests that, after dominating Episode I, the rhyolite magma was either largely exhausted or was excluded from access to the conduit by the flux of dacite. Complete withdrawal of the rhyolite layer from the magma chamber is consistent with postcaldera eruption of the Horseshoe Island dacite on the Katmai caldera floor. Pulses of andesite magma, on the other hand, were able to gain sporadic access to the Novarupta vent until virtually the end of the eruption, leading to ultraproximal emplacement of numerous andesite-dacite pyroclastic fall and flow deposits (Fierstein and others, 1997; Houghton and others, 2004) intercalated within the voluminous sequence of dacite-dominated regional plinian fall deposits (Layers C, D, F, and G).

Compositionally banded pumice (fig. 61) was common throughout the 1912 eruptive sequence following deposition of all-rhyolitic Layer A. In total, such banded pumice makes up 3–5 percent of the trillions of pumice lumps deposited, but its fraction fluctuated greatly during the 60-hour eruption, ranging erratically from 0 to 37 percent in the numerous subunits emplaced (Fierstein and Hildreth, 1992). Banded pumice is especially abundant in Layer B2 and in the middle part of the main VTTS-filling ignimbrite sequence, in both of which it consists principally of mingled rhyolite and andesite. Banded pumice is also abundant in ignimbrite Packages 8 and 9 that ended Episode I and in the thin proximal PDC and scoria-fall units of Episodes II and III, but in these units the mingling is between andesite and dacite. The mingling combinations generally reflect proportions of the magmas coerupting, so the only unit where rhyolite-dacite banded pumice is common is Package 7, which is andesite-poor (figs. 27, 34). The extreme sparsity of rhyolite-dacite banded pumice elsewhere during Episode I, when all three magmas coerupted for many hours, may tell less about the plumbing system than about the physical difficulty of mingling between the high-silica rhyolite and crystal-rich dacite on a syneruptive timescale. The lower viscosity of the hotter andesite may have been the key to unstable conduit flow promoting entrainment of its highly viscous companion magmas.

The compositional ranges of plagioclase and orthopyroxene in the rhyolite are outside the corresponding (wide) ranges of zoning in the dacite and andesite (fig. 12 of Hildreth, 1983). Neither these nor the quartz phenocrysts present in 1912 rhyolite occur in noncontaminated 1912 dacite or andesite. The Zr-SiO_2 relations (fig. 151) also illustrate clearly that 1912 dacite is closely related to the coerupted andesite and is not a hybrid mixture of rhyolite and andesite (though a few dacite samples may contain traces of rhyolite contaminant). Microprobe analyses generally distinguish matrix glass compositions of 1912 rhyolite, dacite, and andesite (Avery, 1992; Fierstein and Hildreth, 1992), but the dacite and rhyolite glass arrays are continuous and overlap slightly (fig. 149).

Plagioclase and orthopyroxene phenocrysts are broadly similar in the andesite and dacite, each exhibiting comparably broad, overlapping ranges of zoning, with normal and reverse zoning coexisting within individual scoria and pumice blocks (Hildreth, 1983; Shearer and others, 1991; Avery, 1992). Zoning ranges of feldspar and pyroxenes in the dacite extend to only slightly more evolved compositions than in the andesite (Hildreth, 1983). The shared complexity suggests intimate preeruptive association of the andesite and dacite and some mixing of their phenocrysts. We envisage such heterogeneities to result from limited convective stirring within a graded or layered body of crystal-rich magma, differentiating by upward percolative loss of interstitial melt and perhaps disturbed on a century or decadal timescale by injection of new andesitic batches from below.

In contrast to the extensive and complex zoning of phenocrysts in the andesite and dacite, the oligoclase and orthopyroxene phenocrysts in the rhyolite are almost unzoned. Unlike the rounded quartz in many pyroclastic rhyolites worldwide, quartz in the 1912 rhyolite consists of dipyramidal euhedra. These features and its nearly aphyric condition (0.5–3 weight percent phenocrysts) suggest that the rhyolite had undergone little cooling and had experienced a simple and probably short history since separating from a crystal-rich residue (or, alternatively, after some strong reheating encounter had resorbed all previous phenocrysts). Once separated, the greater buoyancy and viscosity of the rhyolite may have suppressed stirring at the interface and entrainment of phenocrysts from the crystal-rich residue (plausibly the underlying dacite magma, as discussed below). Such apparently sharp interfaces between density-stratified magmatic layers in reservoirs that fed large, strongly zoned eruptions have been discussed by Fridrich and Mahood (1987), Bacon and Druitt (1988), and Johnson and Fridrich (1990).

We don't know what fraction of the magma stored in the 1912 reservoir actually erupted, but the 13 km^3 of magma that did erupt (if initially all stored jointly) could have been held in a spherical chamber about 3 km in diameter. Alternatively, a squat cylindrical 13-km^3 chamber with height and diameter both 2.55 km would underlie an area only slightly larger than that of the Katmai caldera floor. Because the magma volume erupted was about 2.3 times greater than the volume collapsed, however, it is more likely that the magma reservoir was irregular, perhaps with an elongate protrusion extending from beneath the caldera some unknown distance toward Novarupta or Trident (fig. 158). This might have eventually advanced into a dike or sill along joints or bedding planes in the horizontal Mesozoic sedimentary section. A sill extending to Novarupta from a chamber beneath Mount Katmai (if 10 km long and 1 km in cross width) would at rest contain only 0.1 km^3 of magma if 10 m thick (or 1 km^3 if 100 m thick). During propagation of such a sheet, most of the magma would clearly remain in the main reservoir until the eruption was underway (fig. 158). But a sheet of rhyolite magma should have cooled and crystallized significantly during an advance of 10 km through cold sedimentary wall rocks unless its propagation was rapid.

Dissolved H_2O concentrations measured for rhyolitic melt inclusions trapped in 1912 quartz phenocrysts range from 3.5 to 4.7 weight percent, averaging about 4.3 weight percent (Westrich and others, 1991; Lowenstern, 1993; Wallace, 2005). Such values pertain to rhyolite samples from Layer A, early all-rhyolite ignimbrite, later mixed ignimbrite, and the plug dome, suggesting a limited range in preeruptive rhyolitic water content comparable to the rhyolite's limited ranges in major and trace elements. The very few FTIR measurements of H_2O dissolved in melt inclusions in 1912 dacite are in the range 2.2 to 3.1 weight percent, all three determinations having been for plagioclase phenocrysts from a single lump of plinian Layer C pumice. No reliable values have been measured for 1912 andesite, in which the melt inclusions have generally broken, leaked, or partially crystallized (Lowenstern, 1993). Evidence of Cu-rich bubbles in melt inclusions and of progressive partitioning of S, Cu, and Cl into a preeruptive vapor phase during magmatic evolution suggests fluid saturation of all 1912 magma prior to the eruption (Lowenstern, 1993). Because 4.3 weight percent H_2O is sufficient for water saturation of rhyolite melt at 100–130 MPa, saturation thus implies a reservoir depth of 3–5 km. Embracing the analytical uncertainties and allowing for small contributions of CO_2 and SO_2 in the aqueous fluid phase (Wallace, 2005), these estimates could be expanded to about 80–150 MPa or 3–6 km depth (for rhyolite, dacite, and andesite alike), in acceptable agreement with the phase-equilbrium experiments (fig. 157) recounted above (Coombs and Gardner, 2001; Hammer and others, 2002).

There is little doubt that 1912 andesite and dacite were part of a continuously zoned magma body that underlay Mount Katmai. Large fluctuations in the andesite/dacite proportions that erupted during the 60-hour sequence (figs. 34, 60) may have been governed by their viscosity contrast, by viscous segregation during 10 km of subterranean sheet flow, and by convective dynamics of withdrawal from the main reservoir. Although we clearly favor the model of a zoned reservoir where the rhyolite overlay the andesite-dacite continuum, preeruptive location of the rhyolite is less firmly established. Favoring contiguity and petrogenetic kinship are thermal continuity, Sr-Nd-Pb-O isotopic overlap, concurrent eruption throughout Phase B and part of C (altogether for >24 hours), and phase-equilibrium experiments permissive of preeruptive storage at similar depth. Because of the wide bulk-compositional gap and differences in short-lived isotope ratios, ($^{230}Th/^{232}Th$) and ($^{226}Ra/^{230}Th$), however, it may be useful to summarize and again scrutinize the varied evidence for origin of the rhyolite.

Origin of the Rhyolite Magma

We discuss here the evidence bearing upon the preeruptive site of the 7–8 km^3 of rhyolite magma released in 1912, how it may or may not have been intimately situated along with the andesite-dacite continuum, and when and by what processes it may have been generated. It seems to be generally underappreciated that high-silica rhyolite (75–78% SiO_2), although common enough intracontinentally, is uncommon in island arcs and continental margin arcs. No other late Quaternary example has been reported on the Alaska Peninsula, and we know of only three older remnants there. For most arc volcanoes, the most evolved product is dacite (63–68% SiO_2) or rhyodacite (68–72% SiO_2). Therefore, generation of high-silica rhyolite must entail materials or processes that are otherwise seldom involved or effective in arc magma evolution. Possibilities to consider include partial remelting of silicic plutonic rocks, expulsion of highly evolved interstitial liquid from crystal-rich dacitic mush in a chamber (or a migmatitized domain at any crustal depth), buoyant ascent of such liquid from crystallizing rinds at chamber sidewalls, or protracted settling of a phenocryst suite rich in low-silica phases. In each case, efficient isolation of high-silica rhyolitic melt from abundant complementary crystals and suppression of convective reentrainment would be critical to the process. Considerable time might be required to purge a viscous partial-melt extract of its restite or antecrysts.

Isotopic and chemical affinities of the 7–8 km^3 of high-silica rhyolite to the coerupted dacite and andesite were outlined above. The rhyolite and the intermediate magmas appear all to have assimilated melt from wall rocks that had only modest Sr-isotopic contrast and minimal Pb, Nd, and O isotopic effects. Oxygen-isotope data for 1912 ejecta suggest a normal magmatic fractional-crystallization progression with no detectable contribution from sedimentary or hydrothermally altered rocks (Hildreth, 1987). Likewise, for both matrix glass and whole-rock samples, parallelism of rare-earth patterns, with increasing REE abundance (and Eu deficiency) from andesite to dacite to rhyolite (fig. 153) suggests a simple fractionation sequence with no extraneous contribution to the rhyolite large enough to be chemically identifiable. Very low concentrations of Ni and Cr, each as low as 0–3 ppm, in the nearly aphyric rhyolite demand that crystal-melt fractionation have continued subsequent to any contributions of extraneous partial melt. Although the Sr-isotope data indicate small wall-rock contributions to all the 1912 magmas, the combined Nd, Sr, and O isotope results show unequivocally that the rhyolite magma was not substantially a partial melt of the Mesozoic and older stratified rocks constituting the Alaska Peninsula basement. The Jurassic sandstone and siltstone of the exposed Naknek Formation is not a geochemically viable assimilant, even had the magma been stored that shallowly. Taken altogether, the isotope data favor fresh arc-intrusive rocks as the most likely source of minor partial-melt contributions.

Sr-isotope ratios for the 1912 rhyolite (0.70368–0.70374) are slightly more radiogenic than for coerupted dacite (0.70354–0.70362), but they are wholly overlapped by those for andesite (0.70338–0.70374). The modest Sr content of the rhyolite (58–95 ppm) rendered it sensitive enough to contamination that derivation of only 5–7 percent of its Sr from a source like the Tertiary rhyolite sills exposed nearby (0.7043; 235 ppm Sr; fig. 155; Lowenstern and Mahood, 1991) would suffice to produce a shift of 0.0001 in $^{87}Sr/^{86}Sr$. Because 1912 dacite matrix glass also has low Sr contents (80–101 ppm) like the bulk rhyolite, comparably modest proportions of assimilated Sr would have been required had the contamination process operated on the rhyolitic interstitial melt phase of a dacite mush before its expulsion to form the proximate progenitor of the crystal-poor 1912 rhyolite.

Because the andesite-dacite continuum likewise spanned a significant range (0.00036) in $^{87}Sr/^{86}Sr$, the intermediate magma had apparently also been affected by assimilation comparable to that evident in the rhyolite. Less assimilant is thus required to go from 1912 dacite to 1912 rhyolite than to produce the Sr-isotopic range within the andesite-dacite continuum (fig. 155). The isotopic coherence of the entire 1912 suite, in O, Nd, and Sr, makes it doubtful that the rhyolite could be predominantly a partial melt of any extraneous source rock.

The glass (melt) phase interstitial to the crystal-rich 1912 dacite is itself high-silica rhyolite only slightly less evolved than the nearly aphyric 1912 rhyolite (as determined by microprobe). For example, the average dacite-glass versus bulk-rhyolite compositions (normalized volatile-free) compare as follows (all figures in percent): 1.1 vs 0.9 CaO; 0.25 vs 0.15 MgO; 0.27 vs 0.15 TiO_2; 1.6 vs 1.3 Fe_2O_3; 2.9 vs 3.2 K_2O; and 77.6 vs 78.2 SiO_2. In fact, the ranges overlap for most elements (fig. 149). If most of the melt constituting the rhyolite had once been interstitial liquid that escaped upward from a compacting (and gas-saturated) volume of intermediate crystal mush, such modest compositional differences could have resulted from (1) limited subsequent crystal fractionation of the rhyolite, (2) additions of small wall-rock partial-melt increments to both, as inferred from the Sr-isotope data, (3) internal crystal-liquid reequilibration of the melt-depleted dacite mush, and (4) contamination of the dacite mush by melt and crystals from the contiguous andesite. Note that glass in the dacite pumice would represent, by such a mechanism, the melt phase left behind in the dacite mush after escape of the melt fraction that became the rhyolite.

Whatever proportion originated by upward escape of interstitial melt from crystal mush (whether or not augmented by sidewall ascent of fractionating liquid), and irrespective of the fraction contributed by wall-rock partial melting, the 1912 rhyolite was chemically and isotopically closely related to the zoned crystal-rich dacite magma that erupted behind it (figs. 149, 152–155). Thermally continuous and virtually continuous in melt composition, they were probably physically contiguous as well. Rare quartz phenocrysts found in mineral separates from two dacite samples, the most silicic dacites crushed (Hammer and others, 2002), might be cognate but are more likely to be contaminants, transferred syneruptively or across an intrachamber interface.

As for the phenocryst-poor rhyolite and the 69–77 percent SiO_2 compositional gap, discontinuities in composition and mineralogy are common in caldera-forming pyroclastic eruptions large and small (Hildreth, 1981). They have been explained in terms of density-stratified layered chambers that accumulated incrementally and differentiated internally (Fridrich and Mahood, 1987, Bacon and Druitt, 1988; Johnson and Fridrich, 1990). Derivation of rhyolitic melt from a crystal-rich intermediate parent magma was modelled by Grove and Donnelly-Nolan (1986) in terms of extensive low-pressure crystallization over a small cooling interval—rapid multiphase crystallization along a gently sloping cotectic having shallow thermal curvature in temperature-composition space. Maintaining a compositional gap so established would require efficient expulsion of interstitial rhyolitic melt and subsequent isolation from its intermediate parent, which would then be left behind as a mushy quasi-cumulate. Mechanisms for incremental upward separation of buoyant residual liquid, when successively intercepted new magma batches crystallize within or atop cumulate zones, were proposed by Bacon and Druitt (1988), Grove and others (1997), Philpotts and others (1998), and Sisson and Bacon (1999). Vesiculation caused by rapid crystallization of lenses of andesite-dacite recharge magma could repeatedly hasten upward segregation of gas-saturated residual liquid, thereby intermittently augmenting the overlying zone of thermally buffered, water-saturated, phenocryst-poor rhyolite. Minor assimilation of partial melt from wall rocks need induce little or no crystallization of such doubly buffered rhyolite; moreover, the Sr-isotope data suggest that most assimilation took place before the rhyolite separated from the intermediate mush. We are persuaded that such a model fits best the 1912 reservoir because of the isotopic and chemical affinities among all the 1912 ejecta, their continuity in magmatic temperature and oxygen fugacity, and the need for a huge proximate residue of hot crystal mush from which to have extracted (and maintained aphyric) at least 8 km^3 of high-silica rhyolite.

It was previously suggested that the magmatic focussing implicit in the unusual clustering of several stratovolcanoes might somehow be implicated in generating the unusual high-silica rhyolite at Katmai (Hildreth, 1983, 1987). Among the centers here, however, we now know that only Southwest Katmai has produced magma more evolved than dacite, that Northeast Katmai and the Trident cones and domes have long been dead, and that comparable Quaternary clusters and calderas along the Alaska Peninsula (Miller and others, 1998) lack high-silica rhyolite. Mount Katmai is the only center in the cluster where basalt has ascended to eruptible levels, thereby investing crustal melting with a heat source significantly intensified relative to the flux of crystal-rich andesite typical of the other centers. If thermal "focussing" was indeed a factor in producing the rhyolite, the focus is better thought of as beneath Mount Katmai rather than the cluster as a whole.

Alternatives to a Zoned Reservoir?

Although our earliest data already favored a zoned chamber, the two-chamber idea of Curtis (1968) was initially left open as a possible alternative (Hildreth, 1983, p. 49-50). Subsequently it was stated (Hildreth, 1987) that "for every complexly zoned eruption, it needs to be asked afresh whether the syneruptively mingled compositions reflected (1) addition of a new batch of relatively primitive magma from depth, (2) a single zoned or layered chamber, or (3) mixing of magmas from two or more discrete pods, dikes, or sills in a complex reservoir. Although the continuity in Fe-Ti-oxide temperatures argues in favor of the second scenario for 1912 (Hildreth, 1983), the existing data do not really permit one to exclude a chance encounter of the third kind (Curtis, 1968)."

Two Chambers?

By 1987, the rapid thermal decline of the 1912 vent region had raised doubts that any chamber had ever underlain Novarupta basin, and new work led to the suggestion that structural trends parallel to the local direction of plate convergence might have permitted lateral magma transport from a reservoir beneath Trident (Hildreth, 1987; Wallmann and others, 1990). Having by now established, however, that Falling Mountain and Mount Cerberus are older than 60 ka and compositionally related to Trident (not to Novarupta), that the three cones of Trident existing in 1912 had long been dead (since at least 40 ka), and that magmatic precursors of Southwest Trident were not involved in 1912, there is no longer any basis for supposing that any 1912 magma had been resident under Trident, either. This does not, of course, preclude a transport path of 1912 magma from Mount Katmai to Novarupta beneath long-crystallized parts of the Trident complex.

Sill transport from Katmai to Trident, thence dike transport toward Novarupta (parallel to the plate-motion vector and to vertical jointing in the Naknek Formation) is not unreasonable (Hildreth, 1987; Wallmann and others, 1990).

A New Magma Batch from Depth?

As for the possibility of introducing a new magma batch not already part of the zoned system, a newly arrived "triggering" batch of andesite might be expected to be conspicuous among the ejecta evacuated from a reservoir of this (intermediate) size. Despite persistent search and many analyses, however, no mafic or intermediate 1912 ejecta have been found that do not belong chemically and mineralogically to the very phenocryst-rich andesite-dacite zoned continuum. No phenocryst-poor or decidedly more primitive mafic ejecta have been identified. (The five andesites more mafic than the continuum are comparably crystal-rich, and at least one is certainly olivine-accumulative.)

It was speculated by Eichelberger and Izbekov (2000) that an aphyric rhyolite dike arose from depth (the middle crust?) and "simultaneously" intersected the surface at Novarupta and the andesite-dacite chamber beneath Mount Katmai. Such a deus ex machina would need to have approached the surface without causing surface deformation anywhere along the 10-km-long reach separating the two vents. We can verify none of the surface features attributed by those authors to dike emplacement between Novarupta and Katmai caldera, where the connecting reach is covered by tens of meters of glacial ice and 1912 ejecta. Its strike would have been east-west, at a high angle to vertical jointing in the basement (330°±20°) and to the plate-convergence vector (320°), which closely coincide in this area (Hildreth, 1987). It would be extraordinary for a dike to propagate 10 km in a direction at 50°–60° to the maximum horizontal stress and to the plane of preferred anisotropy in the wall rocks (Wallmann and others, 1990).

For the rhyolite to be an extraneous intruder, the thermal, isotopic, and glass-compositional continuity of the rhyolite with the coerupted dacite would have to be purely coincidental. The phase-equilibrium experimental data and melt-inclusion volatile evidence indicating shallow storage of 1912 rhyolite, dacite, and andesite all at similar depths need not, of course, require differentiation of the aphyric rhyolite within the upper-crustal reservoir where the phase assemblages last equilibrated; but if the rhyolite had risen from some deeper extraneous source, all restite, phenocrysts, or entrained crystal mush would need to have been completely resorbed during rapid ascent.

The Mount Katmai-Trident chain had been magmatically active for at least 140,000 years (Hildreth and others, 2003b) and had erupted ~80 km^3 of magma that ranged from basalt to rhyodacite, predominantly andesite-dacite. It is hard to envisage how a deep-crustal processing domain (MASH zone of Hildreth and Moorbath, 1988; or hot zone of Annen and others, 2006) that had been releasing intermediate magmas for so long and so currently could abruptly generate a large batch of minimum-melt high-silica rhyolite. Because basalt-induced partial melting of amphibole-bearing mafic meta-igneous deep-crustal rocks generally yields intermediate melts (Rapp and Watson, 1995; Sisson and others, 2005; Annen and others, 2006), not high-silica rhyolites, the best candidates for near-primary arc-crustal melts are the uncommon high-temperature crystal-poor dacites and rhyodacites. Moreover, a large volume of rhyolite magma from the deep crust would require an enormous source domain, raising problems of source-rock heterogeneity, hybridization, and source depletion and dehydration over a span of 140,000 years. If a deep felsic pluton were indeed to have been partially remelted, yielding a large batch of rhyolitic magma, it is hard to envisage how such a batch could run the 20–30 km crustal gauntlet of mushy pods and trailings left by the many forerunning basalts and andesites without entrainment or mixing. It appears far more feasible to separate the high-silica rhyolite in the upper crust.

Roof-Zone Remelting?

Extraction of rhyolitic melt from the underlying andesite-dacite reservoir need not have been a one-stage process. During a magmatic history longer than 100,000 years, tonalitic to granodioritic plutons could have crystallized in the upper crust beneath Mount Katmai. Had the 1912 vapor-saturated andesite-dacite chamber developed within or adjacent to such a late Pleistocene pluton, rhyolitic partial melts from it could have augmented the rhyolite melt expelled from the dacite mush. Difficulties with such a model include the apparent absence of inherited zircon or other refractory xenocrysts and the chemical and O-Sr-Nd-Pb isotopic affinities of the 1912 rhyolitic and intermediate magmas. Such affinities appear to limit any excess pluton-derived partial melt in the rhyolite to a modest contribution, as implicit in figure 155. On the other hand, contributions from a Tertiary or early Quaternary arc pluton could have high (^{230}Th/^{232}Th) and equilibrium (^{230}Th/^{238}U), consistent with the near-equilibrium U-Th data for the 1912 rhyolite (fig. 156). Or, if the pluton were of late Pleistocene age like most of Mount Katmai, its partial melts in the presence of refractory zircon could be expected to shift toward the equiline from bulk-rock ^{238}U excesses.

Still closer similarity of partial melts of plutonic wall rock and rhyolitic melt extracted from resident dacitic mush would result if the "granodiorite" had been a crystalline roof zone in the Mount Katmai system itself. Thermal waxing and waning during an intrusive and eruptive history longer than 100,000 years could have produced broadly dacitic (or more silicic) differentiates many times and lodged some of them beneath the edifice as substantial bodies, mushy or solidified. Partial remelting of such forerunner material (Mahood, 1990) during the thermally waxing buildup of the 1912 andesite-dacite chamber might have generated rhyolite melt not unlike that expelled from the dacite itself, reflecting essential "consanguinity" within a broadly comagmatic long-lived system. If it was ~100,000 years old, ^{230}Th ingrowth would have driven such a fossilized zone closer to the

$^{230}Th/^{238}U$ equiline, helping to resolve the U-series uncertainties concerning the 1912 rhyolite.

The process of extracting high-silica-rhyolite partial melts from quartz-bearing granitoid has the added appeal of being consistent with the rarity of such compositions in immature arcs and their abundance in continental interiors. Tertiary plutons sampled in the Katmai area (n=22) are mostly tonalitic (57–65% SiO_2), but a few are more evolved (67–71% SiO_2). Remelting of intrusive remainders of the 22.5-ka rhyolitic eruption (71–72% SiO_2) of Mount Katmai might well have yielded a high-silica melt extract. Nevertheless, the mushy crystal-rich 1912 dacite had done so, too, even without participation of quartz, as its interstitial glass phase contains 76–79 percent SiO_2.

1912 Magma Storage Model

Figure 158 portrays a preeruptive magma-storage model consistent with all the many constraints discussed. The evidence weighs heavily against participation of preexisting Trident magma, and nothing favors more than one magma chamber. The chamber could well have been larger than the 13 km^3 erupted and depicted, and it could have been a bit shallower than shown, but, if the rhyolite had indeed been water-saturated at 4.0–4.5 weight percent, its roof is not likely to have been much deeper. Vertical displacement of Mount Katmai by 1.2±0.1 km demands that the magma body have been at least that thick (in addition to cumulates of unknown but probably significant thickness). Compensation by collapse of only 5.5 km^3 for the 13 km^3 erupted requires that the reservoir have been areally more extensive than the caldera. Magma chambers considerably more extensive than superjacent calderas would appear to be a common pattern, considering the areal spectra of batholiths and calderas (Lipman, 1984). A simple model for 1912 would be a storage zone elongate along the volcanic axis, above the axial zone long subject to distributed magma ascent from deep below. Elongation of the chamber along axis lessens the distance required for sill propagation toward Trident or Novarupta.

Sill injection is favored by the shallow overburden and by the horizontal bedding planes of the Mesozoic marine section (Fenner, 1923; Lowenstern and others, 1991). Advance of a horizontal sill from beneath the volcanic chain toward the off-axis Novarupta depression lessens the local overburden by at least 700 m, not insignificant for magma initially only 3–6 km deep; progressive reduction in vertical stress would favor sill propagation away from the highland edifice. Ultimately, of course, the rhyolitic magma had to crosscut strata to reach the 1912 vent site, but the terminal dike segment could have been short—breaking away from the hypothetical sill either right under lowland Novarupta or somewhere beneath Trident where the plate-motion vector, inferred regional stress field, and jointing in the basement rocks should have favored a north-northwesterly direction of dike propagation (Hildreth, 1987; Wallmann and others, 1990). In this setting, dike transport directly east-west between Mount Katmai and Novarupta would be unlikely, still less so to the extent that outbreak depth influenced vent location, because much of that 10-km path runs beneath terrain that was lower in elevation than the 1912 vent site.

It can be inferred that present-day upper crustal seismicity largely reflects response to hydrothermal processes and not shallow magma storage. The modest batches of andesite-dacite magma typical of the Katmai cluster volcanoes probably ascend from midcrustal reservoirs or deeper, from well below the brittle-ductile transition (at a depth of ~10 km). The only evidence for late Quaternary reservoirs sufficiently large, shallow, and sustained to produce rhyolite, rhyodacite, large-magnitude eruptions, and caldera collapse points exclusively to Mount Katmai. The model depicted (fig. 158) is consistent with the volume and displacement requirements, compositional integrity of the zonation, and eruption of ~3 km^3 of unaccompanied rhyolite magma first. To the extent that the reservoir was elongate toward Trident, the path required for sill propagation shortens. For the rhyolite to have remained phenocryst-poor on departing from the reservoir into cold wall rocks, sill injection had to be fast. Might it have been accomplished during the five days (1–5 June 1912) of preeruptive seismicity? Shallow earthquake swarms that migrate along rift zones in Iceland and Hawaii (from central volcanoes to remote eruption sites) track the propagation of basaltic dikes at rates of 0.1 to 6 km per hour (Brandsdóttir and Einarsson, 1979; Klein and others, 1987). Comparable data are lacking for silicic dikes and sills, but even rhyolite might reasonably be expected to propagate the 10 km to Novarupta in 120 hours (5 days).

Campsite of Geophysical Expedition of 1989 in arcuate graben 1.5 km northeast of Novarupta dome. Typically 8–14 participants at a time (and as many as 29 altogether) used WeatherPort structures and smaller, backpacking tents during 6 weeks of geophysical surveys of the 1912 vent region. West Trident Volcano in background, heavily mantled with 1912 fall deposits. Photo by John Eichelberger.

Chapter 14

Comparisons with Other Historic Eruptions

Every great explosive eruption is unique in some respects. The Novarupta outbreak was extreme in its compositional range and extraordinary in its separation of vent and subsidence sites, but it had much in common with other plinian eruptions. Data compiled in table 3 are provided to compare aspects of the 1912 eruption with those of several other well-characterized explosive eruptions—larger or smaller, mostly historical, some notorious. Many others might have been chosen, but the examples tabulated were selected for variety and because most of the parameters listed are fairly well constrained by data. Significant fractions of ejecta were lost at sea for about half the eruptions listed, and several other well-characterized island eruptions were ignored because the lost fraction is too great.

Of the 24 eruptions listed in table 3, 18 were from arc volcanoes, reflecting principally the rarity of explosive intracontinental activity in the Holocene. That four are from Iceland is a tribute to the detailed work of Sigurdur Thorarinsson. All 24 eruptions were plinian except for Pelée, which is included here for its historical importance in volcanology; several also had phreatomagmatic phases.

Although all the magmas were rich in dissolved volatile components, there is no obvious correlation of plinian explosivity with major-element composition, which ranges from rhyolite to andesite, and even rarely to basalt or phonolite. The range provides a corrective reminder to those of us who have taken for granted the abundance of rhyolites among the ignimbrite-forming eruptions of the intracontinental North American Tertiary and Quaternary. From arc volcanoes generally, high-silica rhyolites like that of 1912 are uncommon, though not rare. The apparent exception, New Zealand (for which Taupo is an example in the table), owes its voluminous Quaternary rhyolites more to quasi-continental rifting than to arc volcanism.

Many of the examples listed in table 3 produced small quantities of ejecta more mafic than the predominant eruptive volume, but only Novarupta, Hekla, and Mazama erupted large subordinate fractions of andesite or basalt. Relatively mafic enclaves were scattered within the dome lavas of Pelée, preclimactic Pinatubo, and postclimactic Novarupta. Among the eruptions listed, fairly homogeneous magma batches include the trachyandesites of Tambora and El Chichón, the andesite of St. Vincent, the dacites of Santa María and Huaynaputina-I, and the rhyolites of Öraefajökull and Taupo. Although they were parts of zoned eruptive sequences, the ~40 km^3 of climactic Mazama rhyodacite (~70.5% SiO_2), the 8 km^3 of Novarupta rhyolite (~77.5% SiO_2), and the 5.5 km^3 of climactic Pinatubo dacite (~64.5% SiO_2) were fairly homogeneous.

Caldera subsidence attended all the eruptions listed that released more than 10 km^3 of magma. Somewhat less voluminous eruptions at Krakatau (9 km^3), Pinatubo (5.5 km^3), and Apoyo (4–5 km^3) also produced calderas, but comparably voluminous eruptions at Huaynaputina (9 km^3), Santa María (4–8 km^3), Quizapu (4 km^3), and Vesuvius (4 km^3) did not. Initiation of subsidence of the 2-km^3 Öskjuvatn caldera in 1875 is attributed to withdrawal of basaltic magma along the rift zone and its eruption 70 km north of Askja, not to the 0.34-km^3 plinian rhyolite eruption at what became the caldera site (Sigurdsson and Sparks, 1981). Despite differences in volume and composition, the withdrawal and collapse at Askja in 1875 are hydraulically the closest analogue we know of for the Katmai Novarupta connection of 1912.

The climactic episodes of most of these plinian eruptions lasted less than a day, and even some of the largest historical events may have lasted little more than two days. At the extreme, we think the great Bishop Tuff eruption lasted less than six days, based on an uninterrupted plinian pumice fall (of varying intensity) and accumulation rates calculated for the comparably dispersed ejecta of well-studied historical plinian deposits (Wilson and Hildreth, 1997; Hildreth and Wilson, 2007). Several of the eruptions listed in table 3, however, consisted of more than one plinian episode, separated by lulls of hours or days, including Tambora, El Chichón, Taupo, and Novarupta. The 1912 sequence at Novarupta spanned 60 hours because it actually consisted of three successive plinian episodes with multihour lulls between.

Many great eruptions commence with a vent-opening outburst, commonly phreatomagmatic, that deposits a small volume of initial ash. The lack of evidence for such a deposit at Novarupta is anomalous and might simply mean that it was scoured from the region around the vent by the blast-like all-rhyolite pyroclastic density currents that accompanied initiation of the main plinian Episode I.

In all or most of the prehistoric, enormous (100–1,000 km^3), caldera-forming, intracontinental ignimbrite eruptions, the volume of ignimbrite exceeds that of its companion fallout, more so where intracaldera tuff can be taken into account.

Table 3. Comparison of some explosive eruptions.

[For the nine eruptions below the line, there is little or no historical record, so the data are based principally on geological investigations. Composition abbreviations: A, andesite; B, basalt; C, comendite; D, dacite; Ph, phonolite; R, rhyolite; Rd, rhyodacite; Tp, tephriphonolite; Ta, trachyandesite; Tr, trachyte. Duration of eruption climactic episode includes neither precursors nor sporadic postclimactic activity; lulls of hours to days between plinian episodes are inferred for Novarupta, El Chichón, Tambora, and Kuwae. Pyroclastic density currents (pdc) include flows, surges, and coignimbrite ashfall (as identified in each report); fall (F) is fallout from column and umbrella cloud, mostly plinian.]

Eruption and year/age	Magma km^3	Composition	Duration hours	Syneruptive deaths	Fall vs pdc	References[a]
Novarupta 1912	13.5	R>D>A	<60	1	F~pdc	8, 9, 15
St. Helens 18 May 1980	0.3	D	9	57	F>pdc	7, 19
Pinatubo 1991	5.5	D>>A>B	9	350	pdc>F	21
El Chichón 1982	0.5	Ta	~155	>1800	F>pdc	29, 35
Quizapu 1932	4	Rd>A	18	0	F>>>pdc	14
Santa María 1902	4–8	D	20	>1500	F	9, 10, 37
St. Vincent 1902	0.8	A	10	>1600	F>pdc	13
Pelée 8 May1902	<0.2	A>D	<1	28,000	pdc>F	11, 12
Tarawera 1886	0.8	B	4	153	F	3, 36
Krakatau 1883	9	D	23–30	36,400	pdc~F	25, 30
Askja 1875	0.34	R>>A>B	17	0	F>>pdc	4, 20, 31, 32
Tambora 1815	33	Ta	~24	>10,000	F>pdc	24, 26, 27
Huaynaputina-I 1600	9	D	20	~1400	F>pdc	1
Öraefajökull 1362	2	Rd	?	>220	F>pdc	34
Vesuvius 79	4	Tp>Ph	19	>3,500	F>pdc	5, 6, 28
Kuwae ~1452	>35	D>>A~B	?	many	pdc>F	23, 40
Baitoushan ~969	~25	C>Tr	?	many	F>pdc	17
Taupo 232	35	R	24–48	0	pdc>F	39
Apoyo-A 2.3 ka	4–5	D	?	?	F~pdc	33
Hekla H_3 2.9 ka	2.2	R>D~A	?	0	F	18
Minoan 3.5 ka	28	D>A	24–48	many	pdc>F	22
Hekla H_4 4.5 ka	~2	R>D>A	24–30	0	F	18
Mazama 7.7 ka	46	Rd>A	24–30	some	pdc~F	2, 41
Bishop Tuff 760 ka	650	R	~130	0	pdc>F	16, 38

[a]**References:** 1. Adams and others, 2001; 2. Bacon and Druitt, 1988; 3. Carey, R.J., and others, 2007; 4. Carey, R.J., and others, 2009; 5. Carey, S., and Sigurdsson, 1987; 6. Cioni and others, 1995; 7. Criswell, 1987; 8. Fierstein and Hildreth, 1992; 9. Fierstein and Nathenson, 1992; 10. Fierstein and Nathenson, 1993; 11. Fisher and Heiken, 1982; 12. Gourgaud and others, 1989; 13. Hay, 1959; 14. Hildreth and Drake, 1992; 15. Hildreth and Fierstein, 2000; 16. Hildreth and Wilson, 2007; 17. Horn and Schmincke, 2000; 18. Larsen and Thorarinsson, 1978; 19. Lipman and Mullineaux, 1981; 20. Macdonald and others, 1987; 21. Newhall and Punongbayan, 1996; 22. Pyle, 1990; 23. Robin and others, 1994; 24. Self and others, 2004; 25. Self and Rampino, 1981; 26. Self and others, 1984; 27. Sigurdsson and Carey, 1989; 28. Sigurdsson and others, 1985; 29. Sigurdsson and others, 1984; 30. Sigurdsson and others, 1991; 31. Sigurdsson and Sparks, 1981; 32. Sparks and others, 1981; 33. Sussman, 1985; 34. Thorarinsson, 1958; 35. Varekamp and others, 1984; 36. Walker and others, 1984; 37. Williams and Self, 1983; 38. Wilson and Hildreth, 1997; 39. Wilson and Walker, 1985; 40. Witter and Self, 2007; 41. Young, 1990.

Among the modest to medium-size historical plinian eruptions tabulated in table 3, however, no such systematic partitioning of the ejecta is apparent. In 14 of the eruptions listed, fallout was volumetrically dominant and, among these, flow deposits were trivial or absent in 6. In 6 of the eruptions listed, deposits of pyroclastic density currents predominated and, among these, massive involvement of seawater or lakewater played a role in 3 (as it did also at Krakatau). In 4 of the 24 listed (Novarupta, Krakatau, Apoyo-A, and Mazama), volumes of fall and flow deposits were roughly equal. As these 4 represent ranges of volume, composition, and eruptive behavior, we recognize no shared pattern that accounts for such partitioning equality. It needs to be recognized, however, that relative proportions of coignimbrite and coplinian ash in many distal fine-ash deposits are poorly known, because such distal ash is seldom scrutinized by microprobe shard-counting like that undertaken by Fierstein on the compositionally heterogeneous ash of 1912.

Deaths caused directly by the eruptions are listed in table 3. Neither direct injuries nor indirect deaths (owing to starvation, debris flows, or forced migration) are included. The death toll clearly correlates more with demography than volcanology. The absence of casualties at Quizapu and Askja reflects their remoteness, as it does for Novarupta, where the only known death attributable to the 1912 eruption was an elderly refugee already suffering from tuberculosis and nephritis who died aboard the *Manning* in Kodiak harbor on the evening of 9 June, after the ashfall had ceased. The list shows that agricultural communities on tropical volcanoes are at elevated risk, as are settlements on or near island volcanoes where tsunami generation can cause widespread devastation. Limitation of death tolls to a tenth of what they might have been during the latter-day eruptions of St. Helens and Pinatubo reflects the wisdom of volcanologists and their influence with the authorities.

The Griggs expedition of 1917, spiffy and dry at Kodiak, just before shipping out to Katmai Bay on 15 June for 10 weeks in the Valley of Ten Thousand Smokes (VTTS). Shown is the entire party, from left to right: Sayre, Maynard, Church, Folsom, Griggs, Hine, Shipley, (Paul) Hagelbarger, Yagashof, and Metroken. Photo courtesy of the National Geographic Society. Folsom and Griggs had together ascended Katmai Pass from the south and discovered the VTTS on 31 July 1916, last full day of that season's explorations.

Chapter 15

Retrospective: Evolution of Ideas about the 1912 Eruption

In the final year of his long life, Fenner (1950) reflected: "A geologist coming into the Katmai country in the years following the eruption found himself faced with phenomena for which no text-book descriptions, or even experience in volcanic regions that had been less recently and less violently active, would have prepared him. He would be surrounded by factual evidence of volcanic processes for which text-books either gave no explanation or gave explanations obviously out of accord with what he saw." Fenner clung too long to imaginative misinterpretations of what he saw, but one can still empathize with his 1919–1923 predicament. Previous pyroclastic eruptions investigated in detail had been few. The leading model at the time was the directed blast of 8 May 1902 from Pelée, but the analogy was misleading for the 1912 ignimbrite, which had no recognizable source vent on any volcanic edifice. The column-collapse scoria flows of 7 May 1902 at St. Vincent would have provided a better model, but their early interpretation was confused with the nuées ardentes of Pelée. A still better model would have been the CE 79 eruption of Vesuvius, but in Fenner's time its interpretation had advanced little since Pliny's eyewitness account. The best model might have been the 1883 eruption of Krakatau, but its vent and most of its ignimbrite were beneath the sea (Simkin and Fiske, 1983).

In retrospect, it seems remarkable that a great eruption that generated international press coverage attracted so few early investigators. Rugged frontiersmen, suitably equipped, were resident in Kodiak, Kanatak, Cold Bay, and Naknek, but none had sufficient curiosity to leave their summertime subsistence pursuits to approach the volcano. Martin sailed to Katmai village in 1912, and Hesse and Horner hiked to Martin Creek in 1913, but clouds, creeks, and the ash blanket kept them from Mount Katmai and the VTTS. In 1915, Griggs, Folsom, and Fulton were stopped by Katmai River from approaching Mount Katmai, but they became the first to view and photograph its beheaded edifice (fig. 100). Only on a second expedition, in 1916, did Griggs climb to the caldera rim and cross Katmai Pass into what he then named the Valley of Ten Thousand Smokes.

Griggs was a respected botanical scholar and an adventurous outdoorsman, whose 1913 studies of revegetation of the ash blanket on both sides of Shelikof Strait prompted the National Geographic Society to select him to lead a 1915 expedition. His Katmai years coincided with a career shift from cytology and taxonomy to colonization dynamics of various marginal environments. Griggs also had a talent for evoking public fascination with nature, helping to strengthen mutually beneficial relations with directors of the National Geographic Society and eventuating in establishment of one of the world's most magnificent and unspoiled wilderness parklands.

Although Griggs was a good field observer, he had little geological training and no previous experience with volcanic phenomena. He was personally in charge of the explorations and was called Professor Griggs or the Director by expedition members. Only in 1919 were geologists invited to participate, and his relations with them that summer were neither close nor smooth. Griggs' geological inexperience is evident in his presumption that Mount Martin was "a new volcano (that) sprung into being at the time of the great explosion" of Mount Katmai (Griggs, 1917; 1922, p. 86). After two seasons of trudging around in the pumice-fall deposits surrounding Katmai River, he first encountered ignimbrite in Mageik Creek and called it "the largest and most striking accumulation of ash observed anywhere."

It is also remarkable that he termed the main VTTS ignimbrite a "great hot mud flow," despite observing that it burned the trees it engulfed (Griggs, 1918c). Griggs had earlier examined the Katmai River debris-flow deposit (figs. 120–122), in which he recognized uncharred trees and bushes; later finding that the main VTTS deposit was rather similar looking may have influenced his "hot mud flow" misinterpretation. From his observations that the "high sand mark" wraps much of the upper VTTS, diminishes gradually downvalley, and left no remnant of a counterpart deposit higher upslope, Griggs (1918c) inferred that the material had "welled quietly up out of the bowels of the earth," nonexplosively. He did observe, however, that the "high sand mark" vanishes in Novarupta basin and that pumice is coarser around Novarupta than elsewhere. Impressed by the fissure fumaroles and high-temperature acid gases, Griggs took the fissures to be magmatic vents, and he inferred existence of a batholithic magma body beneath and coextensive with the VTTS, a notion likely to have come from a geology textbook.

Having viewed Katmai beheaded on his first visit in 1915 and having climbed to its caldera rim in 1916, before he had any inkling of the VTTS and Novarupta, Griggs ever remained attached to his first impression that the pumice falls had issued from Mount Katmai. He rationalized the coarse pumice around Novarupta and the plug dome itself as manifestations of one of many (supposedly fissural) vents distributed throughout the VTTS that had released the ignimbrite. Griggs nonetheless expressed his puzzlement about why the VTTS should extend perpendicular to the volcanic line and why the VTTS and Mount Katmai should have erupted more or less sequentially.

Fenner and Zies recognized that the high-temperature particulate mass had carried no liquid water, leading them to characterize the deposit as an ash flow, sand avalanche, or usually sand flow (Fenner, 1920, 1923). They concurred with Griggs that the ignimbrite did not descend from Mount Katmai, that its source lay somewhere in the VTTS, and that it was overlain by stratified pumice falls (which they all misattributed to Mount Katmai). None of them recognized that the Layer A pumice fall underlies the ignimbrite in Mageik Creek, is ubiquitous on and south of Mount Katmai, but is missing in the VTTS. Failure to distinguish among the several layers of fall deposits (A–H) prevented them from realizing that the ignimbrite was synplinian, emplaced during the fallout sequence, not before it as they inferred from stratigraphic relations in the VTTS alone. The misunderstanding is attributable in part to the fact that (before 1923) all ascents to the caldera rim had been from the south, where one cannot observe either ignimbrite-fallout age relations or the decline in coarseness of the fallout layers away from Novarupta (fig. 105). Locked into convictions that the ignimbrite erupted in the VTTS and most of the fallout at Mount Katmai, they never arrived at a single-vent column-collapse model for the contrasting but concurrent modes of pyroclastic emplacement. Griggs and Fenner acknowledged the coarse pumice falls close to Novarupta, and they granted that Novarupta had been one of many vents, but they never mentioned any attempt to distinguish Novarupta fallout from that attributed to Mount Katmai.

Agreeing with Griggs' interpretation of fissure vents all over the VTTS, Fenner (1920, p. 581) wrote that, concerning the origin of the sand flow, "we suppose that rhyolitic magma, charged with dissolved gases, rose to the surface in the newly formed vents. According to general observation, the usual course for such a magma is either to retain its gases and form a flow of obsidian, or to evolve them with explosive violence and scatter the disrupted particles to a great distance. In this instance, however, it apparently pursued an intermediate course, and produced, by moderately forcible disruption, an outward-spreading and forward-moving torrent of incandescent sand and pumice, each particle of which was surrounded by and partially suspended in gases which it continued to give forth during its impetuous flow." Despite Fenner's idiosyncratic view that magma might reach the surface nonexplosively and evolve most of its gases only subsequently, this is nonetheless one of the first clear formulations of ash-flow emplacement and the cause of its mobility.

When Marshall (1935) introduced the term "ignimbrite" for such deposits, specifically for the many sheets of variably welded rhyolitic tuff in New Zealand, he cited Fenner's VTTS work for both mode of emplacement and manner of cohesion and induration of hot viscous glass shards (sintering, welding), which Fenner had learned in the glass industry. Gilbert's (1938) description of the Bishop Tuff seconded both Fenner and Marshall in citing the importance of continued gas emission during outflow for mobility, welding, cooling, and crystallization of the deposit. Smith (1960) summarized the evolution of ideas about pyroclastic flows, from the abrupt directed blast at Pelée to the extended outpouring at Novarupta. He emphasized the distinction between basal pyroclastic avalanches and overriding billowing ash clouds, and he credited Fenner with recognizing that initial momentum and gravity were supplemented by prolonged internal gas production in accounting for pyroclastic-flow mobility. The column-collapse model for the origin and momentum of pyroclastic flows was implicit in the St. Vincent eruption of 1902, but its appropriateness was not appreciated until Hay's (1959) study, perhaps because it was andesitic, subplinian, and widely conflated with the dome-sourced blasts and block-and-ash flows at Pelée. The first unequivocal description of pumice-rich silicic ignimbrite being generated by collapse of a plinian eruption column was that of Kozu (1934) concerning the 1929 eruption of Komagatake in Hokkaido (Tsuya, 1930).

Neither Marshall nor Gilbert was able to identify vents for their ignimbrite sheets, and, like Fenner, they appealed generally to fissures in the tuff and to burial of vents by the deposits themselves. Citing varied lithic suites in the ignimbrite, Gilbert had inferred "scattered vents," as had Fenner on the basis of fumarolic emissions along the entire VTTS. MacGregor (1952) interpreted extracts from Fenner's many papers to infer that he had really meant to give primacy to Novarupta, but Fenner always referred also to fissure vents throughout the valley, and he always assigned the fallout sequence to Mount Katmai. As late as 1953, many still inferred fissure vents to be distributed across much of the upper VTTS (Williams and others, 1953). Only after Curtis (1968) had isopached the individual fallout layers was it settled that both the plinian phase and the ignimbrite had issued from Novarupta basin.

Fenner's family home was just west of New York City, in eastern New Jersey, where he did his Ph.D. dissertation on basalts of the Watchung Mountains and became intimately familiar with the Mesozoic sills of the Newark Basin (Fenner, 1926, 1929). His dissertation emphasized secondary replacement minerals in the basalt, and he had previously spent 15 years internationally investigating hydrothermal fluids, secondary minerals, and ore deposits. At the Geophysical Laboratory, not long before his first Katmai trip, he had conducted supercritical hydrous experimental work that demonstrated major increases in mineral solubility with temperature (Morey and Fenner, 1917). It seems easy (if perhaps facile) to view Fenner's ideas at Katmai as outgrowths of these previous experiences. Impressed by the horizontal Jurassic strata that extend across the whole VTTS from Mount Katmai, Fenner proposed that (instead of Griggs' batholith) a rhyolite sill had propagated 20 km northwestward from Mount Katmai and, coextensive with the VTTS, had released both the ignimbrite and the fumarolic gases.

More puzzling is why Fenner never even discussed mingling of separate magmas as an alternative for producing the ubiquitous banded pumice, especially as he acknowledged that the adjacent volcanoes were largely andesitic and the juvenile 1912 magma was predominantly rhyolitic. In illustrating a strikingly banded black-and-white pumice clast, he even wrote that it was "formed by the partial mingling of two chemically diverse fluids" (Fenner, 1923, p. 38–39). But he insisted that the black andesitic fluid was not a new magma but a product of fusion by the rhyolite magma of older previously solid lavas and morainal debris. He never mentioned the apparent contradiction between the extended time required for remelting solid andesite and the limited (syneruptive) time indicated by the razor-sharp contacts between black and white pumice bands (fig. 61). The thermal requirements and necessarily inferred physical location of such fusion and mingling lured him into advocating his most imaginative and outrageous hypothesis.

Fenner recognized that lithic ejecta in 1912 fall deposits were far from adequate to account for the Katmai caldera by explosive origin, and he dismissed subsidence as subordinate or speculative (Fenner, 1920, p. 594). He proposed instead that the mountaintop had been assimilated by the new rhyolitic magma and that the andesitic component of the banded pumice represented remelted lavas of the Katmai edifice. Because he presumed separate vents for the ignimbrite down in the VTTS, its abundant banded pumice required a different source of andesitic assimilant, which he assigned to morainal debris in the valley. Such a scenario forced him to envision that the rhyolitic magma had reached the surface nonexplosively, in both the VTTS and the crater of Mount Katmai, and that it had reposed there quiescently long enough to remelt, partially assimilate, and mingle with the surficial andesites before their joint vesiculation and violently explosive fragmentation and dispersal. In dismissing explosivity before the magma had reached the surface, he cited sparsity of lithic ejecta and absence of structural disruption of the basement strata; what he appears to have been dismissing was a magma-chamber explosion, and he evidently had no vision of fragmentation in a conduit that feeds a plume. Fenner made many astute observations, but the principal flaws that drew him into his radical scenario were the two-vent presumption, dismissal of caldera subsidence by magma withdrawal, and ignoring the possibility of simple mingling of coexisting confluent magmas.

Fenner never revealed how he imagined a rhyolitic magma that had decompressed to the surface might have delayed volatile exsolution and plinian dispersal until after it had remelted several cubic kilometers of andesite. He often invoked an intracrater lava lake as an element of his quiescent assimilation process, and he cited fallout clasts of obsidian and revesiculated agglutinate as crusts and "scum" from such a lake; both are actually Episode I welded vent fill that was later reejected (figs. 62, 63). Fenner also cited breadcrusted blocks near Novarupta as evidence of delayed vesiculation, but such blocks are recycled vent-filling agglutinate and fragments of the Episode IV Phantom dome (fig. 54).

Although never quite conceding that rhyolitic magma is generally of lower temperature than andesitic magma (Fenner, 1931, p. 542), Fenner acknowledged repeatedly that a large excess heat supply was required for phenocryst-poor rhyolite to remelt voluminous andesite. For this he appealed to hypothetical and unspecified exothermic reactions within the volatile-rich rhyolite magma while it was supposedly reequilibrating to greatly reduced pressure. "Magmas should not be regarded as inert mixtures, such as are worked with in experimental laboratory melts....During their ascent from great depths, changes of internal equilibrium lead to exothermal reactions, somewhat analogous to the development of the latent heat of crystallization" (Fenner, 1944). He proposed that "the rhyolite liquid was so highly mobile that it soaked into xenoliths and permeated and softened them so thoroughly that they became inflated at the same time as the liquid magma....The view that rhyolitic magmas are essentially highly viscous seems to persist in spite of much evidence to the contrary" (Fenner, 1950, p. 613). He was probably predisposed to favor such nebulous ideas by his experimental work with supercritical fluids (Morey and Fenner, 1917), his studies of hydrothermal and "pneumatolitic" (Fenner, 1933) ore deposits, and contemporary buzz about exothermic reactions (Day and Shepherd, 1913, p. 600). Fenner advocated his model, sometimes aggressively, for the rest of his life (Fenner, 1920, 1923, 1926, 1929, 1931, 1934, 1937, 1938a, 1938b, 1944, 1950). His final statement read: "When relief of pressure gives opportunity for volatiles to escape, they may not be able to take advantage of it immediately. Relief of pressure favors the escape of gases, and therefore reactions proceed in the direction that leads to the formation of gases, but the rearrangements may require time for their development. In a chain of chemical reactions, a bottle-neck may occur which slows the whole series down; but...the speed of reactions increases rapidly with rise of temperature, and, if the net effect in a magma is exothermic, reactions that have begun very slowly proceed more and more rapidly, at a high compound interest rate, as the temperature rises, and finally reach a cataclysmic stage" (Fenner, 1950, p. 609). A century of experimental petrology has proven his radical hypotheses untenable, and a half century of stratigraphic studies at Katmai has proven them unnecessary.

Advances in Volcanology and Progress in Understanding the 1912 Eruption

Understanding of volcanic processes has advanced greatly since Fenner's day, through a combination of observational, theoretical, analytical, experimental, stratigraphic, and granulometric studies. The breakthrough in clarifying the events of 1912 started with the 1953–54 stratigraphic study by Curtis (1968), which demonstrated that virtually all the juvenile eruptive products had issued at Novarupta, not Mount Katmai.

Relationships among conduit flow, magma fragmentation, eruption columns, ignimbrites, and plinian dispersal were progressively clarified in studies by Smith (1960), Walker (1971, 1973), Sparks (1976), Wilson (1976), and hundreds more, culminating in overview volumes by Sparks and others (1997), Gilbert and Sparks (1998), and Freundt and Rosi (1998).

Successive advances in understanding 1912 eruptive processes have included the following:

1. Fundamental was stratigraphic documentation that plinian fallout and ignimbrite emplacement were synchronous and sustained together for more than half a day. We observed as early as 1976 that part of rhyolitic Layer A underlies the ignimbrite and dacitic Layers C-H overlie it, but it required years of detailed regional work to appreciate the synchroneity of zoned Layer B and most of the zoned ignimbrite. Challenges included the rugged glacial terrain and the limited area of overlap between fall and flow distributions, but we were also confronted by the then-reigning supposition that such synchroneity should not exist, based on oversimplified interpretations of the column-collapse model.

2. Progressive compositional changes permitted establishment that various facies in the ignimbrite were deposited contemporaneously, reflecting differences in initial momentum, vertical position in stratified (density-graded) currents, and outflow distance. Cross-stratified bedsets on the ridgecrests, massive flow units in the valleys, and a few thin blast deposits are compositional and temporal counterparts.

3. Compositional zonation of the ignimbrite permitted mapping of nine distinguishable packages, each consisting of a few or several flow units, successively emplaced over an interval of ~16 hours.

4. Progressive downvalley deflation, densification, and increasing intergranular friction caused ignimbrite-depositing currents to change from turbulent to fluidized to sluggish grainflows distally. Laminated ignimbrite and many distal flow units represent not vent-derived currents but features emerging from massive ignimbrite that was shearing to a halt, stalling, and locally remobilizing in pulses.

5. Welding was immediate where ignimbrite was thick, creating valley-axial swales that channelled flow units only hours younger. But crystallization of glassy ignimbrite, which contributed acid gases to the steam-dominated fumaroles, continued for a decade or more.

6. Fine ash settled more slowly than the bulk of each plinian deposit, some of it drifting widely, even westward, in directions independent of the main southeasterly umbrella cloud. Such fines were intrinsic to the primary plinian plume, and some fraction were probably swept back up into the column from the deflation zone around the vent, but only a minor contribution of coignimbrite fines was elutriated from the VTTS-filling ignimbrite sheet. Because there were three plinian episodes, a significant fraction of the slow-settling fine ash of Episode I was deposited with the fallout of Episode II and III.

7. The effectiveness of 1912 dust and acid aerosol on cooling the atmosphere was modest compared to that of many lesser eruptions because the predominant rhyolite, though rich in halogens, was poor in sulfur.

8. Caldera collapse began in the 11th hour of the eruption and may have been largely over by the 38th hour (fig. 60). Seismicity and ejection of phreatic mud layers near the end of Episode III indicate a short resumption of subsidence around the 58th hour; and after the eruption ended, earthquakes of M6.0 in the 66th hour and M6.9 in the 89th hour are hard to explain other than by further wall slumps or caldera-floor settling.

9. The "ten thousand" fumaroles were mostly rooted in the ignimbrite, from which acid gases liberated from the tuff itself leached metals for a decade; soluble surface sublimates were weathered away within subsequent decades. Fumaroles lasted longest on the floor of Katmai caldera and in the arcuately fissured vent area of Novarupta basin, where only wispy subboiling emissions persist today.

10. Magma mingling was syneruptive. Sharp contacts in banded pumice reflect at most hours to minutes of contiguity before vesiculation, fragmentation, and eruptive quench. Some xenoliths were oxidized, but none were partially melted, not even siltstones and granitoids, let alone andesite lavas and boulders.

11. Chemical and petrographic characterization of most lavas from all the nearby volcanoes confirmed the affinity of 1912 andesite and dacite with products of Mount Katmai, not of Trident, Mageik, or even the Southwest Trident lavas and ejecta of 1953–74.

12. The high-silica rhyolite that dominated the 1912 eruption remains a regional rarity of unproven affinity. Its chemical and isotopic similarity to the melt (glass) phase of the coerupted dacite, its thermal and redox continuity with the dacite as shown by Fe-Ti-oxide pairs, and the similar preeruptive storage depths inferred from phase-equilibrium experiments are all consistent with a comagmatic zoned system. If the rhyolite originated elsewhere, the chain of coincidences is remarkable.

13. Cutting the nearly horizontal sedimentary basement strata that form the walls of the VTTS and Katmai

Canyon, there are many dikes and sills, probably all of Miocene or Pliocene age. Magma transport may usually be vertical or nearly so, but sills demonstrate alternatives. From a magma body 3–6 km beneath Katmai caldera, magma transport by a 10-km-long sill to a position beneath Novarupta is more feasible than a dike, which (like many before it; fig. 135) could have ascended through the Katmai edifice itself. Transport from a sill to the surface in Novarupta basin would finally require a dike or pipe-like conduit.

14. Within ~2 km of Novarupta dome, sector-confined fall deposits of local dispersal and thin pyroclastic-density-current deposits are intercalated with the regionally dispersed plinian layers of Episodes II and III. They contribute to overthickened sections of proximal ejecta and reflect preferential partitioning of andesitic and lithic fragments into an annular collar around the emerging jet, producing transient asymmetrical fountaining and column-margin collapse pulses.

15. Geophysical investigations at Novarupta and the VTTS have been bedeviled by shallow crustal heterogeneities. Arrays of seismometers in 1965–67 yielded data believed to indicate magma storage beneath the volcanic axis, but a new deployment in 1987–91 failed to identify the S-wave screening interpreted earlier. P-wave delays were found to be greatest near Katmai Pass, where poorly welded thick ignimbrite fills the former defile. Seismic-refraction and gravity profiles across the central VTTS in 1966–69 were misinterpreted to indicate thin ignimbrite because no account was taken of concealed welded zones. Heat-flow measurements in Novarupta basin showed order-of-magnitude variations in conductive and convective heat flux on a meter scale, owing to vapor transport along faults and fractures, and yielded nothing useful about the deeper thermal regime. A gravity survey of Novarupta basin failed to outline the vent area, owing to limited density contrast between basement siltstone and welded vent fill. An aeromagnetic survey of Novarupta basin was overinterpreted without taking into account great local variations in vent-fill welding, thickness, andesite-dacite-rhyolite proportions, or fumarolic leaching and magnetite deposition.

Can another great eruption occur on the Alaska Peninsula (Fierstein and others, 1998)? Of course. And what might be different next time? The Katmai region is not greatly more populous today than in 1912, but Kodiak and Cook Inlet certainly are. The first airplane in Alaska was a demonstration flight at Fairbanks in 1913; there are thousands of aircraft today. Alaska Volcano Observatory has numerous instruments at Katmai and elsewhere along the peninsula, and satellite surveillance is now routine. Many volcanoes are closely monitored for signs of unrest, and timely warnings would be issued if an eruption were pending. Many Alaskan communities have experienced recent Cook Inlet ashfalls and would be better stockpiled with supplies and better prepared to protect life, lodging, and equipment than in 1912. The 1912 eruption at Novarupta, so unusual in so many ways, has also left us with two exceptional warnings. Don't count your plinian outbursts. Who would have expected three separate plinian eruptions on successive days? And don't count on your next rhyolite to erupt from a volcano at all. It might break out through a ridge of siltstone.

Dave Johnston, USGS volcanologist, who worked with Hildreth and Anita Grunder in the VTTS in 1979 but was fatally swept away by the 18 May 1980 eruption of Mount St. Helens, Washington. Photo taken by Dan Kosco in 1978 at southwest base of Novarupta dome. On the fallen block of flow-banded rhyolite are the initials, carved in 1923, of Clarence Norman Fenner (C.N.F.) and Charles Yori (C.Y.).

Acknowledgments

Our principal sources were field observations during 20 summer expeditions to Katmai, as well as the many earlier publications cited. The National Geographic Society granted access to its photo archives of the Griggs expeditions and provided prints of numerous historical images selected by the authors. The National Park Service made available the field diaries of Jasper Sayre and E.T. Allen, as well as a photo collection by Charles Yori. Our initial fieldwork in 1976 was encouraged by Garniss Curtis and supported by National Science Foundation grants to Ian Carmichael. A dozen seasons on foot were supported by the Geothermal Research and Volcano Hazards Programs of the U.S. Geological Survey, as were Fierstein's two excursions along Shelikof Strait in the good ship *Ten Bears*, piloted by Eric Stirrup and Hank Pennington. Starting in 1996, seven helicopter-based seasons to investigate the eruptive history of the wider Katmai volcanic cluster were supported by the Alaska Volcano Observatory, under the leadership of Terry Keith and Tom Murray. We are especially grateful to Michelle Coombs and Terry Keith, both of that observatory and both old Katmai hands, for detailed and constructive reviews of this lengthy contribution.

Helicopter pilots Paul Walters, Bill Springer, Jim Sink, and Sam Egli ably handled the challenges of howling winds and socked-in passes to bring us back safely (though often green). Equally expert were the fixed-wing pilots of KatmaiAir, led by legendary bush pilot Sonny Peterson. Between stormy sorties, the U.S. Fish and Wildlife Service bunkhouse in King Salmon provided many amenities, thanks to Susan Savage and Donna Dewhurst.

Collaboration with Bruce Houghton and Colin Wilson extended our reach, broadened our perspectives, advanced our understanding of the 1912 eruption, and brightened our morale in wind and rain. The geophysical expeditions of 1989 and 1990, spearheaded by John Eichelberger, brought new dimensions to investigation of the Novarupta vent region and introduced us to the mixed blessings of helicopters. Radioisotopic age determinations by Marvin Lanphere and Andrew Calvert were critical for calibrating the pre-1912 eruptive history of the region. For collaboration, logistical support, field assistance, and comradeship, we are grateful to many colleagues (in roughly chronological order): Larry Jager, Dan Kosco, Tom Miller, Dave Johnston, Anita Grunder, Terry Keith, Tina Neal, Gail Mahood, Jim Riehle, Buck Detterman, Pete Ward, Betsy Yount, John Eichelberger, Jim Papike, Vicki McConnell, John Power, Maura Hanning, Paul Wallace, Bob Symonds, Bea Ritchey, Michelle Coombs, Dave Tucker, Julia Hammer, Nancy Adams, and (perennially) John Paskievitch. Hildreth is especially grateful that Ian Carmichael sent him off to Katmai on the hunch that something might come of it; and he is more than pleased that his own hunch that Fierstein might eventually fill the boots once worn by Dave Johnston was so amply rewarded.

Numerous personnel of the National Park Service were logistically helpful, notably Gil Blinn, Lynn Fuller, George Stroud, Penny and Dennis Knuckles, Bruce Kaye, Bud Rice, Karen Jettmar, Eric Schmidt, Ron Squib, Jim Gavin, Dave Morris, Ralph Furbush, Ray Bane, Hal Grovert, Susan Savage, Janis Meldrum Kozlowski, Rick Potts, Karen Gustin, John Bundy, and Roy Wood. Their service in preserving one of the world's most pristine wilderness parklands is seldom adequately acknowledged. We salute them.

Ian Carmichael (left) and Garniss Curtis, geology professors at the University of California, Berkeley, visiting Three Forks Overlook above the lower Valley of Ten Thousand Smokes in August 2003. Together, they had inspired and supported Hildreth's first expedition to the area in 1976, when he was a Berkeley graduate student.

References

Abbot, C.G., 1913, Do volcanic explosions affect our climate?: National Geographic Magazine, v. 24, p. 181–198.

Abe, K., 1992, Seismicity of the caldera-making eruption of Mount Katmai, Alaska, in 1912: Bulletin of the Seismological Society of America, v. 82, p. 175–191.

Adams, N.K., de Silva, S.L., Self, S., Salas, G., Schubring, S., Permenter, J.L., and Arbesman, K., 2001, The physical volcanology of the 1600 eruption of Huaynaputina, southern Peru: Bulletin of Volcanology, v. 62, p. 493–518.

Adams, N.K., Houghton, B.F., Fagents, S.A., and Hildreth, W., 2006a, The transition from explosive to effusive eruptive regime; the example of the 1912 Novarupta eruption, Alaska: Geological Society of America Bulletin, v. 118, p. 620–634.

Adams, N.K., Houghton, B.F., and Hildreth, W., 2006b, Abrupt transitions during sustained explosive eruptions; examples from the 1912 eruption of Novarupta, Alaska: Bulletin of Volcanology, v. 69, p. 189–206.

Allen, E.T., and Zies, E.G., 1923, A chemical study of the fumaroles of the Katmai region: National Geographic Society, Contributed Technical Papers, Katmai Series, No. 2, p. 75–155.

Andersen, D.J., Lindsley, D.H., and Davidson, P.M., 1993, QUILF—A Pascal program to assess equilibria among Fe-Mg-Mn-Ti oxides, pyroxenes, olivine, and quartz: Computers & Geosciences, v. 19, p. 1333–1350.

Anderson, T., and Flett, J.S., 1903, Report on the eruptions of the Soufrière, in St. Vincent, in 1902, and on a visit to Montagne Pelée, in Martinique—Part I: Philosophical Transactions of the Royal Society of London, Series A, v. 200, p. 353–553.

Annen, C., Blundy, J.D., and Sparks, R.S.J., 2006, The genesis of intermediate and silicic magmas in deep crustal hot zones: Journal of Petrology, v. 47, p. 505–539.

Arculus, R.J., 2003, Use and abuse of the terms calcalkaline and calcalkalic: Journal of Petrology, v. 44, p. 929–935.

Avery, V.F., 1992, A petrogenetic study of the dacite from the 1912 eruption of Novarupta, Katmai National Park, Alaska; implications for magma storage locations: Fairbanks, University of Alaska, M.S. thesis, 177 p.

Bacon, C.R., and Druitt, T.H., 1988, Compositional evolution of the zoned calcalkaline magma chamber of Mount Mazama, Crater Lake, Oregon: Contributions to Mineralogy and Petrology, v. 98, p. 224–256.

Bacon, C.R., and Hirschmann, M.M., 1988, Mg/Mn partitioning as a test for equilibrium between coexisting Fe-Ti oxides: American Mineralogist, v. 73, p. 57–61.

Baedecker, P.A., ed., 1987, Methods for geochemical analysis: U.S. Geological Survey Bulletin 1770. (Paginated by chapter.)

Ballard, S., Carrigan, C.R., and McConnell, V.S., 1991, Shallow conductive-component of heat flow near Novarupta Dome, Katmai, Alaska: Geophysical Research Letters, v. 18, p. 1529–1532.

Berg, E., Kubota, S., and Kienle, J., 1967, Preliminary determination of crustal structure in the Katmai National Monument, Alaska: Bulletin of the Seismological Society of America, v. 57, p. 1367–1392.

Blake, S., and Ivey, G.N., 1986, Magma mixing and the dynamics of withdrawal from stratified reservoirs: Journal of Volcanology and Geothermal Research, v. 27, p. 153–178.

Bordet, P., Marinelli, G., Mittempergher, M., and Tazieff, H., 1963, Contribution à l'étude volcanologique du Katmai et de la Vallée des Dix Mille Fumées: Mémoires de la Société Belge de Géologie, de Paléontologie et d'Hydrologie, No. 7, p. 1–114.

Bowen, N.L., 1913, The melting phenomena of the plagioclase feldspars: American Journal of Science, v. 35, p. 577–599.

Bowen, N.L., 1915, The later stages of the evolution of the igneous rocks: Journal of Geology, v. 23, Supplement, p. 1–89.

Brandsdóttir, B., and Einarsson, P., 1979, Seismic activity associated with the September 1977 deflation of Krafla Volcano in northeastern Iceland: Journal of Volcanology and Geothermal Research, v. 6, p. 197–212.

Branney, M.J., and Kokelaar, P., 2002, Pyroclastic density currents and the sedimentation of ignimbrites: Geological Society of London Memoir 27, 152 p.

Buddington, A.F., and Lindsley, D.F., 1964, Iron-titanium oxide minerals and synthetic equivalents: Journal of Petrology, v. 5, p. 310–357.

Cahalane, V.H., 1959, A biological survey of Katmai National Monument: Smithsonian Institution Publication 4376, 231 p.

Cameron, R.E., 1970, Soil microbial ecology of Valley of 10,000 Smokes, Alaska: Journal of the Arizona Academy of Science, v. 6, no. 1, p. 11–40.

Cameron, W.A., and Larson, G.L., 1992, Baseline inventory of the aquatic resources of Aniakchak National Monument, Alaska: Seattle, National Park Service, Technical Report NPS/PNROSU/NRTR-92/03, 215 p.

Carey, R.J., Houghton, B.F., Sable, J.E., and Wilson, C.J.N., 2007, Contrasting grain size and componentry in complex proximal deposits of the 1886 Tarawera basaltic Plinian eruption: Bulletin of Volcanology, v. 69, p. 903–926.

Carey, R.J., Houghton, B.F., and Thordarson, T., 2009, Abrupt shifts between wet and dry phases of the 1875 eruption of Askja Volcano; microscopic evidence for macroscopic dynamics: Journal of Volcanology and Geothermal Research, v. 184, p. 256–270.

Carey, S., and Sigurdsson, H., 1987, Temporal variations in column height and magma discharge rate during the 79 A.D. eruption of Vesuvius: Geological Society of America Bulletin, v. 99, p. 303–314.

Carey, S., and Sparks, R.S.J., 1986, Quantitative models of the fallout and dispersal of tephra from volcanic eruption columns: Bulletin of Volcanology, v. 48, p. 109–125.

Carrigan, C.R., Schubert, G., and Eichelberger, J.C., 1992, Thermal and dynamical regimes of single- and two-phase magmatic flow in dikes: Journal of Geophysical Research, v. 97, p. 17377–17392.

Choux, C., Druitt, T., and Thomas, N., 2004, Stratification and particle segregation in flowing polydisperse suspensions, with applications to the transport and sedimentation of pyroclastic density currents: Journal of Volcanology and Geothermal Research, v. 138, p. 223–241.

Cioni, R., Civetta, L., Marianelli, P., Metrich, N., Santacroce, R., and Sbrana, A., 1995, Compositional layering and syneruptive mixing of a periodically refilled shallow magma chamber; the AD 79 plinian eruption of Vesuvius: Journal of Petrology, v. 36, no. 3, p. 739–776.

Clocchiatti, R., 1972, Les cristaux de quartz des ponces de la Vallée des Dix Mille Fumées (Katmai, Alaska): Comptes Rendus, Academie des Sciences de Paris, t. 274, p. 3037–3040.

Clocchiatti, R., 1975, Les inclusions vitreuses des cristaux de quartz: Mémoires de la Société Géologique de France, no. 122, 96 p.

Clough, C.T., Maufe, H.B., and Bailey, E.B., 1909, The cauldron-subsidence of Glen Coe and the associated igneous phenomena: Quarterly Journal of the Geological Society of London, v. 65, p. 611–678.

Coombs, M.L., 2001, Experimental and petrologic constraints on magma storage, movement, and interactions at two volcanoes in Katmai National Park, Alaska: Fairbanks, University of Alaska, Ph.D. dissertation, 218 p.

Coombs, M.L., and Gardner, J.E., 2001, Shallow-storage conditions for the rhyolite of the 1912 eruption at Novarupta, Alaska: Geology, v. 29, p. 775–778.

Coombs, M.L., Eichelberger, J.C., and Rutherford, M.J., 2000, Magma storage and mixing conditions for the 1953-1968 eruption of Southwest Trident Volcano, Katmai National Park, Alaska: Contributions to Mineralogy and Petrology, v. 140, p. 99–118.

Cowee, C., Faust Larsen, J., Guetschow, H., Eichelberger, J., Kent, A., and Hutcheon, I., 1999, Melt inclusion textures and volatile compositions from Novarupta dome, Katmai National Park, Alaska: Eos, American Geophysical Union Transactions, v. 80, no. 17, p. S352.

Criswell, C.W., 1987, Chronology and pyroclastic stratigraphy of the May 18, 1980, eruption of Mount St. Helens, Washington: Journal of Geophysical Research, v. 92, p. 10237–10266.

Curtis, G.H., 1968, The stratigraphy of the ejecta from the 1912 eruption of Mount Katmai and Novarupta, Alaska, *in* Coats, R.R., Hay, R.L., and Anderson, C.A., eds., Studies in Volcanology: Geological Society of America Memoir 116, p. 153–210.

Curtis, G., Williams, H., and Juhle, W., 1954, Evidence against assimilation of andesite by rhyolite in the Valley of 10,000 Smokes, Alaska: American Geophysical Union Transactions, v. 35, no. 2, p. 378.

Daly, R.A., 1914, Igneous rocks and their origin: New York, McGraw-Hill, 563 p.

Day, A.L., 1931, Melting temperatures of granite and basalt: Carnegie Institution of Washington Year Book No. 30, p. 75–78.

Day, A.L., and Shepherd, E.S., 1913, Water and volcanic activity: Geological Society of America Bulletin, v. 24, p. 573–606.

Day, R.H., 1995, New information on Kittlitz's murrelet nests: The Condor, v. 97, p. 271–273.

Decker, R.W., 1963, Proposed volcano observatory at Katmai National Monument; a preliminary study: Hanover N.H., Dartmouth College, unpublished report, 54 p. [Katmai National Park archives].

Decker, R.W., 1964, Geophysical investigations in Katmai National Monument, Alaska: American Geophysical Union Transactions, v. 45, p. 124.

Detterman, R.L., Case, J.E., Wilson, F.H., and Yount, M.E., 1987, Geologic map of the Ugashik, Bristol Bay, and western part of Karluk quadrangle, Alaska: U.S. Geological Survey Map I-1685; scale 1:250,000.

Detterman, R.L., Case, J.E., Miller, J.W., Wilson, F.H., and Yount, M.E., 1996, Stratigraphic framework of the Alaska Peninsula: U.S. Geological Survey Bulletin 1969-A, 74 p.

Devine, J.D., Sigurdsson, H., Davis, A.N., and Self, S., 1984, Estimates of sulfur and chlorine yield to the atmosphere from volcanic eruptions and potential climatic effects: Journal of Geophysical Research, v. 89, p. 6309–6325.

Dixon, J.P., and Power, J.A., 2009, The January 2006 volcanic-tectonic earthquake swarm at Mount Martin, Alaska, *in*

Haeussler, P.J., and Galloway, J.P., eds., Studies by the U.S. Geological Survey in Alaska, 2007: U.S. Geological Survey Professional Paper 1760-D, 17 p.

Donnay, J.D.H., 1956, The geological studies of Rolf Werner Juhle (1929-1953)—in memoriam: Journal of Geological Education, v. 4, p. 47–51.

Doukas, M.P., and McGee, K.A., 2007, A compilation of gas-emission data from volcanoes of Cook Inlet (Spurr, Crater Peak, Redoubt, Iliamna, and Augustine) and Alaska Peninsula (Douglas, Fourpeaked, Griggs, Mageik, Martin, Peulik, Ukinrek Maars, and Veniaminof), Alaska, from 1995–2006: U.S. Geological Survey Open-File Report 2007–1400, 13 p.

Druitt, T.H., 1998, Pyroclastic density currents, *in* Gilbert, J.S., and Sparks, R.S.J., eds., The physics of explosive volcanic eruptions: Geological Society of London Special Publication 145, p. 145–182.

Druitt, T.H., Bruni, G., Lettieri, P., and Yates, J.G., 2004, The fluidization behaviour of ignimbrite at high temperature and with mechanical agitation: Geophysical Research Letters, v. 31, L02604, doi: 10.1029/2003GL018593.

Druitt, T.H., Avard, G., Bruni, G., Lettieri, P., and Maez, F., 2007, Gas retention in fine-grained pyroclastic flow materials at high temperatures: Bulletin of Volcanology, v. 69, p. 881–901.

Dumond, D.E., 2005, A Naknek chronicle: Washington, National Park Service, 111 p.

Eichelberger, J., and Hildreth, W., 1986, Research drilling at Katmai, Alaska: Eos (American Geophysical Union Transactions), v. 67, p. 778–780.

Eichelberger, J.C., and Izbekov, P.E., 2000, Eruption of andesite triggered by dyke injection; contrasting cases of Karymsky Volcano, Kamchatka, and Mount Katmai, Alaska: Philosophical Transactions of the Royal Society of London, series A, v. 358, p. 1465–1485.

Eichelberger, J.C., Hildreth, W., and Papike, J.J., 1987, Direct observation of a young igneous system; a proposal for research drilling at Katmai, Alaska: Continental Scientific Drilling Program, 174 p.

Eichelberger, J.C., Hildreth, W., and Papike, J.J., 1989, Direct observation of a young igneous system; a science plan for research drilling at Katmai National Park, Alaska: Continental Scientific Drilling Program, 218 p.

Eichelberger, J.C., Ballard, S., Carrigan, C.R., Goodliffe, A., Hildreth, W., Iwatsubo, E., Kasameyer, P.W., Keith, T.E.C., Kienle, J., Papike, J.J., Pollard, D.D., Stone, D.E., Wallmann, P.C., Ward, P.C., Wilt, M., and Yount, M.E., 1990, Geophysics at Katmai; geophysical expedition to Novarupta Volcano, Katmai National Park, Alaska: Eos (American Geophysical Union Transactions), v. 71, p. 733–735.

Eichelberger, J.C., Hildreth, W., and Papike, J.J., 1991, The Katmai Scientific Drilling Project, surface phase; investigation of an exceptional igneous system: Geophysical Research Letters, v. 18, p. 1513–1516.

Eicher, G.J., Jr., and Rounsefell, G.A., 1957, Effects of lake fertilization by volcanic activity on abundance of salmon: Oceanography, v. 2, no. 2, p. 70–78.

Evermann, B.W., 1914, Alaska fisheries and fur industries in 1913: Washington, Department of Commerce, Bureau of Fisheries Document 797.

Fenner, C.N., 1920, The Katmai region, Alaska, and the great eruption of 1912: Journal of Geology, v. 28, p. 569–606.

Fenner, C.N., 1923, The origin and mode of emplacement of the great tuff deposit of the Valley of Ten Thousand Smokes: National Geographic Society, Contributed Technical Papers, Katmai Series, no.1, p. 1–74.

Fenner, C.N., 1925, Earth movements accompanying the Katmai eruption: Journal of Geology, v. 33, p. 116–139 and 193–223.

Fenner, C.N., 1926, The Katmai magmatic province: Journal of Geology, v. 34, p. 673–772.

Fenner, C.N., 1929, The crystallization of basalts: American Journal of Science, v. 18, p. 225–253.

Fenner, C.N., 1930, Mount Katmai and Mount Mageik: Zeitschrift für Vulkanologie, Band 13, p. 1–24.

Fenner, C.N., 1931, The residual liquids of crystallizing magmas: Mineralogical Magazine, v. 22, p. 539–560.

Fenner, C.N., 1933, Pneumatolytic processes in the formation of minerals and ores, *in* Finch, J.W., and others, eds., The Committee on the Lindgren Volume, Ore deposits of the Western States: American Institute of Mining and Metallurgical Engineers, p. 58–106.

Fenner, C.N., 1934, Some magmatic problems: Journal of the Washington Academy of Sciences, v. 24, p. 113–124.

Fenner, C.N., 1937, A view of magmatic differentiation: Journal of Geology, v. 45, p. 158–168.

Fenner, C.N., 1938a, The phenomena of Falling Mountain: American Journal of Science, v. 35-A, p. 35–48.

Fenner, C.N., 1938b, Contact relations between rhyolite and basalt on Gardiner River, Yellowstone Park: Geological Society of America Bulletin, v. 49, p. 1441–1483.

Fenner, C.N., 1944, Rhyolite-basalt complex on Gardiner River, Yellowstone Park, Wyoming— a discussion: Geological Society of America Bulletin, v. 55, p. 1081–1096.

Fenner, C.N., 1950, The chemical kinetics of the Katmai eruption: American Journal of Science, v. 248, p. 593–627 and 697–725.

Fierstein, J., 2007, Explosive eruptive record in the Katmai region, Alaska Peninsula; an overview: Bulletin of Volcanology, v. 69, p. 469–509.

Fierstein, J., and Hildreth, W., 1992, The plinian eruptions of 1912 at Novarupta, Katmai National Park, Alaska: Bulletin of Volcanology, v. 54, p. 646–684.

Fierstein, J., and Hildreth, W., 2001, Preliminary volcano-hazard assessment for the Katmai Volcanic Cluster, Alaska: U.S. Geological Survey Open-File Report 00–489, 50 p., oversize plate in pocket.

Fierstein, J., and Hildreth, W., 2008, Kaguyak dome field and its Holocene caldera, Alaska Peninsula: Journal of Volcanology and Geothermal Research, v. 177, p. 340–366.

Fierstein, J., and Nathenson, M., 1992, Another look at the calculation of fallout tephra volumes: Bulletin of Volcanology, v. 54, p. 156–167.

Fierstein, J., and Nathenson, M., 1993, Reply to Comment by W.I. Rose: Bulletin of Volcanology, v. 55, p. 375–378.

Fierstein, J., and Wilson, C.J.N., 2005, Assembling an ignimbrite; compositionally defined eruptive packages in the 1912 Valley of Ten Thousand Smokes ignimbrite, Alaska: Geological Society of America Bulletin, v. 117, p. 1094–1107.

Fierstein, J., Houghton, B.F., Wilson, C.J.N., and Hildreth, W., 1997, Complexities of plinian fall deposition at vent; an example from the 1912 Novarupta eruption: Journal of Volcanology and Geothermal Research, v. 76, p. 215–227.

Fierstein, J., Hildreth, W., Hendley, J.W.II, and Stauffer, P.H., 1998, Can another great volcanic eruption happen in Alaska?: U.S. Geological Survey Fact Sheet 075-98, 2 p.

Filson, J., Simkin, T., and Leu, L., 1973, Seismicity of a caldera collapse; Galápagos Islands 1968: Journal of Geophysical Research, v. 78, p. 8591–8622.

Fisher, R.V., 1966, Mechanism of deposition from pyroclastic flows: American Journal of Science, v. 264, p. 350–363.

Fisher, R.V., and Heiken, G., 1982, Mt. Pelée, Martinique; May 8 and 20, 1902, pyroclastic flows and surges: Journal of Volcanology and Geothermal Research, v. 13, p. 339–371.

Fliedner, M.M., and Klemperer, S.L., 2000, Crustal structure transition from oceanic arc to continental arc, eastern Aleutian Islands and Alaska Peninsula: Earth and Planetary Science Letters, v. 179, p. 567–579.

Fouqué, F., 1879, Santorin et ses eruptions: Paris, Masson, 479 p.

Freundt, A., and Bursik, M., 1998, Pyroclastic flow transport mechanisms, *in* Freundt, A., and Rosi, M., eds., From magma to tephra: Elsevier, p. 173–245.

Freundt, A., and Rosi, M., eds., 1998, From magma to tephra: Elsevier, 318 p.

Fridrich, C.J., and Mahood, G.A., 1987, Compositional layers in the zoned magma chamber of the Grizzly Peak Tuff: Geology, v. 15, p. 299–303.

Gedney, L., Matteson, C., and Forbes, R.B., 1970, Seismic refraction profiles of the ash flow in the Valley of Ten Thousand Smokes, Katmai National Monument, Alaska: Journal of Geophysical Research, v. 75, p. 2619–2624.

George, R., Turner, S., Hawkesworth, C., Morris, J., Nye, C., Ryan, J., and Zheng, S.-H., 2003, Melting processes and fluid and sediment transport rates along the Alaska-Aleutian arc from an integrated U-Th-Ra-Be isotope study: Journal of Geophysical Research, v. 108, no. B5, 2252, doi:10.1029/2002JB001916; ECV, p. 6–1 to 6–25.

Gilbert, C.M., 1938, Welded tuff in eastern California: Geological Society of America Bulletin, v. 49, p. 1829–1862.

Gilbert, J.S., and Sparks, R.S.J., eds., 1998, The physics of explosive volcanic eruptions: Geological Society of London Special Publication 145, 186 p.

Girolami, L., Druitt, T.H., Roche, O., and Khrabrykh, Z., 2008, Propagation and hindered settling of laboratory ash flows: Journal of Geophysical Research, v. 113, B02202, doi: 10.1029/2007JB005074.

Goodliffe, A.M., Stone, D.B., Kienle, J., and Kasameyer, P., 1991, The vent of the 1912 Katmai eruption; gravity and magnetic measurements: Geophysical Research Letters, v. 18, p. 1521–1524.

Gourgaud, A., Fichaut, M., and Joron, J.-L., 1989, Magmatology of Mt. Pelée (Martinique, F.W.I.); I. Magma mixing and triggering of the 1902 and 1929 Pelean nuées ardentes: Journal of Volcanology and Geothermal Research, v. 38, p. 143–169.

Griggs, R.F., 1917, The Valley of Ten Thousand Smokes; National Geographic Society explorations in the Katmai district of Alaska: National Geographic Magazine, v. 31, no. 1, p. 12–68.

Griggs, R.F., 1918a, The recovery of vegetation at Kodiak: Ohio Journal of Science, v. 19, p. 1–57.

Griggs, R.F., 1918b, Are the Ten Thousand Smokes real volcanoes?: Ohio Journal of Science, v. 19, p. 97–116.

Griggs, R.F., 1918c, The great hot mud flow of the Valley of Ten Thousand Smokes: Ohio Journal of Science, v. 19, p. 117–142.

Griggs, R.F., 1918d, The character of the eruption as indicated by its effects on vegetation: Ohio Journal of Science, v. 19, p. 173–209.

Griggs, R.F., 1918e, The eruption of Katmai: Nature, v. 101, no. 2547, p. 497–499.

Griggs, R.F., 1918f, The Valley of Ten Thousand Smokes: National Geographic Magazine, v. 33, no. 2, p. 115–169.

Griggs, R.F., 1920, The great Mageik landslide: Ohio Journal of Science, v. 20, p. 325–354.

Griggs, R.F., 1921, Our greatest National Monument: National Geographic Magazine, v. 40, p. 219–292.

Griggs, R.F., 1922, The Valley of Ten Thousand Smokes: Washington, D.C., National Geographic Society, 340 p.

Griggs, R.F., 1933, The colonization of the Katmai ash, a new and inorganic "soil": American Journal of Botany, v. 20, p. 92–113.

Griggs, R.F., 1961, We Two Together: Pacific Grove, California, The Boxwood Press, 320 p.

Grove, T.L., and Donnelly-Nolan, J.M., 1986, The evolution of young silicic lavas at Medicine Lake Volcano, California; implications for the origin of compositional gaps in calc-alkaline series lavas: Contributions to Mineralogy and Petrology, v. 92, p. 281–302.

Grove, T.L., Donnelly-Nolan, J.M., and Housh, T, 1997, Magmatic processes that generated the rhyolite of Glass Mountain, Medicine Lake volcano, California: Contributions to Mineralogy and Petrology, v. 127, p. 205–223.

Grunder, A.L., Laporte, D., and Druitt, T.H., 2005, Experimental and textural investigation of welding; effects of compaction, sintering, and vapor-phase crystallization in the rhyolitic Rattlesnake Tuff: Journal of Volcanology and Geothermal Research, v. 142, p. 89–104.

Hammer, C.U., 1977, Past volcanism revealed by Greenland Ice Sheet impurities: Nature, v. 270, p. 482–486.

Hammer, C.U., Clausen, H.B., and Dansgaard, W., 1980, Greenland ice sheet evidence of post-glacial volcanism and its climatic impact: Nature, v. 288, p. 230–235.

Hammer, J.E., Rutherford, M.J., and Hildreth, W., 2002, Magma storage prior to the 1912 eruption at Novarupta, Alaska: Contributions to Mineralogy and Petrology, v. 144, p. 144–162.

Harker, A., 1909, The natural history of igneous rocks: New York, MacMillan, 384 p.

Hay, R.L., 1959, Formation of the crystal-rich glowing avalanche deposits of St. Vincent, B.W.I.: Journal of Geology, v. 67, p. 540–562.

Hildreth, W., 1981, Gradients in silicic magma chambers—implications for lithospheric magmatism: Journal of Geophysical Research, v. 86, p. 10153–10192.

Hildreth, W., 1983, The compositionally zoned eruption of 1912 in the Valley of Ten Thousand Smokes, Katmai National Park, Alaska: Journal of Volcanology and Geothermal Research, v. 18, p. 1–55.

Hildreth, W., 1987, New perspectives on the eruption of 1912 in the Valley of Ten Thousand Smokes, Katmai National Park, Alaska: Bulletin of Volcanology, v. 49, p. 680–693.

Hildreth, W., 1991, The timing of caldera collapse at Mount Katmai in response to magma withdrawal toward Novarupta: Geophysical Research Letters, v. 18, p. 1541–1544.

Hildreth, W., and Drake, R.E., 1992, Volcan Quizapu, Chilean Andes: Bulletin of Volcanology, v. 54, p. 93–125.

Hildreth, W., and Fierstein, J., 1992, Hydrothermal explosion breccia emplaced during caldera collapse of Mount Katmai, Alaska, on 6 June 1912: Eos, American Geophysical Union Transactions, v. 73, no. 43, p. 635–636.

Hildreth, W., and Fierstein, J., 2000, Katmai volcanic cluster and the great eruption of 1912: Geological Society of America Bulletin [GSA Overview Paper], v. 112, p. 1594–1620.

Hildreth, W., and Fierstein, J., 2003, Geologic map of the Katmai Volcanic Cluster, Katmai National Park, Alaska: U.S. Geological Survey Map I-2778, scale 1:63,360. Includes six 15-minute quadrangles and 43-page pamphlet.

Hildreth, W., and Moorbath, S., 1988, Crustal contributions to arc magmatism in the Andes of central Chile: Contributions to Mineralogy and Petrology, v. 98, p. 455–489.

Hildreth, W., and Wilson, C.J.N., 2007, Compositional zoning of the Bishop Tuff: Journal of Petrology, v. 48, p. 951–999.

Hildreth, W., Fierstein, J., Lanphere, M.A., and Siems, D.F., 1999, Alagogshak volcano; a Pleistocene andesite-dacite stratovolcano in Katmai National Park, *in* Geologic Studies in Alaska by the U.S. Geological Survey 1997: U.S. Geological Survey Professional Paper 1614, p. 105–113.

Hildreth, W., Fierstein, J., Lanphere, M.A., and Siems, D.F., 2000, Mount Mageik, a compound stratovolcano in Katmai National Park, *in* Geologic Studies in Alaska by the U.S. Geological Survey 1998: U.S. Geological Survey Professional Paper 1615, p. 23–34.

Hildreth, W., Fierstein, J., Lanphere, M.A., and Siems, D.F., 2001, Snowy Mountain; a pair of small andesite-dacite stratovolcanoes in Katmai National Park, *in* Geologic Studies in Alaska by the U.S. Geological Survey 1999: U.S. Geological Survey Professional Paper 1633, p. 13–34.

Hildreth, W., Fierstein, J., Lanphere, M.A., and Siems, D.F., 2002, Mount Griggs; a compositionally distinctive Quaternary stratovolcano behind the main volcanic line in Katmai

National Park, *in* Studies by the U.S. Geological Survey in Alaska, 2000: U.S. Geological Survey Professional Paper 1662, p. 87–112.

Hildreth, W., Fierstein, J., Lanphere, M.A., and Siems, D.F., 2003a, Trident Volcano; four contiguous stratocones adjacent to Katmai Pass, Alaska Peninsula, *in* Studies by the U.S. Geological Survey in Alaska, 2001: U.S. Geological Survey Professional Paper 1678, p. 1–28.

Hildreth, W., Lanphere, M.A., and Fierstein, J., 2003b, Geochronology and eruptive history of the Katmai Volcanic Cluster, Alaska Peninsula: Earth and Planetary Science Letters, v. 214, p. 93–114.

Hildreth, W., Fierstein, J., Siems, D.F., Budahn, J.R., and Ruíz, J., 2004, Rear-arc vs arc-front volcanoes in the Katmai reach of the Alaska Peninsula; a critical appraisal of across-arc compositional variation: Contributions to Mineralogy and Petrology, v. 147, p. 243–275.

Hildreth, W., Fierstein, J., and Calvert, A.T., 2007, Blue Mountain and The Gas Rocks; rear-arc dome clusters on the Alaska Peninsula: U.S. Geological Survey Professional Paper 1739-A, p. 1–27.

Holmes, A., 1916, The origin of igneous rocks [Essay-review on The later stages of the evolution of the igneous rocks, by N.L. Bowen]: Science Progress, v. 11, p. 67–73.

Horn, S., and Schmincke, H.-U., 2000, Volatile emission during the eruption of Baitoushan Volcano (China/North Korea) ca. 969 AD: Bulletin of Volcanology, v. 61, p. 537–555.

Houghton, B.F., Wilson, C.J.N., Fierstein, J., and Hildreth, W., 2004, Complex proximal deposition during the plinian eruptions of 1912 at Novarupta, Alaska: Bulletin of Volcanology, v. 66, p. 95–133.

Hubbard, B.R., 1932, Mush You Malemutes!: New York, The America Press, 179 p.

Hubbard, B.R., 1935, Cradle of the Storms: New York, Dodd, Mead and Co., 285 p.

Humphreys, W.J., 1913, Volcanic dust and other factors in the production of climatic changes and their possible relation to ice ages: Mount Weather Observatory Bulletin, v. 6, p. 1–34.

Hussey, J.A., 1971, Embattled Katmai—a history of Katmai National Monument: San Francisco, National Park Service, 457 p.

Iddings, J.P., 1892, The origin of igneous rocks: Bulletin of the Philosophical Society of Washington, v. 12, p. 89–213.

Iddings, J.P., 1909, 1913, Igneous rocks: New York, John Wiley, 2 volumes, 454 p, 685 p.

Iddings, J.P., 1914, The problem of volcanism: New Haven, Yale University Press, 273 p.

Iverson, R.M., and Vallance, J.W., 2001, New views of granular mass flows: Geology, v. 29, p. 115–118.

Jaggar, T.A., 1927, Eruption of Mageik in Alaska: The Volcano Letter, no. 147, Hawaiian Volcano Observatory [Implausible eruption report; see Hildreth and others, 2000]

Johnson, C.M., and Fridrich, C.J., 1990, Non-monotonic chemical and O, Sr, Nd, and Pb isotope zonations and heterogeneity in the mafic-to-silicic-composition magma chamber of the Grizzly Peak Tuff, Colorado: Contributions to Mineralogy and Petrology, v. 105, p. 677–690.

Jolly, A.D., and McNutt, S.R., 1999, Seismicity at the volcanoes of Katmai National Park, Alaska, July 1995-December 1997: Journal of Volcanology and Geothermal Research, v. 93, p. 173–190.

Jolly, A.D., Moran, S.C., McNutt, S.R., and Stone, D.B., 2007, Three-dimensional P-wave velocity structure derived from local earthquakes at the Katmai group of volcanoes, Alaska: Journal of Volcanology and Geothermal Research, v. 159, p. 326–342.

Jones, P.D., Wigley, T.M.L., and Kelly, P.M., 1982, Variations in surface air temperatures, Part 1. Northern Hemisphere, 1881-1980: Monthly Weather Review, v. 110, p. 59–70.

Kanamori, H., Mori, J., Hauksson, E., Heaton, T.H., Hutton, L.K., and Jones, L.M., 1993, Determination of earthquake energy release and M_L using TERRAscope: Bulletin of the Seismological Society of America, v. 83, p. 330–346.

Keith, T.E.C., 1991a, Fossil and active fumaroles in the 1912 eruptive deposits, Valley of Ten Thousand Smokes, Alaska: Journal of Volcanology and Geothermal Research, v. 45, p. 227–254.

Keith, T.E.C., 1991b, Argillic alteration in the Novarupta vent region, Katmai National Park, Alaska: Geophysical Research Letters, v. 18, p. 1549–1552.

Keith, T.E.C., Thompson, J.M., Hutchinson, R.A., and White, L.D., 1992, Geochemistry of waters in the Valley of Ten Thousand Smokes region, Alaska: Journal of Volcanology and Geothermal Research, v. 49, p. 209–231.

Keller, A.S., and Reiser, H.N., 1959, Geology of the Mount Katmai area, Alaska: U.S. Geological Survey Bulletin 1058-G, p. 261–298.

Kienle, J., 1970, Gravity traverses in the Valley of Ten Thousand Smokes, Katmai National Monument, Alaska: Journal of Geophysical Research, v. 75, p. 6641–6649.

Kienle, J., 1991, Depth of the ash flow deposit in the Valley of Ten Thousand Smokes, Katmai National Park, Alaska: Geophysical Research Letters, v. 18, p. 1533–1536.

Kienle, J., Swanson, S.E., and Pulpan, H., 1983, Magmatism and subduction in the eastern Aleutian arc, *in* Shimozuru, D., and Yokoyama, I., eds., Arc volcanism; physics and tectonics: Tokyo, Terra Scientific Publishing Company, p. 191–224.

Kimball, H.H., 1913, The effect upon atmospheric transparency of the eruption of Katmai volcano: Monthly Weather Review (Weather Bureau, U.S. Dept. of Agriculture), v. 41, p. 153–159.

Klein, F.W., Koyanagi, R.Y., Nakata, J.S., and Tanigawa, W.R., 1987, The seismicity of Kilauea's magma system, *in* Decker, R.W., Wright, T.L., and Stauffer, P.H., eds., Volcanism in Hawaii: U.S. Geological Survey Professional Paper 1350, p. 1019–1185.

Kleinman, J.W., and Iwatsubo, E.Y., 1991, A geodetic network in the Novarupta area, Katmai National Park, Alaska: Geophysical Research Letters, v. 18, p. 1517–1519.

Kleinman, J.W., Iwatsubo, E.Y., Power, J.A., and Endo, E.T., 1997, Geodetic studies in the Novarupta area, Katmai National Park, Alaska, 1990 to 1995, *in* Dumoulin, J.A., and Gray, J.E., eds., Geologic studies in Alaska by the U.S. Geological Survey, 1995: U.S. Geological Survey Professional Paper 1574, p. 83–92.

Kodosky, L.G., 1989, Surface mercury geochemistry as a guide to volcanic vent structure and zones of high heat flow in the Valley of Ten Thousand Smokes, Katmai National Park, Alaska: Journal of Volcanology and Geothermal Research, v. 38, p. 227–242.

Kosco, D.G., 1981, Characteristics of andesitic to dacitic volcanism at Katmai National Park, Alaska: Berkeley, University of California, Ph.D. dissertation, 249 p.

Kozu, S., 1934, The great activity of Komagatake (Japan) in 1929: Tschermaks Mineralogische und Petrographische Mitteilungen, v. 45, p. 133–174.

Kubota, S., and Berg, E., 1967, Evidence for magma in the Katmai volcanic range: Bulletin Volcanologique, v. 31, p. 175–214.

Lacis, A., Hansen, J., and Sato, M., 1992, Climate forcing by stratospheric aerosols: Geophysical Research Letters, v. 19, no. 15, p. 1607–1610.

Lacroix, A., 1904, La Montagne Pelée et ses eruptions: Paris, Masson, 662 p., 30 plates.

Larsen, G., and Thorarinsson, S., 1978, H_4 and other acid Hekla tephras: Jökull, v. 27, p. 28–46.

LeBas, M.J., LeMaitre, R.W., Streckeisen, A., and Zanettin, B., 1986, A chemical classification of volcanic rocks based on the total alkali–silica diagram: Journal of Petrology, v. 27, p. 745–750.

Lindgren, W., 1913, Mineral Deposits: New York, McGraw-Hill, 883 p.

Lipman, P.W., 1984, The roots of ash flow calderas in western North America; windows into the tops of granitic batholiths: Journal of Geophysical Research, v. 89, p. 8801–8841.

Lipman, P.W., and Mullineaux, D.R., eds., 1981, The 1980 eruptions of Mount St. Helens, Washington: U.S. Geological Survey Professional Paper 1250, 844 p.

Lovering, T.S., 1957, Halogen-acid alteration of ash at Fumarole No. 1, Valley of Ten Thousand Smokes, Alaska: Geological Society of America Bulletin, v. 68, p. 1585–1604.

Lowenstern, J.B., 1993, Evidence for a copper-bearing fluid in magma erupted at the Valley of Ten Thousand Smokes, Alaska: Contributions to Mineralogy and Petrology, v. 114, p. 409–421.

Lowenstern, J.B., and Mahood, G.A., 1991, Petrogenesis of high-silica rhyolite on the Alaska Peninsula: Geophysical Research Letters, v. 18, p. 1565–1568.

Lowenstern, J.B., Wallmann, P.C., and Pollard, D.D., 1991, The West Mageik Lake sill complex as an analogue for magma transport during the 1912 eruption at the Valley of Ten Thousand Smokes, Alaska: Geophysical Research Letters, v. 18. p. 1569–1572.

Lu, Z., Fatland, R., Wyss, M., Li, S., Eichelberger, J., Dean, K., and Freymueller, J., 1997, Deformation of New Trident volcano measured by ERS-1 SAR interferometry, Katmai National Park, Alaska: Geophysical Research Letters, v. 24, no. 6, p. 695–698.

Lyons, W.B., Mayewski, P.A., Spencer, M.J., and Twickler, M.S., 1990, A northern hemisphere volcanic chemistry record (1869-1984) and climatic implications using a South Greenland ice core: Annals of Glaciology, v. 14, p. 176–182.

Macdonald, R., Sparks, R.S.J., Sigurdsson, H., Mattey, D.P., McGarvie, D.W., and Smith, R.L., 1987, The 1875 eruption of Askja volcano, Iceland; combined fractional crystallization and selective contamination in the generation of rhyolitic magma: Mineralogical Magazine, v. 51, p. 183–202.

MacGregor, A,G., 1952, Eruptive mechanisms—Mt. Pelée, the Soufrière of St. Vincent (West Indies) and the Valley of Ten Thousand Smokes (Alaska): Bulletin of Volcanology, series 2, v. 12, p. 49–74.

Mahood, G.A., 1990, Second reply to comment on "Evidence for long residence times of rhyolite magma in the Long Valley magma system; the isotopic record in the precaldera rhyolite lavas of Glass Mountain": Earth and Planetary Science Letters, v. 99, p. 395–399.

Marshall, P., 1935, Acid rocks of the Taupo-Rotorua volcanic district: Royal Society of New Zealand Transactions, v. 64, part 3, p. 323–366.

Martin, G.C., 1913, The recent eruption of Katmai volcano in Alaska: National Geographic Magazine, v. 24, p. 131–181.

Martin, G.C., 1921, Preliminary report on petroleum in Alaska: U.S. Geological Survey Bulletin 719, 83 p.

Martin, G.C., 1926, The Mesozoic stratigraphy of Alaska: U.S. Geological Survey Bulletin 776, 493 p.

Matumoto, T., 1971, Seismic body waves observed in the vicinity of Mount Katmai, Alaska, and evidence for the existence of molten chambers: Geological Society of America Bulletin, v. 82, p. 2905–2920.

Matumoto, T., and Ward, P.L., 1967, Microearthquake study of Mount Katmai and vicinity, Alaska: Journal of Geophysical Research, v. 72, p. 2557–2568.

Miller, T.P., McGimsey, R.G., Richter, D.H., Riehle, J.R., Nye, C.J., Yount, M.E., and Dumoulin, J.A., 1998, Catalog of the historically active volcanoes of Alaska: U.S. Geological Survey Open File Report 98–582, 104 p.

Miyashiro, A., 1974, Volcanic rock series in island arcs and active continental margins:American Journal of Science, v. 274, p. 321–355.

Moore, J.C., and 11 others, 1991, EDGE deep seismic reflection transect of the eastern Aleutian arc-trench layered lower crust reveals underplating and continental growth: Geology, v. 19, p. 420–424.

Moran, S.C., 2003, Multiple seismogenic processes for high-frequency earthquakes at Katmai National Park, Alaska; evidence from stress tensor inversions of fault-plane solutions: Bulletin of the Seismological Society of America, v. 93, p. 94–108.

Morey, G.W., and Fenner, C.N., 1917, The ternary system H_2O-K_2SiO_3-SiO_2: Journal of the American Chemical Society, v. 39, no. 6, p. 1173–1229.

Mori, J., White, R.A., Harlow, D.H., Okubo, P., Power, J.A., Hoblitt, R.P., Laguerta, E.P., Lanuza, A., and Bautista, B.C., 1996, Volcanic earthquakes following the 1991 climactic eruption of Mount Pinatubo; strong seismicity during a waning eruption, in Newhall, C.G., and Punongbayan, R.S., eds., 1996, Fire and Mud—eruptions and lahars of Mount Pinatubo, Philippines: Seattle, University of Washington Press, p. 339–350.

Motyka, R.J., 1977, Katmai caldera; glacier growth, lake rise, and geothermal activity, in Short Notes on Alaskan Geology—1977: Alaska Division of Geological and Geophysical Surveys, Geologic Report 55, p. 17–21.

Motyka, R.J., 1978, Surveillance of Katmai caldera and crater lake, Alaska: 1977: Final report on U.S. National Park Service purchase order PX 9100-7-1009: Geophysical Institute,University of Alaska, Fairbanks, 19 p.

Mullen, M.W., 1987, Petrology and provenance of sandstones of the Naknek Formation, Alaska Peninsula, in Hamilton, T.D., and Galloway, J.P., eds., Geological studies in Alaska by the U.S. Geological Survey during 1986: U.S. Geological Survey Circular 998, p. 86–90.

Muller, E.H., and Coulter, H.W., 1957a, Incipient glacier development within Katmai caldera, Alaska: Journal of Glaciology, v. 3, p. 13–17.

Muller, E.H., and Coulter, H.W., 1957b, The Knife Creek Glaciers of Katmai National Monument, Alaska: Journal of Glaciology, v. 3, p. 116–122.

Muller, E.H., Juhle, W., and Coulter, H.W., 1954, Current volcanic activity in Katmai National Monument: Science, v. 119, p. 319–321.

National Park Service, 1975, Interview with eyewitnesses of Katmai eruption; transcript of taped interview with Harry and Jennie Kaiakokonok at King Salmon on 29 April 1975 by Park Ranger Michael J. Tollefson: Katmai National Park archives, 28 p.

Newhall, C.G., and Punongbayan, R.S., eds., 1996, Fire and mud—eruptions and lahars of Mount Pinatubo, Philippines: Seattle, University of Washington Press, 1,126 p.

Newhall, C.G., Daag, A.S., Delfin, F.G., Jr., Hoblitt, R.P., McGeehin, J., Pallister, J.S., Regalado, M.T.M., Rubin, M., Tubianosa, B.S., Tamayo, R.A., Jr., and Umbal, J.V., 1996, Eruptive history of Mount Pinatubo, in Newhall, C.G., and Punongbayan, R.S., eds., 1996, Fire and mud—eruptions and lahars of Mount Pinatubo, Philippines: Seattle, University of Washington Press, p. 165–195.

Newman, S., and Lowenstern, J.B., 2002, VolatileCalc; a silicate melt-H2O-CO2 solution model written in Visual Basic for Excel: Computer Geoscience, v. 28, p. 597–604.

Page, R. A., Biswas, N. N., Lahr, J. C., and Pulpan, H., 1991, Seismicity of continental Alaska, in Slemmons, D. B., Engdahl, E. R., Zoback, M. D., and Blackwell, D. D., eds., Neotectonics of North America: Boulder, Colorado, Geological Society of America, Decade Map Volume l, chapter 4.

Palais, J.M., and Sigurdsson, H., 1989, Petrologic evidence of volatile emissions from major historic and pre-historic volcanic eruptions, in Kidson, J.W., ed., Understanding climate change: American Geophysical Union Monograph 52, p. 31–53.

Papike, J.J., 1992, The Valley of Ten Thousand Smokes, Katmai, Alaska; a unique geochemistry laboratory: Geochimica et Cosmochimica Acta, v. 56, p. 1429–1449.

Papike, J.J., Keith, T.E.C., Spilde, M.N., Galbreath, K.C., Shearer, C.K., and Laul, J.C., 1991a, Geochemistry and mineralogy of fumarolic deposits, Valley of Ten Thousand Smokes, Alaska; bulk chemical and mineralogical evolution of dacite-rich protolith: American Mineralogist, v. 76, p. 1662–1673.

Papike, J.J., Keith, T.E.C., Spilde, M.N., Shearer, C.K., Galbreath, K.C., and Laul, J.C., 1991b, Major and trace element mass flux in fumarolic deposits, Valley of Ten Thousand Smokes, Alaska; rhyolite-rich protolith: Geophysical Research Letters, v. 18, p. 1545–1548.

Payne, A.L., and Eichelberger, J.C., 2009, Ti-in-quartz geothermometry of magmas from Katmai, Alaska; a combined cathodoluminescence/electron microprobe study: 6th Annual Workshop on Japan-Kamchatka-Alaska Subduction Processes, Scientific Program and Abstracts, University of Alaska, Fairbanks (June 22-26, 2009), p. 120.

Peacock, M.A., 1931, Classification of igneous rock series: Journal of Geology, v. 39, p. 54–67.

Perry, K.W., 1912, Extract from report of Capt. K.W. Perry, U.S.R.S., on board the Revenue Cutter "Manning": National Geographic Magazine, v. 23, no. 8 (August 1912), p. 824–832.

Philpotts, A.R., Shi, J., and Brustman, C., 1998, Role of plagioclase crystal chains in the differentiation of partly crystallized basaltic magma: Nature, v. 395, p. 343–346.

Pinney, D.S., and Beget, J.E., 1991, Late Pleistocene volcanic deposits near the Valley of Ten Thousand Smokes, Katmai National Park, Alaska, in Reger, R.D., ed., Short notes on Alaskan geology 1991: Alaska Division of Geological & Geophysical Surveys Professional Report 111, p. 45–53.

Plafker, G., Moore, J.C., and Winkler, G.R., 1994, Geology of the southern Alaska margin, in Plafker, G., and Berg, H.C., eds., The Geology of Alaska: The Geology of North America, v. G-1, Geological Society of America, chapter 12.

Poreda, R.J., and Craig, H., 1989, Helium isotope ratios in circum-Pacific volcanic arcs: Nature,v. 338, p. 473–478.

Prejean, S, Haney, M., Pesicek, J., and Thurber, C., 2008, Seismicity and structure of the Katmai volcanic cluster, Alaska, revealed: American Geophysical Union, Fall Meeting, abstract V54A-07.

Punongbayan, R.S., Newhall, C.G., and Hoblitt, R.P., 1996, Photographic record of rapid geomorphic change at Mount Pinatubo, 1991-94, in Newhall, C.G., and Punongbayan, R.S., eds., Fire and Mud—eruptions and lahars of Mount Pinatubo, Philippines: Seattle, University of Washington Press, p. 21–66.

Pyle, D.M., 1989, The thickness, volume and grainsize of tephra fall deposits: Bulletin of Volcanology, v. 51, p. 1–15.

Pyle, D.M., 1990, New estimates for the volume of the Minoan eruption, *in* Hardy, D.A., ed., Thera and the Aegean World III: London, The Thera Foundation, v. 2, p. 113–121,

Ramdohr, P., 1962, Erzmikroskopische Untersuchungen an Magnetiten der Exhalationen im Valley of the 10000 Smokes: Neues Jahrbuch fur Mineralogie, Monatshefte 1962, p. 49–59.

Rampino, M.R., 1988, Volcanic winters: Annual Reviews of Earth and Planetary Sciences, v. 16, p. 73–99.

Rampino, M.R., and Self, S., 1984, Sulphur-rich volcanic eruptions and stratospheric aerosols: Nature, v. 310, p. 677–679.

Rapp, R.P., and Watson, E.B., 1995, Dehydration melting of metabasalt at 8-32 kbar; implications for continental growth and crust-mantle recycling: Journal of Petrology, v. 36, p. 891–931.

Ray, D.K., 1967, Geochemistry and petrology of the Mt. Trident andesites, Katmai National Monument, Alaska. Fairbanks, University of Alaska, Ph.D. dissertation, 198 p.

Reagan, M.K., Sims, K.W.W., Erich, J., Thomas, R.B., Cheng, H., Edwards, R.L., Layne, G., and Ball, L., 2003, Timescales of differentiation from mafic parents to rhyolite in North American continental arcs: Journal of Petrology, v. 44, p. 1703–1726.

Reck, H., 1928, Deutung der vulkanischen Geschichte und der Calderabildung auf der Insel La Palma: Zeitschrift für Vulkanologie, Band 11, p. 217-243.

Reck, H., 1936, Santorin—der Werdegang eines Inselvulkans und sein Ausbruch 1925–1928: Berlin, Verlag von Dietrich Reimer, 3 volumes.

Riehle, J.R., Detterman, R.L., Yount, M.E., and Miller, J.W., 1993, Geologic map of the Mount Katmai quadrangle and adjacent parts of the Naknek and Afognak quadrangles, Alaska: U.S. Geological Survey Map I-2204; scale 1:250,000.

Rigg, G.B., 1914, The effects of the Katmai eruption on marine vegetation: Science, v. 40, no. 1032, p. 509–513.

Robin, C., Monzier, M., and Eissen, J.-P., 1994, Formation of the mid-fifteenth century Kuwae caldera (Vanuatu) by an initial hydroclastic and subsequent ignimbritic eruption: Bulletin of Volcanology, v. 56, p. 170–183.

Sayre, J.D., and Hagelbarger, P.R., 1919, A study of temperatures in the Valley of Ten Thousand Smokes: Ohio Journal of Science, v. 19, p. 249–278.

Sbar, M.L., and Matumoto, T., 1972, Refraction profiles in the Valley of Ten Thousand Smokes, Katmai, Alaska: Bulletin Volcanologique, v. 35, p. 335–349.

Scott, W.E., Hoblitt, R.P., Torres, R.C., Self, S., Martinez, Ma. M.L., and Nillos, T., Jr., 1996, Pyroclastic flows of the June 15, 1991, climactic eruption of Mount Pinatubo, *in* Newhall, C.G., and Punongbayan, R.S., eds., Fire and mud—eruptions and lahars of Mount Pinatubo, Philippines: Seattle, University of Washington Press, p. 545–570.

Self, S., and Rampino, M.R., 1981, The 1883 eruption of Krakatau: Nature, v. 294, p. 699–704.

Self, S., Rampino, M.R., Newton, M.S., and Wolff, J.A., 1984, Volcanological study of the great Tambora eruption of 1815: Geology, v. 12, p. 659–663.

Self, S., Gertisser, R., Thordarson, T., Rampino, M.R., and Wolff, J.A., 2004, Magma volume, volatile emissions, and stratospheric aerosols from the 1815 eruption of Tambora: Geophysical Research Letters, v. 31, L20608, doi:10.1029/2004GL020925.

Shearer, C.K., Papike, J.J., and Spilde, M.N., 1991, Pyroxene/melt trace element behavior; a study of pyroxenes from the Valley of Ten Thousand Smokes, Alaska: Geophysical Research Letters, v. 18, p. 1557–1560.

Sheppard, D.S., Janik, C.J., and Keith, T.E.C., 1992, A comparison of gas geochemistry of fumaroles in the 1912 ash-flow sheet and on active stratovolcanoes, Katmai National Park, Alaska: Journal of Volcanology and Geothermal Research, v. 53, p. 185–197.

Shipley, J.W., 1920, Some chemical observations on the volcanic emanations and incrustations in the Valley of 10,000 Smokes, Katmai, Alaska: American Journal of Science, v. 50, p. 141–153.

Sigurdsson, H., 1990, Evidence of volcanic loading of the atmosphere and climate response: Paleogeography, Paleoclimatology, and Paleoecology, v. 89, p. 277–289.

Sigurdsson, H., and Carey, S., 1989, Plinian and co-ignimbrite tephra fall from the 1815 eruption of Tambora volano: Bulletin of Volcanology, v. 51, p. 243–270.

Sigurdsson, H., and Sparks, R.S.J., 1981, Petrology of rhyolitic and mixed magma ejecta from the 1875 eruption of Askja, Iceland: Journal of Petrology, v. 22, p. 41–84.

Sigurdsson, H., Carey, S., and Espindola, J.M., 1984, The 1982 eruptions of El Chichón volcano, Mexico; stratigraphy of pyroclastic deposits: Journal of Volcanology and Geothermal Research, v. 23, p. 11–37.

Sigurdsson, H., Carey, S., Cornell, W., and Pescatore, T., 1985, The eruption of Vesuvius in 79 A.D.: National Geographic Research, v. 1 (3), p. 332–387.

Sigurdsson, H., Carey, S., and Mandeville, C., 1991, Krakatau: National Geographic Research & Exploration, v. 7(3), p. 310–327.

Simkin, T., and Fiske, R.S., 1983, Krakatau 1883—the volcanic eruption and its effects: Washington, Smithsonian Institution Press, 464 p.

Simkin, T., and Howard, K.A., 1970, Caldera collapse in the Galápagos Islands, 1968: Science, v. 169, p. 429–437.

Sisson, T.W., and Bacon, C.R., 1999, Gas-driven filter pressing in magmas: Geology, v. 27, p. 613–616.

Sisson, T.W., Ratajeski, K., Hankins, W.B., and Glazner, A.F., 2005, Voluminous granitic magmas from common basaltic sources: Contributions to Mineralogy and Petrology, v. 148, p. 635–661.

Smith, N.R., and Griggs, R.F., 1932, The microflora of the ash of Katmai volcano with especial reference to nitrogen fixing bacteria: Soil Science, v. 34, p. 365–373.

Smith, R.L., 1960, Ash flows: Geological Society of America Bulletin, v. 71, p. 795–842.

Snyder, G.L., 1954, Eruption of Trident volcano, Katmai National Monument, Alaska: U.S. Geological Survey Circular 318, 7 p.

Sparks, R.S.J., 1976, Grain size variations in ignimbrites and implications for the transport of pyroclastic flows: Sedimentology, v. 3, p. 147–188.

Sparks, R.S.J., Self, S., and Walker, G.P.L., 1973, Products of ignimbrite eruptions: Geology, v. 1, p. 115–118.

Sparks, R.S.J., Wilson, L., and Hulme, G., 1978, Theoretical modeling of the generation, movement, and emplacement of pyroclastic flows by column collapse: Journal of Geophysical Research, v. 83, p. 1727–1739.

Sparks, R.S.J., Wilson, L., and Sigurdsson, H., 1981, The pyroclastic deposits of the 1875 eruption of Askja, Iceland: Philosophical Transactions of the Royal Society of London, Series A, v. 299, p. 241–273.

Sparks, R.S.J., Bursik, M.I., Carey, S.N., Gilbert, J.S., Glaze, L.S., Sigurdsson, H., and Woods, A.W., 1997, Volcanic plumes: John Wiley and Sons, 574 p.

Sparks, R.S.J., Tait, S.R., and Yanev, Y., 1999, Dense welding caused by volatile resorption: Journal of the Geological Society, London, v. 156, p. 217–225.

Spera, F.J., Yuen, D.A., Greer, J.C., and Sewell, G., 1986, Dynamics of magma withdrawal from stratified magma chambers: Geology, v. 14, p. 723–726.

Sun, S.-s., and McDonough, W.F., 1989, Chemical and isotopic systematics of oceanic basalts; implications for mantle composition and processes, *in* Saunders, A.D., and Norry, M.J., eds., Magmatism in the Ocean Basins: Geological Society Special Publication 42, p. 313–345.

Sussman, D., 1985, Apoyo caldera, Nicaragua; a major Quaternary silicic eruptive center: Journal of Volcanology and Geothermal Research, v. 24, p. 249–282.

Symonds, R.B., Rose, R.I., and Reed, M.H., 1988, Contribution of Cl- and F-bearing gases to the atmosphere by volcanoes: Nature, v. 334, p. 415–418.

Thorarinsson, S., 1958, The Öraefajökull eruption of 1362: Reykjavik, Acta Naturalia Islandica, v. II, no. 2, 99 p.

Trop, J.M., Szuch, D.A., Rioux, M., and Blodgett, R.B., 2005, Sedimentology and provenance of the Upper Jurassic Naknek Formation, Talkeetna Mountains, Alaska; bearings on the accretionary tectonic history of the Wrangellia composite terrane: Geological Society of America Bulletin, v. 117, p. 570–588.

Tsuya, H., 1930, The volcano Komagatake, Hokkaido, its geology, activity, and petrography: Bulletin of the Earthquake Research Institute, Tokyo Imperial University, v. 8, p. 238–270.

Turner, S., Sandiford, M., Reagan, M., Hawkesworth, C., and Hildreth, W., 2010, Origins of large-volume, compositionally zoned volcanic eruptions; new constraints from U-series isotopes and numerical thermal modeling for the 1912 Katmai-Novarupta eruption: Journal of Geophysical Research, v. 115, B12201, doi:10.1029/2009JB007195.

Valentine, G., 1987, Stratified flow in pyroclastic surges: Bulletin of Volcanology, v. 49, p. 616–630.

Varekamp, J.C., Luhr, J.F., and Prestegaard, K.L., 1984, The 1982 eruptions of El Chichón volcano (Chiapas, Mexico); character of the eruptions, ash-fall deposits, and gas phase: Journal of Volcanology and Geothermal Research, v. 23, p. 39–68.

Verbeek, R.D.M., 1885, Krakatau: Batavia, 495 p.

Volz, F.E., 1975a, Distribution of turbidity after the 1912 Katmai eruption in Alaska: Journal of Geophysical Research, v. 80, p. 2643–2648.

Volz, F.E., 1975b, Burden of volcanic dust and nuclear debris after injection into the stratosphere at 40°–58°N: Journal of Geophysical Research, v. 80, p. 2649–2652.

Walker, G.P.L., 1971, Grain-size characteristics of pyroclastic deposits: Journal of Geology, v. 79, p. 696–714.

Walker, G.P.L., 1973, Explosive volcanic eruptions—a new classification scheme: Geologische Rundschau, v. 62, p. 431–446.

Walker, G.P.L., Heming, R.F., and Wilson, C.J.N., 1980, Low-aspect ratio ignimbrites: Nature, v. 283, p. 286–287.

Walker, G.P.L., Self, S., and Froggatt, P.C., 1981, The ground layer of the Taupo ignimbrite; a striking example of sedimentation from a pyroclastic flow: Journal of Volcanology and Geothermal Research, v. 10, p. 1–11.

Walker, G.P.L., Self. S., and Wilson, L., 1984, Tarawera 1886, New Zealand—a basaltic plinian fissure eruption: Journal of Volcanology and Geothermal Research, v. 21, p. 61–78.

Wallace, P.J., 2005, Volatiles in subduction zone magmas; concentrations and fluxes based on melt inclusion and volcanic gas data: Journal of Volcanology and Geothermal Research, v. 140, p. 217–240.

Wallmann, P.C., Pollard, D.D., Hildreth, W., and Eichelberger, J.C., 1990, New structural limits on magma chamber locations at the Valley of Ten Thousand Smokes, Katmai National Park, Alaska: Geology, v. 18, p. 1240–1243.

Ward, P.L., and Matumoto, T., 1967, A summary of volcanic and seismic activity in Katmai National Monument, Alaska: Bulletin Volcanologique, v. 31, p. 107–129.

Ward, P.L., Pitt, A.M., and Endo, E., 1991, Seismic evidence for magma in the vicinity of Mt. Katmai, Alaska: Geophysical Research Letters, v. 18, p. 1537–1540.

Westrich, H.R., Eichelberger, J.C., and Hervig, R.L., 1991, Degassing of the 1912 Katmai magmas: Geophysical Research Letters, v. 18, p. 1561–1564.

Wilcox, R.E., 1999, The idea of magma mixing; history of a struggle for acceptance: Journal of Geology, v. 107, p. 421–432.

Williams, H., 1941, Calderas and their origin: University of California Publications in Geological Sciences, v. 25, no. 6, p. 239–346.

Williams, H., Curtis, G., and Juhle, W., 1953, Mount Katmai and the Valley of Ten Thousand Smokes, Alaska (a new interpretation of the great eruptions of 1912): Proceedings of the 8th Pacific Science Congress (Quezon City, Philippines), v. II, p. 129.

Williams, S.N., and Self, S., 1983, The October 1902 plinian eruption of Santa Maria volcano, Guatemala: Journal of Volcanology and Geothermal Research, v. 16, p. 33–56.

Wilson, C.J.N., 1980, The role of fluidization in the emplacement of pyroclastic flows; an experimental approach: Journal of Volcanology and Geothermal Research, v. 8, p. 231–249.

Wilson, C.J.N., 1984, The role of fluidization in the emplacement of pyroclastic flows, 2. experimental results and their interpretation: Journal of Volcanology and Geothermal Research, v. 20, p. 55–84.

Wilson, C.J.N., and Hildreth, W., 1997, The Bishop Tuff; new insights from eruptive stratigraphy: Journal of Geology, v. 105, p. 407–439.

Wilson, C.J.N., and Walker, G.P.L., 1982, Ignimbrite depositional facies; the anatomy of a pyroclastic flow: Journal of the Geological Society of London, v. 139, p. 581–592.

Wilson, C.J.N., and Walker, G.P.L., 1985, The Taupo eruption, New Zealand, I. General aspects:Philosphical Transactions of the Royal Society of London, Series A, v. 314, p. 199–228.

Wilson, L., 1976, Explosive volcanic eruptions III. Plinian eruption columns: Geophysical Journal of the Royal Astronomical Society, v. 45, p. 543–556.

Wilson, L., Sparks, R.S.J., and Walker, G.P.L., 1980, Explosive volcanic eruptions IV. The control of magma properties and conduit geometry on eruption column behavior: Geophysical Journal of the Royal Astronomical Society, v. 63, p. 117–148.

Wilson, L., and Walker, G.P.L., 1987, Explosive volcanic eruptions—VI. Ejecta dispersal in plinian eruptions; the control of eruption conditions and atmospheric properties:Geophysical Journal of the Royal Astronomical Society, v. 89, p. 657–679.

Witter, J.B., and Self, S., 2007, The Kuwae (Vanuatu) eruption of AD 1452; potential magnitude and volatile release: Bulletin of Volcanology, v. 69, p. 301–318.

Wright, F.E., 1952, Memorial to Clarence Norman Fenner: Proceedings Volume of the Geological Society of America for 1951, p. 103–107.

Yoder, H.S., Jr., 1992, Norman L. Bowen (1887-1956), MIT Class of 1912, first predoctoral fellow of the Geophysical Laboratory: Earth Sciences History, v. 11, no. 1, p. 45–55.

Yoder, H.S., Jr., 1993, Timetable of petrology: Journal of Geological Education, v. 41, p. 447–489.

Yoder, H.S., Jr., 2004, The Geophysical Laboratory: Volume III of Centennial History of the Carnegie Institution of Washington, Cambridge University Press, 270 p.

Young, S.R., 1990, Physical volcanology of Holocene airfall deposits from Mt. Mazama, Crater Lake, Oregon: University of Lancaster (U.K.), Ph.D. dissertation, 299 p.

Zies, E.G., 1924a, The fumarolic incrustations in the Valley of Ten Thousand Smokes: National Geographic Society, Contributed Technical Papers, Katmai Series, No. 3, p. 157–179.

Zies, E.G., 1924b, Hot springs of the Valley of Ten Thousand Smokes: Journal of Geology, v. 32, p. 303–310.

Zies, E.G., 1929, The Valley of Ten Thousand Smokes; I. The fumarolic incrustations and their bearing on ore deposition; II. The acid gases contributed to the sea during volcanic activity: National Geographic Society, Contributed Technical Papers, Katmai Series, No. 4, p. 1–79.

Appendixes A–B–C

Appendix A. Major-element chemical analyses of eruptive products, Novarupta 1912.

[The 10 major oxides (reported in weight percent) are normalized to H_2O-free totals of 99.6 weight percent (allowing 0.4 weight percent for trace oxides and halogens); determinations by wavelength-dispersive x-ray fluorescence in U.S. Geological Survey laboratory at Lakewood, Colorado; supervised by J.E. Taggart. Precision and accuracy are discussed by Baedecker (1987) and Bacon and Druitt (1988). FeO* is total iron calculated as FeO. "Original total" is the volatile-free sum of the 10 oxides, as analyzed, before normalization, with total iron calculated as Fe_2O_3. LOI is weight loss on ignition at 900°C. —, not analyzed.]

Sample	SiO_2	TiO_2	Al_2O_3	FeO*	MnO	MgO	CaO	Na_2O	K_2O	P_2O_5	LOI	Original total
						Layer A						
K-88	77.36	0.14	12.32	1.24	0.04	0.10	0.94	4.27	3.15	0.05	1.30	97.84
K-1000-a	77.48	0.14	12.18	1.30	0.04	0.18	0.85	4.21	3.16	0.05	1.69	97.31
K-1001	77.43	0.16	12.16	1.30	0.05	0.14	0.87	4.36	3.12	—	1.45	98.27
K-1065	77.53	0.16	12.16	1.28	0.05	0.13	0.85	4.29	3.14	—	1.64	98.27
K-1079-a	77.46	0.14	12.20	1.26	0.04	0.17	0.86	4.26	3.15	0.05	1.36	97.98
K-1190	77.45	0.17	12.18	1.29	0.05	0.13	0.85	4.33	3.14	—	1.51	98.12
						Layer B						
K-90-a	77.43	0.16	12.21	1.29	0.05	0.14	0.89	4.29	3.13	—	1.50	97.89
K-90-c	64.50	0.65	15.97	5.23	0.12	2.19	5.07	4.06	1.65	0.15	0.55	99.14
K-827-a	77.44	0.17	12.16	1.32	0.05	0.16	0.86	4.28	3.14	—	1.70	98.26
K-827-c	67.97	0.51	15.06	4.21	0.09	1.67	3.88	3.96	2.16	0.11	0.79	99.21
K-827-d	59.35	0.71	16.88	6.72	0.13	3.64	7.36	3.40	1.26	0.15	0.18	99.69
K-1008-c	63.31	0.66	16.05	5.81	0.12	2.63	5.48	3.77	1.63	0.14	0.50	99.27
K-1061-a	77.31	0.17	12.19	1.38	0.05	0.16	0.88	4.33	3.12	—	1.56	98.04
K-1061-c	60.26	0.70	16.57	6.64	0.13	3.52	6.79	3.50	1.35	0.15	0.35	99.17
K-1068-a	77.24	0.17	12.27	1.35	0.05	0.17	0.91	4.30	3.13	—	1.40	98.25
K-1068-a	77.37	0.14	12.25	1.29	0.04	0.19	0.88	4.25	3.14	0.05	1.62	97.58
K-1073	77.54	0.16	12.14	1.29	0.05	0.12	0.86	4.30	3.14	—	1.39	97.62
K-1074-a	77.38	0.14	12.22	1.29	0.04	0.17	0.92	4.25	3.14	0.05	1.23	97.82
K-1158-c	65.59	0.65	15.59	5.04	0.11	1.97	4.65	4.10	1.75	0.14	0.70	99.02
K-1160-a	77.11	0.16	12.56	1.32	0.05	0.13	0.86	4.28	3.12	—	1.25	98.30
K-1160-c	65.25	0.59	15.68	4.96	0.11	2.18	4.99	3.83	1.88	0.13	0.58	99.07
K-1194-a	77.36	0.17	12.27	1.30	0.05	0.13	0.88	4.31	3.12	—	1.19	98.23
K-1199-c	66.45	0.60	15.43	4.70	0.11	1.85	4.33	4.09	1.91	0.14	0.72	98.78
K-1199-d	60.10	0.70	16.56	6.74	0.13	3.59	6.78	3.52	1.32	0.14	0.25	99.26
K-1200-c	61.36	0.70	16.34	6.41	0.13	3.11	6.31	3.62	1.46	0.15	0.37	99.34
K-2466	59.11	0.71	16.82	6.67	0.13	3.73	7.29	3.63	1.29	0.22	0.34	98.39
K-2466-b	59.64	0.73	16.70	6.76	0.13	3.59	7.16	3.41	1.31	0.17	0.34	98.58
K-2488-a	61.53	0.72	16.22	6.30	0.13	3.12	6.21	3.66	1.50	0.20	1.10	98.62
K-2488-b	60.88	0.70	16.67	6.26	0.13	3.15	6.57	3.62	1.42	0.19	1.05	98.56
K-2488-c	61.49	0.72	16.19	6.32	0.13	3.16	6.19	3.66	1.52	0.22	1.15	98.03
K-2488-d	61.38	0.70	16.59	6.15	0.13	3.00	6.44	3.55	1.49	0.18	1.12	97.51
						Layer C						
K-45	64.57	0.71	15.65	5.37	0.12	2.14	4.97	4.20	1.74	0.14	—	96.83
K-45-glass	75.24	0.39	12.71	1.94	0.04	0.59	1.57	4.11	2.93	0.06	—	97.40
K-91	65.29	0.64	15.85	5.02	0.11	1.98	4.71	4.14	1.71	0.15	0.57	99.31
K-1270-a	65.81	0.64	15.45	5.08	0.11	1.91	4.57	4.09	1.78	0.14	2.09	97.32
K-1270-b	65.55	0.64	15.65	5.01	0.11	1.96	4.65	4.12	1.76	0.13	2.05	97.39
K-1270-c	65.54	0.65	15.59	5.06	0.11	1.97	4.69	4.09	1.75	0.14	2.32	97.10
K-1270-d	60.14	0.71	16.73	6.61	0.12	3.32	7.03	3.46	1.35	0.14	1.19	98.22
K-1270-e	63.17	0.68	15.92	5.95	0.12	2.62	5.63	3.75	1.60	0.14	1.93	97.59
K-1270-f	65.54	0.62	15.64	5.04	0.11	2.00	4.71	4.04	1.75	0.13	2.39	97.41
K-1271-a	65.51	0.64	15.74	5.03	0.11	1.89	4.68	4.10	1.76	0.14	2.09	97.46
K-1271-b	61.90	0.68	16.47	6.11	0.12	2.92	6.24	3.55	1.49	0.13	1.49	98.00
K-1271-c	65.31	0.65	15.69	5.09	0.11	2.02	4.78	4.08	1.73	0.14	2.04	97.76
K-1271-d	65.66	0.65	15.47	5.12	0.11	1.98	4.63	4.07	1.77	0.14	2.14	97.23

Appendix A. Major-element chemical analyses of eruptive products, Novarupta 1912.—Continued

Sample	SiO_2	TiO_2	Al_2O_3	FeO*	MnO	MgO	CaO	Na_2O	K_2O	P_2O_5	LOI	Original total
K-1271-e	65.62	0.64	15.69	5.00	0.11	1.91	4.69	4.07	1.74	0.14	2.01	97.15
K-1271-f	65.12	0.65	15.54	5.34	0.11	2.12	4.86	4.01	1.72	0.14	2.06	97.43
K-1271-g	60.63	0.70	16.76	6.46	0.12	3.13	6.80	3.48	1.38	0.14	1.26	98.08
K-1271-h	65.73	0.68	15.61	5.04	0.11	1.90	4.46	4.11	1.81	0.15	2.09	96.97
K-1271-i	65.58	0.67	15.60	5.07	0.12	1.91	4.61	4.11	1.77	0.18	2.13	96.77
K-2002-a	61.53	0.72	16.37	6.28	0.12	3.00	6.22	3.71	1.46	0.17	1.43	97.93
K-2002-b	63.78	0.63	15.92	5.46	0.11	2.47	5.45	3.97	1.63	0.16	0.57	98.84
K-2620-a	65.28	0.65	15.83	5.10	0.11	1.97	4.69	3.99	1.75	0.21	1.95	97.50
K-2620-b	65.57	0.69	15.72	5.01	0.11	1.98	4.56	3.93	1.81	0.22	2.04	96.92
Layer D												
K-39-r	64.94	0.68	15.84	5.31	0.12	1.92	4.85	3.98	1.82	0.12	—	96.27
K-55	64.83	0.70	15.90	5.26	0.12	2.16	5.11	3.66	1.71	0.14	—	96.51
K-94-a	77.48	0.16	12.22	1.26	0.05	0.13	0.84	4.30	3.14	—	1.36	98.59
K-94-b	65.10	0.65	15.77	5.14	0.12	2.03	4.76	4.19	1.68	0.15	0.67	99.14
K-94-c	65.47	0.64	15.77	4.92	0.11	1.93	4.68	4.20	1.73	0.15	0.61	99.18
K-1083-a	77.45	0.17	12.18	1.30	0.05	0.13	0.85	4.31	3.15	0.00	1.43	98.12
K-1083-c	65.57	0.65	15.69	4.99	0.11	1.92	4.62	4.16	1.74	0.15	0.52	99.04
K-1268-a	65.29	0.66	15.78	5.16	0.11	1.97	4.78	3.99	1.72	0.14	2.09	97.17
K-1268-b	59.92	0.72	16.55	6.84	0.13	3.45	7.11	3.42	1.30	0.14	1.26	98.07
K-1268-c	62.45	0.67	16.25	6.01	0.12	2.78	6.04	3.58	1.55	0.14	1.66	97.45
K-1268-d	65.27	0.62	15.88	5.06	0.11	1.97	4.85	3.99	1.71	0.14	2.06	97.21
K-1268-e	65.06	0.66	15.75	5.30	0.11	2.02	4.84	4.03	1.69	0.14	1.96	97.37
K-1268-f	62.04	0.70	16.13	6.37	0.12	3.00	6.17	3.41	1.52	0.14	2.11	96.97
K-1269-a	65.25	0.66	15.70	5.14	0.11	2.02	4.84	4.01	1.72	0.14	2.07	97.09
K-1269-b	65.59	0.64	15.60	5.08	0.11	1.96	4.72	3.98	1.79	0.13	2.23	97.04
K-1269-c	66.54	0.59	15.40	4.69	0.10	1.84	4.46	3.96	1.90	0.12	2.26	97.00
K-1269-d	65.16	0.64	15.75	5.14	0.11	2.06	4.86	4.02	1.73	0.13	2.27	97.37
K-1269-e	65.64	0.64	15.64	4.99	0.10	1.91	4.71	4.05	1.77	0.14	2.10	97.42
K-1269-f	65.79	0.64	15.50	5.03	0.11	1.92	4.64	4.07	1.77	0.13	2.12	97.04
Layer F												
K-53	64.47	0.71	15.61	5.40	0.12	2.26	5.15	3.99	1.72	0.16	—	96.80
K-631-b	64.74	0.68	15.88	5.29	0.12	2.09	4.81	4.18	1.65	0.16	1.17	98.46
K-631-c	66.41	0.60	15.44	4.74	0.11	1.84	4.40	4.06	1.86	0.14	1.44	98.08
K-1267-a	63.10	0.67	16.13	5.83	0.12	2.63	5.73	3.65	1.59	0.13	1.65	97.54
K-1267-b	65.43	0.64	15.74	5.09	0.11	1.96	4.71	4.02	1.74	0.14	1.85	97.42
K-1267-c	62.95	0.67	16.04	5.91	0.12	2.74	5.84	3.63	1.56	0.13	1.63	97.46
K-1267-d	64.02	0.67	15.98	5.55	0.11	2.34	5.37	3.81	1.61	0.14	2.04	97.23
K-1267-e	63.91	0.66	15.88	5.66	0.11	2.47	5.46	3.65	1.68	0.13	1.77	97.24
K-1267-f	63.73	0.67	16.01	5.66	0.11	2.46	5.44	3.72	1.66	0.13	1.57	97.68
K-1267-g	65.52	0.67	15.69	5.07	0.11	1.95	4.56	4.08	1.78	0.16	1.94	97.13
Layer G												
K-23	65.32	0.65	15.80	5.04	0.11	1.73	4.62	4.32	1.87	0.14	—	97.71
K-54	65.59	0.67	15.58	5.11	0.12	1.99	4.67	3.98	1.74	0.16	2.27	96.94
K-101	62.82	0.72	16.16	5.74	0.13	2.65	5.68	4.00	1.57	0.13	—	98.28
K-107-d	61.01	0.73	16.60	6.53	0.13	3.19	6.43	3.42	1.40	0.15	3.49	95.98
K-1266-a	63.64	0.62	15.78	5.66	0.11	2.57	5.62	3.75	1.71	0.13	0.30	99.07
K-1266-b	65.62	0.64	15.60	5.02	0.11	1.94	4.67	4.12	1.73	0.14	0.49	98.35
K-1266-c	65.13	0.64	15.78	5.10	0.11	1.99	4.86	4.15	1.70	0.14	0.39	99.10
K-1266-d	64.09	0.66	15.54	5.70	0.12	2.54	5.26	3.87	1.68	0.14	0.36	98.68
K-1266-e	63.77	0.66	15.99	5.50	0.11	2.28	5.40	4.13	1.60	0.14	0.34	99.02

Appendix A. Major-element chemical analyses of eruptive products, Novarupta 1912.—Continued

Sample	SiO_2	TiO_2	Al_2O_3	FeO*	MnO	MgO	CaO	Na_2O	K_2O	P_2O_5	LOI	Original total
K-1266-f	63.42	0.65	15.75	5.76	0.11	2.69	5.71	3.71	1.69	0.13	0.38	98.63
					Novarupta lava dome							
K-16	76.86	0.17	12.53	1.43	0.05	0.02	0.95	4.35	3.21	0.02	—	97.68
K-17	56.54	0.81	17.37	8.34	0.15	4.13	8.06	3.19	0.90	0.10	0.05	99.19
K-18	65.06	0.71	15.50	5.37	0.12	1.95	4.85	4.22	1.68	0.15	—	99.76
K-22	76.54	0.23	12.52	1.52	0.06	0.08	1.14	4.38	3.10	0.04	—	99.02
K-622	77.44	0.15	12.13	1.28	0.04	0.18	0.89	4.29	3.14	0.05	0.36	98.52
K-623	55.55	0.75	17.88	7.72	0.14	4.31	9.22	2.97	0.92	0.13	0.40	99.15
K-2004	57.14	0.72	16.87	7.69	0.15	4.61	7.62	3.71	0.92	0.16	0.05	99.18
RLS-214	76.52	0.21	12.50	1.53	0.07	0.34	1.16	4.32	2.94	0.02	—	99.70
					Pyroclastic flows							
K-7A-r	77.66	0.13	12.34	1.24	0.04	0.12	0.83	4.00	3.20	0.03	2.55	96.07
K-7A-s	77.83	0.20	11.97	1.36	0.05	0.10	0.84	4.03	3.19	0.03	—	97.20
K-11-a-r	77.42	0.14	12.30	1.25	0.04	0.12	0.86	4.27	3.16	0.03	2.69	96.36
K-11-a-s	77.47	0.21	11.90	1.41	0.05	0.10	0.92	4.34	3.17	0.03	—	96.60
K-12-c	66.57	0.64	15.38	4.68	0.11	1.72	4.21	4.24	1.87	0.18	1.54	97.79
K-13-c	66.13	0.65	15.43	4.87	0.11	1.77	4.38	4.27	1.81	0.17	1.39	97.30
K-14-c	65.92	0.61	15.53	4.84	0.11	2.03	4.63	3.79	1.96	0.17	1.70	97.33
K-15-c	67.18	0.60	15.14	4.54	0.10	1.72	4.28	3.90	2.00	0.13	1.35	97.98
K-37-c	67.22	0.61	14.92	4.55	0.10	1.77	4.33	3.90	2.06	0.12	—	97.30
K-42-r	77.41	0.15	12.35	1.23	0.04	0.11	0.83	4.24	3.21	0.03	2.60	95.99
K-42-s	77.19	0.21	12.44	1.26	0.06	0.10	0.88	4.26	3.17	0.03	—	96.37
K-44-r	77.49	0.13	12.30	1.24	0.04	0.12	0.85	4.22	3.17	0.03	3.23	97.17
K-44-s	77.14	0.21	12.48	1.28	0.05	0.11	0.94	4.16	3.19	0.03	—	95.75
K-57	64.85	0.70	15.61	5.17	0.12	2.04	4.97	4.26	1.76	0.13	—	97.37
K-60	61.35	0.74	16.24	6.34	0.13	2.97	6.26	3.91	1.53	0.15	—	99.93
K-63-c	65.37	0.63	15.64	5.04	0.11	2.20	4.70	3.87	1.88	0.17	1.55	97.44
K-64-c	65.27	0.68	15.75	5.12	0.11	2.09	4.83	3.71	1.87	0.16	2.24	97.36
K-66-c	64.94	0.67	15.57	5.43	0.12	2.26	4.96	3.79	1.72	0.13	1.10	98.30
K-66-d	61.97	0.70	16.24	6.21	0.13	2.93	6.18	3.60	1.49	0.15	0.74	98.85
K-74-c	68.62	0.54	14.76	4.08	0.10	1.48	3.85	3.88	2.18	0.12	1.15	98.27
K-75-c	65.89	0.65	15.53	4.95	0.11	1.86	4.57	4.13	1.75	0.16	1.45	98.10
K-75-d	59.27	0.75	16.31	6.88	0.14	3.88	7.30	3.63	1.32	0.12	—	97.89
K-75-d-glass	66.81	0.72	15.61	4.40	0.10	1.56	3.91	4.17	2.15	0.17	—	98.72
K-75-e	66.56	0.64	15.31	4.80	0.11	1.72	4.31	4.19	1.80	0.15	1.39	97.90
K-76	65.35	0.66	15.62	5.16	0.12	1.93	4.77	4.12	1.71	0.15	1.62	97.84
K-78	65.52	0.70	15.40	5.32	0.12	2.07	4.70	3.89	1.72	0.15	1.65	97.68
K-82-c	65.61	0.65	15.24	5.24	0.12	2.16	4.73	3.87	1.83	0.14	1.44	97.83
K-82-d	60.44	0.73	16.41	6.72	0.14	3.40	6.82	3.44	1.37	0.15	1.10	98.21
K-100-a	77.72	0.18	11.95	1.30	0.05	0.11	0.85	4.25	3.16	0.03	2.53	96.91
K-100-c	64.02	0.68	15.69	5.57	0.12	2.51	5.31	3.93	1.61	0.15	1.24	98.08
K-100-d	57.93	0.72	16.59	7.27	0.14	4.62	7.80	3.22	1.17	0.14	0.49	99.04
K-108-c	65.86	0.66	15.39	5.06	0.12	1.94	4.60	4.06	1.76	0.15	1.30	97.99
K-108-d	64.93	0.67	15.62	5.29	0.12	2.09	5.03	4.02	1.68	0.15	1.41	97.87
K-113-c	65.35	0.67	15.54	5.22	0.12	2.06	4.77	4.00	1.73	0.15	1.02	98.21
K-113-d	60.00	0.74	16.61	6.77	0.14	3.40	6.94	3.55	1.29	0.15	—	99.23
K-119D-r	59.00	0.73	17.09	7.01	0.14	3.50	7.40	3.30	1.31	0.12	—	97.99
K-121	77.70	0.13	12.11	1.25	0.04	0.16	0.86	4.11	3.20	0.05	2.80	96.27
K-122	77.64	0.13	12.11	1.28	0.04	0.17	0.85	4.17	3.16	0.05	2.57	96.21
K-122-d	58.69	0.72	17.02	6.97	0.14	3.65	7.35	3.64	1.30	0.13	—	98.62

Appendix A. Major-element chemical analyses of eruptive products, Novarupta 1912.—Continued

Sample	SiO_2	TiO_2	Al_2O_3	FeO*	MnO	MgO	CaO	Na_2O	K_2O	P_2O_5	LOI	Original total
K-123-d-r	58.69	0.71	16.92	7.05	0.15	3.80	7.31	3.55	1.29	0.13	—	97.99
K-131-d-r	58.72	0.72	16.86	7.08	0.14	3.71	7.44	3.49	1.31	0.12	—	97.63
K-135-c	66.04	0.67	15.38	5.00	0.12	1.83	4.48	4.17	1.77	0.15	1.52	97.59
K-135-d	59.85	0.73	16.73	6.76	0.13	3.54	7.14	3.27	1.30	0.15	0.73	98.44
K-138-r	77.30	0.15	12.41	1.25	0.04	0.13	0.86	4.29	3.16	0.02	3.21	95.48
K-138-s	77.05	0.21	12.41	1.25	0.06	0.13	0.96	4.28	3.24	0.02	—	95.39
K-138-glass	77.25	0.17	12.34	1.27	0.05	0.36	0.81	4.08	3.24	0.03	—	97.65
K-139-r	77.37	0.15	12.34	1.20	0.04	0.11	0.86	4.29	3.19	0.03	3.04	96.03
K-139-s	77.22	0.21	12.33	1.25	0.06	0.09	0.93	4.20	3.28	0.03	—	95.56
K-162	64.52	0.66	16.11	5.16	0.10	2.27	4.79	4.08	1.78	0.13	—	98.24
K-182	65.52	0.67	15.77	5.03	0.09	2.11	4.56	3.91	1.80	0.13	—	97.70
K-184	59.47	0.72	16.77	6.81	0.12	3.56	7.03	3.63	1.35	0.15	—	99.70
K-200	62.52	0.70	16.06	5.98	0.12	2.79	5.90	3.86	1.52	0.14	1.29	97.98
K-202	64.78	0.67	15.86	5.23	0.11	2.08	4.75	4.20	1.77	0.14	—	98.09
K-203	60.63	0.73	16.57	6.63	0.13	3.25	6.64	3.49	1.39	0.14	1.22	98.57
K-581	59.69	0.72	16.56	6.84	0.13	3.46	7.13	3.61	1.31	0.14	0.43	98.62
K-1363	77.39	0.14	12.21	1.30	0.04	0.18	0.87	4.25	3.15	0.05	2.16	97.04
K-1364	77.46	0.14	12.21	1.27	0.04	0.16	0.86	4.25	3.15	0.05	2.16	97.08
K-1365	77.54	0.14	12.22	1.26	0.04	0.19	0.86	4.12	3.19	0.05	3.02	96.21
K-1365-r	77.55	0.18	12.11	1.28	0.05	0.12	0.83	4.30	3.15	0.02	2.60	96.96
K-1365-s	77.49	0.18	12.09	1.28	0.05	0.14	0.84	4.36	3.17	—	2.75	96.40
K-1365-t	77.61	0.18	11.99	1.28	0.05	0.11	0.84	4.35	3.16	0.03	2.59	96.58
K-1365-u	77.37	0.17	12.24	1.26	0.05	0.12	0.85	4.32	3.16	0.06	2.76	96.17
K-1490	67.72	0.57	15.10	4.37	0.10	1.59	3.91	4.08	2.03	0.13	1.61	97.65
K-1495	65.73	0.65	15.67	5.01	0.11	1.88	4.53	4.05	1.81	0.16	1.00	97.89
K-1511-c	60.08	0.74	16.77	6.78	0.13	3.28	6.76	3.53	1.36	0.16	0.86	97.98
K-1518-a	67.45	0.59	14.92	4.54	0.10	1.81	4.05	3.90	2.10	0.14	1.39	97.46
K-1518-b	59.20	0.72	16.73	6.90	0.13	3.90	7.23	3.36	1.29	0.15	0.87	98.25
K-1518-c	60.04	0.73	16.91	6.65	0.13	3.31	6.85	3.46	1.36	0.15	0.66	98.37
K-1518-d	59.99	0.74	16.55	6.91	0.13	3.58	6.75	3.41	1.38	0.15	0.78	98.11
K-1518-e	59.24	0.73	16.84	6.95	0.14	3.71	7.20	3.37	1.27	0.15	0.60	98.19
K-1519-b	59.23	0.70	16.46	6.96	0.14	4.36	6.99	3.32	1.29	0.14	0.95	98.03
K-1525	77.19	0.18	12.28	1.30	0.05	0.19	0.87	4.32	3.17	0.05	2.52	96.51
K-2582-a	66.09	0.59	15.44	4.71	0.11	2.03	4.60	3.91	1.94	0.19	1.60	97.99
K-2582-b	65.49	0.66	15.57	5.01	0.11	1.98	4.62	4.20	1.76	0.20	1.29	97.91
K-2582-c	66.73	0.61	15.16	4.67	0.11	1.78	4.15	4.21	1.97	0.20	1.54	97.19
K-2584-b	77.27	0.17	12.35	1.35	0.06	0.15	0.92	4.16	3.11	0.05	2.14	96.76
K-2588-c	77.18	0.18	12.45	1.36	0.05	0.14	0.94	4.15	3.10	0.05	2.14	96.89
K-2614-a	59.19	0.73	16.75	6.95	0.14	3.61	7.22	3.45	1.35	0.20	1.15	98.11
K-2614-b	59.47	0.70	16.44	7.03	0.14	3.81	6.96	3.45	1.40	0.20	1.46	98.14
K-2614-c	59.31	0.72	16.58	7.21	0.15	3.46	6.97	3.48	1.52	0.19	1.54	97.90
K-2614-d	58.80	0.71	16.20	7.34	0.15	4.16	7.26	3.38	1.40	0.19	1.45	97.73
K-2616-a	59.69	0.73	16.62	6.89	0.14	3.61	6.85	3.54	1.33	0.20	1.33	98.27
K-2616-b	59.50	0.73	16.57	6.88	0.14	3.66	7.07	3.53	1.33	0.20	1.11	98.60
K-2617-a	63.85	0.67	16.17	5.54	0.12	2.46	5.36	3.61	1.61	0.21	1.76	97.33
K-2617-b	77.60	0.17	12.27	1.28	0.05	0.12	0.80	4.03	3.18	0.09	3.00	95.75
K-2618-a	65.12	0.67	15.57	5.20	0.12	2.09	4.71	4.19	1.73	0.20	1.60	97.89
K-2618-b	64.28	0.68	15.76	5.43	0.12	2.24	5.06	4.19	1.63	0.20	1.54	97.93
K-2618-c	66.28	0.59	15.40	4.72	0.10	1.88	4.42	4.04	1.99	0.19	1.51	97.68
K-2618-d	59.57	0.73	16.75	6.84	0.13	3.49	7.02	3.53	1.33	0.20	1.02	98.14
K-2619	60.65	0.73	16.58	6.46	0.13	3.14	6.71	3.61	1.38	0.21	0.91	98.54

Appendix A. Major-element chemical analyses of eruptive products, Novarupta 1912.—Continued

Sample	SiO_2	TiO_2	Al_2O_3	FeO*	MnO	MgO	CaO	Na_2O	K_2O	P_2O_5	LOI	Original total
K-2621-a	61.08	0.72	16.43	6.50	0.13	3.19	6.31	3.60	1.43	0.21	0.99	98.82
K-2621-b	65.94	0.63	15.67	4.81	0.11	1.84	4.46	4.14	1.78	0.20	2.24	97.27
K-2621-c	60.52	0.71	16.79	6.46	0.13	3.26	6.63	3.54	1.37	0.19	1.39	97.26
K-2622-a	65.52	0.66	15.62	5.03	0.11	1.88	4.54	4.28	1.75	0.20	1.34	98.20
K-2622-b	63.41	0.68	16.03	5.62	0.12	2.41	5.47	4.08	1.56	0.21	1.41	98.16
K-2623	65.20	0.64	15.86	5.15	0.11	2.19	4.84	3.60	1.80	0.19	1.86	97.31
K-2624	60.42	0.70	16.37	6.45	0.13	3.55	6.81	3.60	1.38	0.19	0.94	98.58
K-2625	67.11	0.56	15.13	4.33	0.10	1.76	4.26	4.09	2.08	0.18	1.08	98.10
K-2626-a	65.12	0.68	15.79	5.27	0.11	2.16	4.75	3.78	1.71	0.22	2.01	97.12
K-2626-b	60.86	0.75	16.49	6.72	0.13	3.30	6.42	3.31	1.41	0.21	1.25	97.87
K-2626-c	59.92	0.72	17.15	6.66	0.13	3.36	6.91	3.23	1.32	0.20	1.41	97.58
K-2626-d	64.06	0.70	15.99	5.59	0.12	2.35	5.16	3.80	1.62	0.22	1.60	97.17
K-2627-a	60.88	0.72	16.49	6.48	0.13	3.17	6.62	3.49	1.40	0.21	1.50	97.84
K-2627-b	59.34	0.74	16.53	6.94	0.14	3.73	7.17	3.52	1.28	0.20	0.94	98.19
K-2649-a	60.30	0.71	16.72	6.53	0.13	3.21	6.77	3.63	1.39	0.21	1.26	98.28
K-2649-b	61.22	0.71	16.37	6.34	0.13	3.10	6.39	3.70	1.44	0.20	0.87	98.59
K-2649-c	59.68	0.74	16.76	6.86	0.14	3.39	6.97	3.52	1.31	0.21	0.33	98.62
K-2650-a	67.44	0.58	14.60	4.67	0.10	1.93	4.17	3.81	2.11	0.19	1.60	98.22
K-2650-b	66.02	0.63	15.46	4.87	0.11	1.87	4.44	4.15	1.84	0.20	1.50	97.90
K-2651-a	66.31	0.65	15.48	4.82	0.11	1.81	4.39	3.89	1.94	0.20	1.90	97.18
K-2651-b	66.23	0.62	15.33	4.91	0.10	1.93	4.38	3.79	2.12	0.18	1.74	97.45
K-2651-c	60.25	0.72	16.78	6.61	0.13	3.29	6.76	3.49	1.37	0.20	1.41	97.37
K-2651-d	65.10	0.66	15.71	5.24	0.11	2.10	4.82	3.81	1.86	0.20	2.09	97.00
K-2651-e	61.05	0.72	16.49	6.36	0.13	3.16	6.46	3.56	1.47	0.18	1.57	97.24
K-2652-a	60.00	0.73	16.77	6.55	0.13	3.38	6.93	3.59	1.32	0.20	0.79	98.61
K-2652-b	59.68	0.72	16.58	6.78	0.14	3.56	7.10	3.57	1.28	0.20	0.34	99.14
K-2652-c	65.98	0.64	15.65	4.74	0.11	1.86	4.49	4.07	1.84	0.20	1.92	97.37
K-2652-d	65.70	0.65	15.56	5.01	0.11	1.93	4.56	4.07	1.79	0.20	1.99	97.32
K-2653-a	59.61	0.73	16.74	6.83	0.13	3.52	7.16	3.40	1.26	0.21	0.44	98.75
K-2653-b	59.47	0.74	16.70	6.90	0.14	3.65	7.12	3.38	1.27	0.21	0.21	98.98
K-2653-c	59.57	0.74	16.72	6.71	0.13	3.66	7.19	3.37	1.30	0.21	0.38	98.31
K-2653-d	60.72	0.73	16.41	6.61	0.13	3.32	6.61	3.45	1.40	0.21	0.33	98.90
K-2654-a	64.96	0.68	15.78	5.23*	0.11	2.06	4.81	4.06	1.69	0.21	1.53	97.82
K-2654-b	64.87	0.66	15.93	5.20	0.11	2.10	4.86	3.99	1.66	0.22	3.20	96.27
K-2654-c	60.53	0.73	16.72	6.41	0.13	3.14	6.66	3.71	1.38	0.20	0.45	98.90
K-2654-d	59.96	0.72	16.77	6.65	0.13	3.36	6.97	3.51	1.33	0.20	0.45	99.17
K-2654-e	59.22	0.73	16.92	6.81	0.13	3.56	7.27	3.49	1.27	0.20	0.47	98.89
K-2654-f	64.99	0.66	15.88	5.13	0.12	2.05	4.80	4.09	1.67	0.21	4.36	94.71
						Novarupta scoria falls						
K-542	63.72	0.69	15.88	5.73	0.12	2.51	5.45	3.76	1.60	0.14	1.13	98.47
K-543	50.41	0.69	14.03	10.13	0.17	11.43	10.12	2.02	0.50	0.09	0.20	99.37
K-721	59.98	0.70	17.00	6.53	0.12	3.11	7.15	3.53	1.34	0.14	0.43	99.63
K-851	54.63	0.77	17.05	8.24	0.14	5.39	9.44	2.94	0.87	0.13	0.30	99.91
K-1594-a	61.63	0.71	16.47	6.13	0.12	2.96	6.25	3.69	1.48	0.17	0.87	98.58
K-1594-b	61.62	0.71	16.16	6.37	0.13	3.10	6.16	3.70	1.48	0.17	0.81	98.60
K-1595-a	62.28	0.70	16.33	6.03	0.12	2.84	5.96	3.62	1.53	0.18	1.05	98.19
K-1595-b	62.32	0.69	16.21	5.99	0.12	2.95	6.02	3.58	1.54	0.17	1.17	98.28
K-1595-c	61.44	0.70	16.70	6.11	0.12	2.99	6.40	3.52	1.44	0.18	1.19	98.39
K-1596-a	65.08	0.68	15.79	5.18	0.11	1.94	4.75	4.19	1.71	0.17	0.95	98.41
K-1596-b	61.73	0.75	15.96	6.57	0.13	3.11	6.04	3.61	1.52	0.18	1.02	98.58
K-1596-c	63.64	0.66	16.21	5.41	0.11	2.40	5.42	3.92	1.64	0.17	0.87	98.28

Appendix A. Major-element chemical analyses of eruptive products, Novarupta 1912.—Continued

Sample	SiO_2	TiO_2	Al_2O_3	FeO*	MnO	MgO	CaO	Na_2O	K_2O	P_2O_5	LOI	Original total
K-1596-d	61.19	0.72	16.33	6.29	0.13	3.20	6.50	3.63	1.45	0.17	0.61	98.81
K-1597-a	65.67	0.64	15.83	4.85	0.11	1.81	4.50	4.29	1.74	0.16	1.21	98.12
K-1597-b	65.77	0.67	15.65	4.92	0.11	1.86	4.46	4.21	1.77	0.17	1.35	97.98
K-1597-c	65.19	0.64	15.92	4.95	0.11	1.93	4.72	4.27	1.71	0.16	1.23	98.25
K-1597-d	65.11	0.65	15.82	5.10	0.11	2.05	4.77	4.08	1.75	0.17	1.09	98.21
K-2005	61.82	0.71	16.14	6.23	0.13	3.01	6.16	3.75	1.49	0.17	0.52	98.76
K-2012-a	61.81	0.69	16.56	5.98	0.12	2.87	6.16	3.75	1.49	0.17	0.80	98.62
K-2012-b	65.41	0.66	15.82	4.95	0.11	1.90	4.62	4.25	1.70	0.17	0.99	98.22
K-2013-a	61.65	0.72	16.37	6.21	0.12	2.98	6.16	3.72	1.49	0.18	0.91	98.55
K-2013-b	61.14	0.71	16.65	6.22	0.12	2.93	6.53	3.70	1.43	0.17	0.74	98.71
K-2013-c	61.84	0.71	16.22	6.36	0.13	3.08	6.03	3.56	1.48	0.18	0.99	98.25
K-2013-d	65.66	0.63	15.80	4.87	0.10	1.87	4.68	4.07	1.76	0.16	1.32	97.69
K-2013-e	65.37	0.65	15.69	5.01	0.11	1.92	4.67	4.29	1.72	0.17	1.08	98.42
K-2015-a	61.76	0.72	16.25	6.28	0.13	2.95	6.17	3.68	1.49	0.17	0.99	98.05
K-2015-b	65.32	0.65	15.82	5.00	0.11	1.93	4.68	4.21	1.70	0.17	1.21	98.19
K-2015-c	65.60	0.64	15.64	4.97	0.11	1.93	4.54	4.26	1.75	0.16	1.10	98.07
					Blocks from Phantom Dome							
K-20V	64.76	0.66	15.89	5.29	0.11	2.09	4.96	4.06	1.64	0.14	0.24	99.05
K-513-v	64.34	0.56	15.66	5.30	0.11	2.58	5.43	3.82	1.69	0.10	0.29	99.22
K-2022-a	65.57	0.65	15.61	5.00	0.11	1.95	4.62	4.17	1.75	0.15	0.66	98.89
K-2022-b	64.94	0.65	15.78	5.31	0.11	2.17	4.87	3.88	1.74	0.15	2.10	97.23
K-2022-c	66.65	0.69	15.40	4.38	0.11	1.99	4.37	4.15	1.78	0.09	0.79	98.33
K-2023-a	62.24	0.72	16.16	6.03	0.13	2.89	5.89	3.89	1.49	0.15	< 0.01	99.22
K-2023-b	64.05	0.66	15.86	5.50	0.12	2.36	5.24	3.98	1.69	0.14	0.20	99.21
K-2023-c	63.89	0.69	15.82	5.64	0.12	2.42	5.20	4.02	1.64	0.16	0.26	98.84

Appendix B. Trace-element chemical analyses of eruptive products, Novarupta 1912.

[Determinations (in parts per million) by energy-dispersive x-ray fluorescence in U.S. Geological Survey laboratory at Lakewood, Colorado; D.F. Siems, analyst. Precision and accuracy are discussed by Baedecker (1987) and Bacon and Druitt (1988). —, not analyzed. Additional data are given by Turner and others (2010).]

Sample	Rb	Sr	Y	Zr	Ba	Ni	Cu	Zn	Cr
				Layer A					
K-1000-a	58	70	45	155	906	1	—	56	—
K-1001	65	64	42	153	877	2	<2	44	11
K-1065	67	64	44	155	863	<2	<2	42	10
K-1079-a	56	71	44	151	915	1	—	58	—
K-1190	64	62	42	153	938	3	<2	43	9
				Layer B					
K-90-a	65	64	42	153	915	4	3	41	11
K-90-c	34	261	36	154	538	4	9	64	12
K-827-a	66	64	43	151	842	4	3	42	11
K-827-c	48	207	30	152	613	5	10	57	13
K-827-d	27	307	24	114	348	7	13	73	30
K-1008-c	36	272	29	140	371	0	9	74	17
K-1061-a	67	66	44	155	814	2	2	43	11
K-1061-c	28	301	30	121	326	4	16	78	27
K-1068-a	59	73	44	156	909	1	—	63	—
K-1073	68	64	44	155	821	3	2	42	13
K-1074-a	59	73	44	151	882	1	—	61	—
K-1158-c	37	242	31	166	482	4	6	64	11
K-1160-c	43	247	30	150	519	5	7	60	18
K-1194-a	64	64	43	155	982	3	<2	41	9
K-1199-c	42	234	32	168	487	4	6	64	15
K-1199-d	28	301	30	123	350	6	13	79	37
K-1200-c	34	285	27	135	401	4	14	73	23
K-2466	27	301	23	117	446	—	—	—	—
K-2466-b	25	304	24	117	445	—	—	—	—
K-2488-a	31	276	26	136	508	—	—	—	—
K-2488-b	29	290	26	133	485	—	—	—	—
K-2488-c	24	269	27	138	487	—	—	—	—
K-2488-d	29	280	23	126	505	—	—	—	—
				Layer C					
K-91	36	248	31	165	508	4	7	59	10
K-1270-a	42	238	37	166	598	2	21	68	—
K-1270-b	38	246	35	171	582	2	15	65	—
K-1270-c	38	252	34	167	559	4	25	73	—
K-1270-d	33	309	31	130	452	4	29	80	13
K-1270-e	42	271	33	155	520	3	23	79	—
K-1270-f	36	244	37	161	586	1	23	72	—
K-1271-a	38	251	33	164	583	4	14	72	—
K-1271-b	34	284	30	143	489	6	32	79	—
K-1271-c	38	256	36	166	567	4	18	74	—

Appendix B. Trace-element chemical analyses of eruptive products, Novarupta 1912.—Continued

Sample	Rb	Sr	Y	Zr	Ba	Ni	Cu	Zn	Cr
K-1271-d	36	249	36	170	567	8	23	77	—
K-1271-e	40	250	35	166	564	5	20	66	—
K-1271-f	35	251	36	159	549	6	22	75	—
K-1271-g	30	299	29	130	473	8	31	84	3
K-1271-h	31	229	33	167	—	—	—	—	—
K-1271-i	35	236	30	163	571	—	—	—	—
K-2002-a	32	273	29	134	—	—	—	—	—
K-2002-b	29	254	32	139	—	—	—	—	—
K-2620-a	36	244	31	161	568	3	22	82	8
K-2620-b	37	236	32	167	481	4	14	90	8
				Layer D					
K-39-r	36	243	30	161	578	2	12	54	5
K-55	38	238	30	166	574	4	18	70	15
K-94-a	66	63	44	155	1040	3	3	40	9
K-94-b	36	253	32	160	543	2	5	64	8
K-94-c	37	243	30	160	608	4	8	59	9
K-1083-a	67	64	45	155	675	3	<2	47	10
K-1083-c	38	245	31	162	526	5	7	63	8
K-1268-a	37	249	36	162	565	1	58	64	—
K-1268-b	31	313	31	125	440	2	64	72	1
K-1268-c	38	274	34	145	484	1	66	74	3
K-1268-d	38	254	38	165	571	1	50	69	0
K-1268-e	36	253	37	162	560	2	47	75	0
K-1268-f	36	280	33	138	487	1	69	80	10
K-1269-a	39	249	34	159	545	4	23	71	—
K-1269-b	42	251	36	167	581	4	16	72	—
K-1269-c	40	238	34	158	605	5	23	75	—
K-1269-d	36	256	34	163	564	9	23	76	—
K-1269-e	37	251	38	167	572	3	23	71	—
K-1269-f	38	243	37	170	577	4	21	74	—
				Layer F					
K-53	35	251	29	154	526	8	29	64	9
K-631-b	34	254	36	153	432	<2	8	72	10
K-631-c	39	227	32	151	467	5	9	65	10
K-1267-a	40	278	35	152	525	2	25	74	—
K-1267-b	38	250	34	163	569	1	28	69	—
K-1267-c	34	274	33	148	543	1	25	79	—
K-1267-d	36	263	33	154	529	2	34	75	—
K-1267-e	42	265	32	157	534	2	25	74	—
K-1267-f	38	260	32	155	533	3	29	75	—
K-1267-g	33	234	31	164	—	—	—	—	—
				Layer G					
K-23	38	259	32	161	554	10	14	61	11

Appendix B. Trace-element chemical analyses of eruptive products, Novarupta 1912.—Continued

Sample	Rb	Sr	Y	Zr	Ba	Ni	Cu	Zn	Cr
K-54	35	252	29	161	537	8	15	61	5
K-101	33	273	28	151	514	—	—	—	—
K-107-d	29	279	32	137	535	16	13	74	18
K-1266-a	42	273	34	151	587	1	7	68	—
K-1266-b	36	245	36	168	582	2	2	73	—
K-1266-c	35	257	34	164	563	1	3	73	—
K-1266-d	36	254	33	158	555	2	5	76	1
K-1266-e	36	270	35	160	546	1	3	74	—
K-1266-f	42	266	34	151	562	3	9	73	—
Novarupta lava dome									
K-16	60	75	43	158	964	—	—	—	—
K-17	18	315	22	102	339	—	—	—	—
K-18	33	257	32	164	558	14	14	51	—
K-18-s	39	262	33	163	568	6	16	58	12
K-22	69	96	42	160	917	—	4	41	9
K-622	60	76	43	159	971	—	—	—	—
K-623	19	359	21	101	311	—	—	—	—
K-623-r	17	369	19	93	—	—	—	—	—
K-2004	16	333	22	86	—	—	—	—	—
RLS-214	66	86	48	157	952	—	—	31	—
Pyroclastic flows									
K-7A-r	79	91	48	165	925	2	3	44	6
K-11-a-r	69	84	48	162	924	1	1	37	6
K-12-c	31	217	29	158	608	—	—	—	—
K-13-c	33	230	33	174	625	—	—	—	—
K-14-c	38	239	30	160	649	—	—	—	—
K-15-c	50	245	36	176	648	6	11	50	7
K-37-c	50	218	32	183	691	4	22	56	5
K-42-r	63	75	41	156	915	—	8	44	4
K-44-r	63	76	42	158	944	—	—	—	—
K-57	36	258	30	160	548	4	10	55	6
K-60	31	290	25	141	523	13	26	55	—
K-63-c	34	238	30	156	625	—	—	—	—
K-64-c	37	229	31	157	511	7	18	139	11
K-66-c	40	255	30	161	562	5	2	60	12
K-66-d	32	282	27	142	499	7	9	65	21
K-74-c	47	213	29	165	694	3	6	49	4
K-75-c	34	249	32	165	580	8	16	60	7
K-75-d	25	306	25	124	431	—	—	—	—
K-75-e	38	248	32	170	573	2	16	53	5
K-76	35	258	32	164	569	6	17	64	8
K-78	33	246	34	162	565	6	14	61	18
K-82-c	37	246	30	165	615	6	20	53	6

Appendix B. Trace-element chemical analyses of eruptive products, Novarupta 1912.—Continued

Sample	Rb	Sr	Y	Zr	Ba	Ni	Cu	Zn	Cr
K-82-d	35	299	27	127	462	11	26	84	24
K-100-a	64	77	41	154	903	—	1	43	4
K-100-c	34	267	30	152	506	4	15	60	14
K-100-d	24	307	23	104	319	27	33	74	54
K-108-c	40	250	30	165	577	7	22	61	7
K-108-d	32	261	29	156	520	8	17	66	8
K-113-c	42	246	32	167	572	4	9	63	11
K-113-d	26	290	25	128	454	18	18	63	—
K-119D-r	25	308	25	119	420	—	—	—	—
K-121	62	66	42	150	861	1	-	45	—
K-122	54	64	39	146	866	1	—	61	—
K-122-d	22	308	25	124	436	—	—	—	—
K-122-d-r	27	315	25	119	—	—	—	—	—
K-123-d-r	30	299	26	131	447	—	—	—	—
K-131-d-r	26	322	25	120	425	—	—	—	—
K-135-c	38	243	34	164	585	5	16	61	10
K-135-d	29	307	27	120	417	9	35	71	32
K-139-r	65	73	39	158	914	2	5	28	4
K-162	41	252	36	170	597	5	24	55	6
K-182	28	242	29	166	592	12	14	41	—
K-184	28	301	26	138	500	12	10	54	—
K-200	29	272	29	148	511	—	—	—	—
K-202	33	242	30	164	581	13	14	54	—
K-203	29	294	27	138	477	—	—	—	—
K-581	24	308	24	128	454	—	—	—	—
K-1363	56	66	44	150	917	1	—	61	—
K-1364	52	68	39	141	878	1	—	70	—
K-1365	54	67	41	147	900	1	—	58	—
K-1365-r	63	70	39	155	919	2	3	40	7
K-1365-u	58	66	39	154	952	—	—	—	—
K-1490	35	205	33	161	—	—	—	—	—
K-1495	35	247	36	167	—	—	—	—	—
K-1518-a	42	213	35	176	—	—	—	—	—
K-1525	58	69	46	156	—	—	—	—	—
K-2582-a	38	234	30	166	661	—	—	—	—
K-2582-b	31	242	31	163	602	—	—	—	—
K-2582-c	36	216	34	171	642	—	—	—	—
K-2584-b	63	78	39	157	946	—	—	—	—
K-2588-c	59	77	43	159	952	—	—	—	—
K-2614-a	29	303	24	118	382	6	23	109	29
K-2614-b	29	292	25	120	436	9	24	93	35
K-2614-c	29	298	30	119	452	8	27	113	26
K-2614-d	28	292	23	115	394	7	26	96	55

Appendix B. Trace-element chemical analyses of eruptive products, Novarupta 1912.—Continued

Sample	Rb	Sr	Y	Zr	Ba	Ni	Cu	Zn	Cr
K-2616-a	28	300	29	121	514	12	27	84	23
K-2616-b	28	304	29	121	446	11	25	87	28
K-2617-a	32	259	34	146	464	5	15	67	13
K-2617-b	65	61	43	154	846	3	3	44	9
K-2618-a	38	247	32	165	541	5	17	63	11
K-2618-b	33	254	35	153	665	2	21	62	10
K-2618-c	44	232	30	166	656	3	18	58	11
K-2618-d	28	307	30	122	445	5	30	76	22
K-2619	29	297	31	128	464	3	28	70	22
K-2621-a	31	287	32	134	580	7	26	72	16
K-2621-b	39	253	30	164	797	4	19	47	8
K-2621-c	29	293	30	126	497	6	37	75	24
K-2622-a	37	240	32	165	611	3	20	60	9
K-2622-b	32	270	34	148	637	2	10	61	13
K-2623	40	232	30	160	631	8	11	61	12
K-2624	28	294	29	127	520	11	23	76	30
K-2625	46	223	31	171	621	<2	5	56	12
K-2626-a	36	241	31	159	525	2	19	74	12
K-2626-b	29	289	31	129	456	7	25	77	20
K-2626-c	28	302	24	120	442	8	29	76	22
K-2626-d	32	260	35	149	469	2	19	64	14
K-2627-a	30	287	25	127	497	6	24	91	21
K-2627-b	27	300	24	117	431	6	26	77	32
K-2649-a	30	300	26	123	385	4	25	90	20
K-2649-b	31	286	31	132	498	7	29	75	20
K-2649-c	28	307	29	121	429	5	24	74	19
K-2650-a	50	214	32	171	541	3	15	59	14
K-2650-b	42	232	31	172	727	3	17	45	9
K-2651-a	39	235	33	173	501	4	13	61	11
K-2651-b	43	223	31	164	522	5	8	69	11
K-2651-c	29	298	30	126	406	7	12	92	26
K-2651-d	38	242	31	159	509	4	7	77	11
K-2651-e	29	285	30	130	406	7	9	96	19
K-2652-a	29	302	26	126	401	5	7	73	25
K-2652-b	28	303	25	121	455	7	8	79	25
K-2652-c	37	233	31	162	440	2	5	64	9
K-2652-d	38	235	32	164	493	5	8	63	10
K-2653-a	28	313	25	123	506	7	13	80	26
K-2653-b	27	308	30	124	500	7	8	76	29
K-2653-c	27	305	30	121	425	5	6	77	32
K-2653-d	31	296	31	131	542	4	14	74	22
K-2654-a	36	245	31	159	503	5	19	64	10

Appendix B. Trace-element chemical analyses of eruptive products, Novarupta 1912.—Continued

Sample	Rb	Sr	Y	Zr	Ba	Ni	Cu	Zn	Cr
K-2654-b	37	254	31	156	532	3	14	51	9
K-2654-c	29	301	31	130	537	6	21	74	19
K-2654-d	28	310	32	127	508	6	21	70	23
K-2654-e	27	315	25	118	458	8	26	75	28
K-2654-f	37	253	31	162	607	5	15	50	8
Novarupta scoria falls									
K-542	31	263	31	155	525	—	—	—	—
K-543	13	299	16	66	186	—	—	—	—
K-721	29	309	31	125	445	3	61	75	5
K-851	23	341	25	88	268	11	9	73	61
K-851-r	16	349	21	87	—	—	—	—	—
K-1594-a	27	286	29	133	—	—	—	—	—
K-1594-b	31	271	30	136	—	—	—	—	—
K-1595-a	32	273	30	142	—	—	—	—	—
K-1595-c	28	289	27	132	—	—	—	—	—
K-1596-a	30	245	33	157	—	—	—	—	—
K-1596-c	33	258	29	152	—	—	—	—	—
K-1596-d	31	286	29	137	—	—	—	—	—
K-1597-a	33	242	31	160	—	—	—	—	—
K-1597-b	35	236	31	161	—	—	—	—	—
K-1597-c	30	249	33	147	—	—	—	—	
K-1597-d	36	242	35	160	—	—	—	—	—
K-2005	29	270	29	133	—	—	—	—	—
K-2012-a	31	276	30	137	—	—	—	—	—
K-2012-b	29	250	30	161	—	—	—	—	—
K-2013-a	27	279	30	132	—	—	—	—	—
K-2013-b	26	278	24	131	—	—	—	—	—
K-2013-c	29	277	27	135	—	—	—	—	—
K-2013-d	31	242	31	155	—	—	—	—	—
K-2013-e	32	242	33	160	—	—	—	—	—
K-2015-a	30	271	27	131	—	—	—	—	—
K-2015-b	28	240	30	159	—	—	—	—	—
K-2015-c	32	231	32	157	—	—	—	—	—
Blocks from Phantom dome									
K-20V	29	263	29	162	569	—	—	—	—
K-513-v	33	256	29	133	577	—	—	—	—
K-2022-a	30	233	31	163	—	—	—	—	—
K-2022-b	34	246	32	150	—	—	—	—	—
K-2022-c	30	233	31	166	—	—	—	—	—
K-2023-a	25	279	26	143	—	—	—	—	—
K-2023-b	31	256	32	157	—	—	—	—	—
K-2023-c	33	254	30	150	—	—	—	—	—

Appendix C. Trace-element analyses of 1912 eruptive products by Instrumental neutron activation analysis (INAA).

[Determinations (in parts per million) by INAA in U.S. Geological Survey laboratories. —, not analyzed. Methods and precision discussed by Baedecker (1987).]

Sample	Ba	Co	Cr	Cs	Hf	Rb	Sb	Ta	Th	U	Zn	Zr	Sc	La	Ce	Nd	Sm	Eu	Gd	Tb	Tm	Yb	Lu
Layer A																							
K-88-r	897	0.6	0.6	2.40	4.80	62	—	0.40	5.1	2.30	42	185	7.2	19.0	40.0	24	4.90	0.780	5.9	0.86	0.58	4.8	0.69
K-88-s	984	0.7	1.4	2.60	5.00	65	—	0.49	5.6	2.10	45	248	7.4	19.0	41.0	21	5.30	0.780	5.7	0.95	0.70	4.6	0.69
Layer D																							
K-39-r	508	10.6	7.9	1.50	4.20	37	—	0.27	2.9	1.40	72	170	15.9	13.0	26.0	19	3.50	0.980	4.9	0.67	0.44	3.4	0.51
K-39-s	600	11.3	8.2	1.60	4.40	41	—	0.35	3.3	1.80	75	200	16.7	13.0	27.0	18	3.80	1.000	4.5	0.79	0.49	3.3	0.50
K-55	634	116.0	8.2	1.53	4.90	38.9	0.47	0.71	3.4	1.52	77.3	154	17.3	13.2	31.5	18.5	5.04	1.100	5.5	0.84	0.59	3.8	0.59
Layer F																							
K-53	566	68.6	12.2	1.36	4.49	49.1	0.52	0.48	3.1	1.40	79.0	—	20.4	12.6	29.9	16.6	4.84	1.120	4.7	0.84	0.57	3.7	0.54
Layer G																							
K-23	570	10.5	7.5	1.40	4.10	36	—	0.41	3.1	1.50	72.0	120	16.5	13.0	27.0	16	3.60	1.020	3.1	0.71	0.46	3.4	0.50
Novarupta lava dome																							
K-16-r	914	0.6	0.9	2.60	4.90	64	—	0.42	5.4	2.30	46	161	7.4	19.0	40.0	22.0	4.90	0.790	6.8	1.04	0.59	4.7	0.67
K-16-s	995	0.6	2.5	2.60	5.00	69	—	0.49	5.6	3.00	44	155	7.3	19.0	40.0	22.0	5.30	0.800	6.2	1.01	0.70	4.7	0.69
K-17	360	23.6	27.9	0.72	2.50	23	0.32	0.16	1.4	0.63	84	140	29.3	6.9	14.0	8.6	2.92	0.903	—	0.54	—	2.5	0.38
K-622	991	0.7	3.5	2.57	5.12	59.9	0.89	0.41	5.2	2.29	40.4	141	7.6	18.2	41.4	23.3	5.56	0.855	6.5	1.11	0.73	4.8	0.72
K-623	354	22.5	43.5	0.68	2.41	19.9	0.36	0.19	1.9	0.79	83.0	115	31.3	7.6	16.7	11.0	3.05	0.932	3.5	0.56	0.36	2.3	0.35
RLS-214	940	1.6	1.6	2.30	4.90	64	1.20	0.44	5.0	2.10	37	178	8.4	19.0	39.0	20	5.60	0.780	6.1	1.14	0.55	4.6	0.67
Pyroclastic flows																							
K-60	505	35.0	16.7	2.00	3.48	36	2.15	0.33	2.9	1.20	88.8	110	21.0	10.8	23.2	14	3.98	0.970	—	0.70	—	2.9	0.42
K-113-d	470	92.1	27.0	1.10	3.10	27	0.36	0.56	2.8	1.30	72	120	23.6	10.2	21.7	14	3.76	0.915	—	0.69	—	2.8	0.41
K-119D-r	469	18.1	36.0	1.10	3.20	28	—	0.30	2.8	1.20	80	180	23.8	10.0	21.0	14	3.20	0.890	3.6	0.59	0.43	2.6	0.39
K-119D-s	467	17.8	35.2	1.00	3.10	30	—	0.27	2.9	1.50	84	170	24.1	10.0	20.0	16	3.20	0.910	4.2	0.60	0.41	2.7	0.40
K-122-d	462	19.3	33.2	1.10	3.20	34	—	0.26	3.0	1.40	86	220	24.3	11.0	22.0	16	3.20	0.930	4.1	0.55	0.44	2.8	0.41
K-123-d-r	522	17.8	28.3	1.10	3.40	28	—	0.30	3.0	1.50	85	180	23.6	11.0	23.0	18	3.20	0.910	4.0	0.59	0.39	2.8	0.41
K-123-d-s	501	18.0	27.6	1.40	3.50	20	—	0.40	2.9	1.30	90	110	25.0	11.0	24.0	14	3.80	0.970	4.1	0.62	0.47	2.8	0.40
K-123-d-t	487	16.6	24.4	1.00	3.20	30	0.70	0.20	2.8	1.40	83	120	22.3	11.0	23.0	17	3.70	0.890	3.9	0.56	0.43	2.8	0.42
K-131-d-r	483	18.7	36.8	1.10	3.00	29	—	0.22	2.7	1.30	106	100	24.2	10.0	21.0	15	3.10	0.900	3.5	0.55	0.43	2.7	0.38
K-131-d-s	483	18.5	35.6	1.20	3.00	36	—	0.28	2.7	2.00	102	140	24.1	10.0	21.0	21	3.30	0.920	4.3	0.60	0.39	2.8	0.41
K-162	618	11.0	10.2	1.50	4.30	39	—	0.42	3.2	1.80	74	160	16.1	13.0	26.0	17	3.60	1.020	5.1	0.73	0.49	3.3	0.49
K-182	562	10.2	6.5	1.40	4.30	37	—	0.31	3.1	1.70	67.0	150	16.1	12.0	25.0	22	3.60	0.920	5.0	0.72	0.48	3.3	0.50
K-184	492	18.6	36.2	1.20	3.30	34	—	0.30	2.9	1.50	85.0	110	24.8	11.0	23.0	19	3.30	0.950	3.9	0.62	0.35	2.7	0.42

Appendix C. Trace-element analyses of 1912 eruptive products by Instrumental neutron activation analysis (INAA).—Continued

Sample	Ba	Co	Cr	Cs	Hf	Rb	Sb	Ta	Th	U	Zn	Zr	Sc	La	Ce	Nd	Sm	Eu	Gd	Tb	Tm	Yb	Lu
K-200	522	14.0	22.0	1.20	3.72	33	0.46	0.25	3.0	1.40	71.7	140	21.1	11.2	24.3	15	4.13	0.978	—	0.69	—	3.0	0.47
K-202	565	11.8	6.5	1.40	4.20	39	—	0.36	3.1	1.60	79.0	160	17.2	13.0	27.0	19	3.90	1.030	5.1	0.71	0.47	3.5	0.53
K-203	460	18.0	28.0	1.10	3.40	31	0.39	0.22	3.1	1.20	80	190	22.3	11.0	22.6	14	3.86	0.920	—	0.64	—	2.9	0.42
K-581	482	18.1	30.0	1.10	3.34	27.0	0.34	0.20	2.7	1.26	86.5	99.5	25.4	9.9	23.3	13.9	3.77	1.020	4.1	0.65	0.44	2.8	0.43
K-1365	878	0.5	<0.29	2.45	4.94	60.9	0.93	0.41	5.1	2.44	39	156	7.1	18.1	41.3	23.9	5.94	0.800	6.1	1.04	0.71	4.9	0.72
K-1490	711	9.45	6.59	1.64	4.75	43.10	0.62	0.32	3.78	2.01	66.1	172	15.8	14.3	32.8	19.9	5.10	1.05	5.46	0.88	0.62	3.91	0.6
K-1490-glass	904	3.40	3.27	2.35	6.49	61.10	0.76	0.43	5.45	2.77	36.5	229	8.52	17.5	40.9	23.4	6.11	0.86	6.35	1	0.75	4.73	0.72
Novarupta scoria falls																							
K-542	599	13.8	10.5	1.38	4.43	35.0	0.47	0.29	3.3	1.52	77.5	145	21.0	12.3	30.1	18.5	4.77	1.120	5.0	0.81	0.58	3.6	0.53
K-543	197	54.3	230.0	0.36	1.57	10.1	—	0.08	1.0	0.39	68.9	54.7	45.5	4.9	11.2	7.47	2.35	0.787	2.8	0.43	0.26	1.6	0.24
K-721	440	17.9	25.6	1.10	3.26	31	0.39	0.24	2.9	1.20	72	120	23.2	10.5	22.0	12	3.78	0.942	—	0.68	—	2.7	0.41
K-851	270	28.0	93.6	0.69	2.29	24	<0.40	0.19	1.9	0.73	79	89	33.5	7.2	16.1	10	2.96	0.844	—	0.55	—	2.0	0.31
Blocks from Phantom Dome																							
K-20V	580	11.4	9.8	1.20	4.14	34	0.53	0.27	3.1	1.50	73.0	140	17.9	12.0	26.2	15	4.52	1.070	—	0.77	—	3.5	0.51
K-513-v	610	14.0	23.0	1.40	3.62	40	0.57	0.26	3.2	1.50	73	230	20.7	12.0	26.0	18	4.34	0.990	—	0.74	—	3.5	0.54

Mid-evening sunlight bathes the Late Pleistocene dacite domes Falling Mountain and Mount Cerberus, both draped by 1912 pumice falls. The sun-lit domes frame the entrance to Katmai Pass. View southeastward toward snow-streaked West Trident (peak 5605) in cloud. In fair weather, fieldwork often extended as late as 11 p.m. during Alaskan summers. Photo taken in August 1982.

Produced in the Menlo Park Publishing Service Center, California
Manuscript approved for publication March 22, 2012
Text edited by Peter H. Stauffer
Layout and design by Judy Weathers